The Polymerase Chain Reaction (PCR) *for* Human Viral Diagnosis

Edited by
Jonathan P. Clewley

CRC Press
Boca Raton Ann Arbor London Tokyo

Library of Congress Cataloging-in-Publication Data

The Polymerase chain reaction (PCR) for human viral diagnosis / editor, Jonathan P. Clewley.
     p.   cm.
    Includes bibliographical references and index.
    ISBN 0-8493-4833-1
    1. Virus diseases--Diagnosis. 2. Polymerase chain reaction--Diagnostic use. 3. Virus diseases--Genetic aspects. I.
    Clewley, Jonathan P.
    [DNLM: 1. Virus Diseases--diagnosis. 2. Polymerase Chain Reaction. 3. Viruses--isolation & purification. WC
    500 P783 1995]
    RC114.5.P58   1995
    616.9′25075--dc20
    DNLM/DLC
    for Library of Congress
                                                        94-12662
                                                          CIP

No claim to original U.S. Government works
International Standard Book Number 0-8493-4833-1
Library of Congress Card Number 94-12662
Printed in the United States of America   2 3 4 5 6 7 8 9 0
Printed on acid-free paper

The Polymerase Chain Reaction (PCR) *for* Human Viral Diagnosis

FOREWORD

This work is intended to address the present and potential role of the polymerase chain reaction (PCR) for the detection and diagnosis of human viral infections. PCR has been demonstrated to be a very sensitive method for revealing the presence of otherwise undetectable quantities of viral nucleic acid, either DNA or RNA. However, the application of PCR to viral diagnosis is not straightforward.

The significance of the presence of a few copies of a viral genome in a healthy individual depends on the nature of the replication cycle of the virus. For instance, viruses which cause a cytolytic infection (e.g. respiratory viruses and parvovirus B19) may be detectable, in the presence of neutralizing antibody, some time after clinical illness has resolved, and may not be relevant to the health of the host. Other viruses which can establish a lysogenic or latent state (e.g. herpesviruses and papillomaviruses) may have future pathological consequences. However, the significance and importance of a few copies of a virus genome in an individual without detectable serological markers of infection may be questionable. PCR is not a predictor of infectivity as the amplicons may come from defective, non-infectious virus.

This volume is not a comprehensive review of all human viruses for which PCR tests have been described; it aims to illustrate the uses of PCR in human virology for a range of representative human viral infections. A search of MedLine in March 1993 using PCR and human viruses as keywords retrieved over 1300 articles. Some of those papers are descriptions of the establishment of PCR tests for particular viruses. Others put PCR to good use as a means to an end not easily otherwise attainable. It is hoped that this volume illustrates some of those good uses of PCR for the detection and characterization of viruses of pathogenic consequence for humans.

After a general introduction to PCR, there follows a discussion of its use in the diagnostic laboratory. Usually, this has involved visualization of the PCR amplicon on a gel, but PCR has a valuable role to play in increasing the sensitivity of *in situ* hybridization tests, particularly for those viruses that establish a persistent or latent infection. This development, which combines histology with DNA amplification, is described by pioneers of the technique. Other constributions address the detection of specific diseases (including enteric, hepatic, and respiratory virus infections) or specific virus groups (including retroviruses and flaviviruses).

Although this is not intended to be a methods book, details of primers are given, and methodological details can be gleaned from several of the chapters. It is hoped that this blend will appeal both to clinicians and to molecular biologists.

THE EDITOR

Jonathan P. Clewley is a Clinical Scientist in the Public Health Laboratory Service's Virus Reference Division at the Central Public Health Laboratory in the United Kingdom.

Dr. Clewley graduated in 1972 from the University of Sussex, with a B.Sc. degree in the biological sciences and obtained his Ph.D. degree in 1976 from the University of Warwick. After completing his doctoral degree, Dr. Clewley continued his training as a post-doctoral fellow at the University of Alabama in Birmingham. During this period, he worked on the characterization of the genomes of single-stranded RNA viruses by oligonucleotide fingerprinting. He returned to the UK in 1978 and worked on the Kirsten murine sarcoma/leukemia viruses at the University of Warwick, before taking up his present post in 1982. At the Virus Reference Division he has been concerned with the application of molecular biology to clinical, diagnostic virology. Since the first description of PCR in 1985 he has established laboratory facilities for the diagnostic and research use of PCR in virology, and has helped to set up several practical tests for the detection of viral genomes in clinical specimens. He has particular research interests in the human parvovirus B19; in retroviruses, in particular HIV; and single-stranded RNA viruses, including hepatitis C and the picornaviruses.

Dr. Clewley is a member of the Society for General Microbiology and the American Society for Virology. His general interests include virology, molecular biology, molecular evolution, and structural biology.

CONTRIBUTORS

Peter Balfe
Department of Medical Microbiology
University College and Middlesex School of
Medicine
London, England

Janet Beneke
Department of Laboratory Medicine and
Pathology
Hennepin County Medical Center
Minneapolis, Minnesota

Hervé Bourhy
Unité de la Rage
Institut Pasteur
Paris, France

David W. G. Brown
PHLS Virus Reference Division
Central Public Health Laboratory
London, England

William F. Carman
Institute of Virology
University of Glasgow
Glasgow, Scotland

Jonathan P. Clewley
PHLS Virus Reference Division
Central Public Health Laboratory
London, England

Bernard J. Cohen
PHLS Virus Reference Division
Central Public Health Laboratory
London, England

Janet E. Embretson
ViroMed Laboratories, Inc.
Minneapolis, Minnesota

Timothy Forsey
National Institute for Biological Standards and
Control
Herts, England

Deepak A. Gadkari
Deputy Director
National AIDS Research Institute
Pune, India

Patricia E. Gibson
PHLS Virus Reference Division
Central Public Health Laboratory
London, England

Jonathan Green
PHLS Virus Reference Division
Central Public Health Laboratory
London, England

Ashley T. Haase
Department of Microbiology
University of Minnesota Medical School
Minneapolis, Minnesota

Timo Hyypiä
Department of Virology
University of Turku
Turku, Finland

Gregor W. Leckie
Probe Diagnostics Business Unit
Abbott Diagnostics Division
Abbott Laboratories
North Chicago, Illinois

Helen H. Lee
Probe Diagnostics Business Unit
Abbott Diagnostics Division
Abbott Laboratories
Abbott Park, Illinois

Elizabeth A. B. McCruden
Institute of Virology
University of Glasgow
Glasgow, Scotland

John P. Norcott
PHLS Virus Reference Division
Central Public Health Laboratory
London, England

Elisabeth Puchhammer-Stöckl
Institute of Virology
University of Vienna
Vienna, Austria

Ernest F. Retzel
Department of Microbiology
University of Minnesota Medical School
Minneapolis, Minnesota

Betty H. Robertson
Hepatitis Branch
Division of Viral and Rickettsial Diseases
Centers for Disease Control
Atlanta, Georgia

James S. Robertson
National Institute for Biological Standards and
   Control
Herts, England

Débora Sacramento
Secao de raiva e encefalomyelite
Instituto Biologico
Sao Paulo, Brazil

Glyn Stanway
Department of Biology
University of Essex
Colchester, England

Katherine A. Staskus
Department of Microbiology
University of Minnesota Medical School
Minneapolis, Minnesota

Priscilla Swanson
Transfusion Diagnostics
Abbott Laboratories
Abbott Park, Illinois

Noël Tordo
Laboratoire des Lyssavirus
Institut Pasteur
Paris, France

TABLE OF CONTENTS

Chapter 1

Introduction to PCR for Viral Diagnosis

Jonathan P. Clewley

TABLE OF CONTENTS

I. INTRODUCTION

The polymerase chain reaction (PCR) first appeared in 1985 as a method for the prenatal diagnosis of sickle cell anemia.[1] It was based on the remarkable insight of Kary Mullis, who realized that repetition of a DNA extension reaction bounded by two synthetic oligonucleotide primers would generate a large quantity of any specified DNA sequence.[2] Like most brilliant ideas, it has generated controversy over the question of who thought of it first,[3] with some claiming that it was described in the 1970s.[4] However the American courts judged otherwise, and it seems clear that Mullis independently devised and established PCR.

Initially, PCR required the addition of thermolabile *Escherchia coli* DNA polymerase at each cycle, limiting its application, although one of the earliest papers describing the detection of human papilloma virus (HPV) DNA by PCR used an automated pipetting device for the addition of the polymerase.[5] Briefly exhibited by the manufacturer, the machine was not, at the time, likely to be of use for routine PCR because of the problems of contamination. Automated workstations will, however, play a major role in genome sequencing initiatives, and may pave the way for future automated diagnostic PCR sequencing machines *(vide infra)*. The innovations that delivered PCR onto the benches of scientists who were not prepared to add enzyme at each cycle were the purification of thermostable polymerase from *Thermus aquaticus (Taq)* and the invention of thermal cycling heating blocks, both of which came from Mullis and scientists at the Cetus and Perkin Elmer corporations. PCR has since been developed to the extent that there are numerous variations on the basic method. These variations are beyond the scope of this book, but they are essentially addressed to the detection, cloning, and characterization of any (unknown) DNA sequence. Many PCR innovations are published in *PCR Methods and Applications, Nucleic Acids Research,* and *BioTechniques*. PCR and its methodology have been reviewed in both books and articles,[6-13] and its clinical applications have been reviewed by numerous authors.[14-23]

II. PCR CONDITIONS AND PARAMETERS

The essential ingredients of a chain reaction to amplify DNA are a thermostable polymerase, buffer, magnesium ions, nucleotides, primers, co-solvents, sample, and an overlay of mineral oil or wax to prevent evaporation. *Taq* is the most widely used polymerase,[7] but it lacks a $3' \rightarrow 5'$ exonuclease or proofreading activity. There are other enzymes available, some of which have properties that make them better suited than *Taq* for certain purposes.[10,12] For instance, *Pfu* from *Pyrococcus furiosus* and Vent™* from *Thermococcus litoralis* both have a higher copying fidelity than *Taq* and are thus better choices for sequencing or protein expression studies where nucleotide misincorporation is not tolerable.[24] These two enzymes, and some others, are also more thermostable than *Taq,* and may be of use for DNA templates with extreme GC contents (e.g., herpes simiae B virus). Batches of polymerase from different manufacturers may differ in their properties due to the particular purification process and exact polypeptide length of the enzyme. Two buffer formulations are widely used. One consists of 50 m*M* KCl, 10 m*M* Tris, pH 8.3 and the other of 67 m*M* Tris, pH 8.8, 16.6 m*M* $(NH_4)_2SO_4$. The pH of the reaction needs to be about 7.0; therefore the buffers are made up at room temperature to a higher pH to allow for changes at elevated temperature. The magnesium ion concentration is usually 1.5 m*M*, but this may need to be optimized. The Stoffel fragment of *Taq,* which lacks $5' \rightarrow 3'$ exonuclease activity, has activity over a broader range of magnesium ion concentrations than the native enzyme. The Stoffel fragment may thus be useful for amplifying more than one target DNA sequence in the same reaction (multiplex PCR). Nucleotides are included at 40 to 200 μ*M* each. The lower amount is used when it is important to minimize base misincorporation. As nucleotides bind magnesium ions, the concentrations of these two components need to be manipulated in concert. It is also important not to add too much enzyme as an excess will increase the production of unwanted, nonspecific reaction products.

A. PRIMERS

The primers are synthetic oligonucleotides on average 20 bases long. Longers primers (about 24 bases) increase the specificity of the reaction. Shorter primers (about 16 bases) may allow nontarget amplification. It is important to avoid amplification primer artifacts (primer dimers), and this can be achieved by choosing primers that do not have any obvious secondary structure and do not hybridize to one another. Thus, the primers should not, if avoidable, have more than three consecutive identical bases or any complementary regions that may anneal or form hairpins. They should have an average base composition (A + T = G + C).

DNA synthesis is initiated from the 3' end of the primer, and base mismatch there may cause amplification to fail. Experimental studies[25] have shown that a 3' T residue mismatch is more likely to be tolerated than a G or C mismatch. There are several computer programs available for designing primers, though these do not guarantee the success of any reaction, which must be determined empirically. *Oligo*[26,27] is probably the most comprehensive program for analyzing primers and probes. Moreover, failure of a computer program to pick or validate a primer should not prevent it being tried experimentally; if a specific target sequence has to amplified, then the flanking sequences must be used, even if they appear unsuitable.

Experimentation with the reaction conditions (e.g., co-solvents, single-stranded binding protein, hot start, as described below) may well yield the desired product. The concentration of the primers should be the lowest that gives consistent amplification, although they are present usually in excess of the amount of amplicon that can be synthesized.[28]

Multiple alignment programs are very useful for choosing primers that will amplify genomes that are very variable (e.g., HIV) or for finding conserved regions of the genomes of a diverse group of viruses (e.g., the 5' noncoding region [NCR] of the picornaviruses). Primers can be modified at the 5' end if necessary.[7,10,11] A restriction enzyme site can be designed into the 5' end to facilitate cloning of the PCR product, or an M13 universal sequence inserted to allow sequencing with a standard primer. The incorporation of biotin into one of the primers allows capture, by streptavidin, of either the whole product, for detection purposes, or just one single strand for sequencing. The complementary (nonbiotinylated) strand can also be recovered for sequencing.

B. THE AMPLICON

For diagnostic purposes the size of the product of the PCR, the amplicon, is not critical and may range from, for example, 61 bp for hepatitis C virus[29] to 591 for B19[30] or larger. Small amplicons may be more difficult to differentiate from primer artifacts, if these occur. For sequencing studies, amplicons of about 400 to 600 bp may be best as these can be sequenced in both directions with a primer from either end to give maximum information. However, larger amplicons (e.g., the *env* gene of HIV-1) can be sequenced with several primers.[31] It is not clear whether there is an optimal location of nested primers with respect to the internal pair. That is, whether it is better to locate them far away from (more than 100 bp) or near

---

* New England Biolabs.

to (less than 20 bp) the internal primers. By locating the outer primers farther away the possibility of interference between the primers carried over from the first round amplification is reduced, as is the likelihood of the synthesis of intermediate amplicons primed by a combination of the first- and second-round primers. Also, by having a large outer amplicon and a smaller inner amplicon, the reaction conditions (extension time, denaturation temperature) can be biased toward the production of only the inner amplicon in the second round of amplification. However, a single internal primer can be used with one of the two outer primers in a hemi-nested PCR.[8] If nested PCR is being used, the primers in the outer reaction should be at a lower concentration (e.g., 5 pmol or less per 50 µl) than the primers in the inner reaction (e.g., 25 pmol per 50 µl). This is an adaptation of "booster PCR",[32] and it improves the efficiency of the reaction as well as minimizing the synthesis in the second round of amplicons primed by a combination of primers carried over from the first round. However, one PCR strategy exploits the multiple bands formed from a combination of the inner and outer primers.[33] Attempts are being made to include the first and second round primers in one reaction mixture, so that nested PCR can be completed in one tube.[8,34]

This has been termed "drop in-drop out" PCR,[8] and is based on the idea that the reaction conditions and primer sequences can be manipulated so that only the outer primers initiate DNA synthesis during the first part of the cycling; a subsequent change of conditions brings the inner primers into play and produces the desired final target amplicon. The first experimental description of the use of this approach was for hepatitis C virus (HCV) RNA.[35] cDNA was synthesized from RNA and amplified in a reaction containing three primers (hemi-nesting): 20 cycles of 94 and 70°C for 30 s each allowed amplification from the outer 31-mer primers. The thermal profile was then changed to 40 cycles of 94, 56, and 72°C for 30 s each to allow amplification from an internal 20-mer primer. The sensitivity of this assay was ten molecules of *in vitro* transcribed RNA. Although promising, this approach still needs more development, for instance the combination of reverse transcription and DNA polymerization in one tube without loss of sensitivity. Also, it needs to be clearly demonstrated for such assays that the same degree of sensitivity cannot be achieved merely by increasing the number of cycles with the inner primers.

C. CO-SOLVENTS

Co-solvents[7] are compounds added to the reaction to improve the amplification of genomes with complex secondary structure (high GC content) and to decrease spurious amplification. These include detergents (Tween 20 and NP40, or Triton X-100*), DMSO,[36] formamide,[37] glycerol, tetramethyl ammonium chloride (TMAC),[38] and reducing agents (2-mercaptoethanol or dithiothreitol). The detergents help keep the polymerase in solution. The other additives either reduce the melting temperature of the target or stabilize the enzyme. High-molecular-weight (MW) carriers like nuclease-free bovine serum albumin (BSA) or gelatin (at 0.001%) are also thought to improve enzyme stability. However, BSA may precipitate at elevated temperatures causing the *Taq* to co-precipitate, and it may be noted that gelatin has been dropped from an original buffer formulation for PCR (Perkin Elmer). It is best to keep the reaction as simple as possible at first, and try the controlled addition of co-solvents only if there are problems with either lack of specificity or lack of amplification. As soon as the reaction ingredients are combined, which is usually done at room temperature, the *Taq* polymerase will start polymerizing whatever suitable template is present, whether the intended target or not. Polymerases require only a base-paired 3′-hydroxyl group and a single-stranded (ss) template to initiate DNA synthesis. This template can be provided by the primers alone if they are not carefully designed, or by any slightly degraded DNA such as would be expected to be present if a whole-cell lysate were added to the PCR mixture (e.g., for amplification of HIV DNA). Unwanted DNA polymerization at room temperature can be minimized by including *E. coli* ssDNA binding protein, which will bind to the primers and any other ssDNA, preventing the *Taq* from binding and initiating DNA synthesis.[39,40] When the temperature is raised at the start of the PCR, the ss binding protein is denatured and amplification occurs normally.

The hot-start technique can also be used to prevent spurious amplification.[28,41] This involves separating the *Taq* from the nucleotides, primers, and magnesium ions with an impermeable wax barrier (e.g., Ampliwax™**).[42] The wax melts when the PCR is started, allowing all the ingredients to mix. Spurious priming can also be minimized by "touchdown" PCR,[43] in which the annealing temperature of the reaction is decreased by, for example, 1°C every 2 cycles from 65 to 55°C, at which temperature the reaction is continued.

---

*   Triton is a registered trademark of Union Carbide Chemical and Plastics Company.
** Perkin Elmer Corporation, Norwalk, CT.

D. SAMPLE PREPARATION

The sample may be added in any form that is not inhibitory to the reaction, and although this sometimes requires extraction and purification of the DNA, it does not always do so. Purified lysed cells, e.g., lymphocytes used for the diagnostic detection of HIV infection, can be directly combined with the reaction mixture after digestion with proteinase K.[44] Heat treatment may be sufficient to allow serum to be added directly.[45] Some samples that are otherwise inhibitory to PCR can be added after treatment with 25% Chelex,*[46] and this is probably the simplest direct method of sample preparation for PCR, although it may not be suitable for RNA because of the high pH of the Chelex. If possible, phenol extractions should be avoided as they involve the generation of aerosols and transfer from tube to tube, which can cause sample cross-contamination. Residual phenol in the nucleic acid preparation may also degrade the enzyme. Denaturation of the sample in a chaotropic agent (potassium iodide or guanidinium isothiocyanate) and binding of the nucleic acid to a silica matrix[47,48] is a convenient method for purifying DNA or RNA prior to amplification. The addition of too much DNA (e.g., 1 μg of chromosomal DNA) will decrease the specificity and inhibit the reaction; 100 to 200 ng of chromosomal DNA is optimal in a 50-μl reaction. Also, the addition of excess RNA has been reported to inhibit amplification.[49] If high-MW DNA is being added to the reaction, the efficiency of amplification can be increased by digesting the DNA with a restriction enzyme that cuts outside of the target sequence.[50] Alternatively it can be extensively boiled;[51-53] this will reduce the melting temperature of the target and allow the primers easier access.

If RNA is to be amplified, it must be transcribed to cDNA before amplification, and this can be achieved with either murine or avian reverse transcriptase (RTase).[54] Combining the RTase and *Taq* in one tube, although possible,[55] is not recommended as the enzymes have different optimal divalent cation requirements and the RTase may inhibit the *Taq,* necessitating the use of a large amount of the enzyme.[56] It is also possible to use a thermostable enzyme possessing both RTase and DNA polymerase activities (e.g., *Tth* polymerase) although the divalent cation Mn^{2+} has to be replaced by Mg^{2+} to switch from RTase to DNA polymerase activity.[8,57] Ampliwax beads can be used to separate the initial RTase reaction from the subsequent *Taq* reaction, obviating the need to transfer manually. However, it is sometimes useful to prepare cDNA with random primers and add an aliquot to several different PCRs, and also to store some cDNA for any necessary further analysis.

E. TEMPERATURE CYCLING

The target DNA is denatured at 94 to 98°C to allow the primers to bind to the separated single strands when the temperature is lowered. After the first 5 to 10 cycles, the denaturation temperature can be reduced to 90 to 94°C as the PCR amplicon will usually have a lower melting temperature than the input DNA.[58] This lower denaturation temperature will also be less harmful to the enzyme. The denaturation temperature is held for up to 1 min, depending on the thermal characteristics of the heating block and the reaction tubes. Heat transfer to the tube contents occurs more quickly if small reaction volumes and thin-walled tubes are used, allowing shorter denaturation times. Fast heat transfer can be achieved by using capillary tubes and hot-air thermal cyclers. The temperature used for annealing the primers to the target is between 50 and 60°C for most amplifications.[27] The higher temperature ensures greater specificity of the PCR, i.e., it reduces the number of spurious amplification products, but may not always allow annealing of generic primers with some base mismatches, as used for amplification of all the members of a virus genus[59,60] (e.g., hepadnaviruses; retroviruses; primers in the 5′ NCR of the picornaviruses). The annealing temperature is also often for 1 min, although this too can be reduced if rapid heat transfer is being achieved. The extension temperature is 72°C and is held for 1 min for most amplicons (up to 1 kb). For larger amplicons (up to 5 kb) an initial 3-min extension period can be used for the first 10 cycles, with a subsequent increase of 10 to 30 s at each cycle for the remainder of the amplification.[61] It is important to allow sufficient time for the enzyme to fully complete the synthesis of one strand. This reduces the possibility of recombination occurring between distinct sequence amplicons.[62] However, the annealing and extension steps can be combined using a two-temperature PCR (e.g., 94 and 63°C).[8] This is fast, but it may not be as sensitive as a three-temperature PCR, and it is probably not ideal for diagnostic purposes.[62]

For a single-round PCR, 35 cycles are often used, as maximum amplification of a discrete amplicon has usually occurred by this time. Some batches of *Taq* will remain active for about 60 cycles, but higher-MW products (aggregates and multimers) often accumulate on lengthy cycling.[63] Some protocols include a 5- to 10-min final extension step to complete synthesis of all the amplicons and polish (blunt)

---

* Trademark of Bio-Rad Laboratories.

their ends. It is doubtful whether this is necessary. At the end of the amplification, the reaction can be allowed to cool to room temperature, where it can be kept indefinitely until the tubes are opened. There is little point in cooling to 4°C a reaction that has been heated to 94°C 35 times. *Taq,* like most polymerases, adds an extra A residue to the 3′ ends of amplicons, but does not remove it as it lacks a 3′ → 5′ exonuclease activity. Removal of this nontarget base addition may be necessary for cloning procedures and must be done with another enzyme, e.g., the Klenow fragment of *E. coli* or T4 DNA polymerase, both of which possess a 3′ → 5′ exonuclease activity.

F. KINETICS OF THE REACTION

Early descriptions of PCR often called it an exponential reaction of the type 2^n, where n is the number of cycles, such that after 20 cycles there would be a greater-than-10^6 increase in the target sequence. In practice, this extent of amplification was not found to occur, and the use of 30 to 40 cycles became widespread. The need for the extra cycles was explained by attributing a less-than-100% efficiency to the amplification so that at 80% efficiency only a few more than 10^5 copies of the target are synthesized. This is based on the yield of the reaction being $(1 + e)^n$, where e is the "efficiency" of the reaction from 0 to 1. A typical PCR reaction might be exponential for the first few cycles (for instance, 13) and then reach a plateau or become linear, entering what Mullis has termed the "anemic mode".[28] Several factors contribute to this anemia,[53] the most important being synthesis of nontarget amplicons (e.g., primer dimers or "ugly little fragments"[28]) and competition between them and the intended target for *Taq* and other reaction ingredients. The *Taq* can also become limiting by progressive heat denaturation. In addition, incomplete amplicon denaturation coupled with efficient reannealing of newly synthesized amplicons will interfere with the binding of the primers to single-strand targets.

Quantification of the amount of a viral genome that is present in a clinical specimen (viral load) can be useful for assessing the significance of its presence if it is a common pathogen (e.g., a respiratory virus), and for determining the effect of antiviral therapy, if available (e.g., HCV; herpesviruses; HIV). For reasons of reproducibility and sensitivity, most PCRs were designed to achieve maximum amplification independent of the amount of starting target. Quantification is difficult because of the factors interfering with exponential amplification discussed above, and claims to have achieved it have been looked on with skepticism. The problem of quantification of PCR is compounded by this difficulty in achieving reproducible reaction kinetics, and because a model quantitative PCR behaves differently when confronted with a real clinical specimen, compared with a test or artificial DNA template. There are three main approaches to quantification: a dilution series of the specimen or extracted DNA/RNA to establish a cutoff for amplification, which is assumed to represent one molecule; amplification of an external standard in a separate tube (e.g., HLA or β-globin genes) to determine the reaction rate; and amplification of an internal standard in the same tube (an artificial template, distinguishable from the real one, but resembling it as much as possible) also to determine the reaction rate. This is the quantitative-competitive PCR approach. Ferre has argued that optimization of the reaction coupled with procedures to eliminate nonspecific amplification, such as hot start, boiling the DNA sample for up to 2 h, and/or use of single-stranded binding protein, may effectively allow quantification of any PCR, provided it is adequately validated.[53] Development of a rigorous mathematical model for the enzymology of PCR will help clarify the kinetics of the reaction and how quantification can best be achieved.[9,28,53]

G. CLONING AMPLICONS

PCR amplicons have often proved difficult to clone,[64] because of the 3′ A overhang and because of the presence of large quantities of primers and nucleotides in the amplification mixture. Vectors with a single T overhang have been constructed to take advantage of the A overhang for cloning of amplicons,[65] and are commercially available (e.g., from Invitron, Promega, and Novagen). Otherwise, the amplicon must be purified away from the residual reaction components, blunt-ended, and cloned by standard methods[66] (kits for this are available from several manufactures, including Stratagene and Pharmacia). Alternatively, a polymerase with a 3′ → 5′ exonuclease activity (e.g., Vent) that generates blunt ends can be used.[67] If the primers are modified so that they include a restriction site, the amplicon is digested with the appropriate enzyme and cloned into a similarly digested vector.[68] However, if the restriction site used is too close to the 5′ end of the primer, it will not be cut efficiently by the enzyme. Staggered ends can also be generated by using primers containing uracil residues, which can be digested after amplification by uracil *N*-glycosylase (UNG), leaving the amplicon in a suitable form for direct cloning[69] (kit available from Life Technologies). Amplicons that are to be cloned should come from reactions where only one target sequence was present during amplification, because of the possibility of

DNA recombination during PCR.[62] For example, HIV exists as a quasispecies of genome sequences, and Meyerhans and colleagues[62] observed that recombinants accounted for 5.4% of *tat* gene amplicons after PCR. Such recombinants will give misleading sequence data, and may not maintain an open reading frame.

H. DETECTION OF THE AMPLICON

An aliquot of the reaction mixture is commonly examined by electrophoresis on agarose gels. NuSieve* agarose (a 3 to 4% gel of the 3:1 composite) is very good for resolution of small (less than 1 kb) amplicons. However, the visualization of an ethidium-bromide-stained band of the expected size is not proof that it is the correct target. If there is ever only one band on the gel, it can be relied on diagnostically once it has been verified by hybridization, restriction enzyme digestion, or sequencing. Some workers argue that a nested PCR provides the equivalent specificity of a hybridization step. Certainly nested PCR should reduce the number of any spurious bands, but it is still necessary to ascertain the sequence of a visible band when the PCR is initially set up.

Hybridization may be performed after Southern blotting of a gel on which the PCR products were run, using either a synthetic oligonucleotide probe or a cloned DNA equivalent to the intended amplicon. (Very large cloned probes are sometimes unsuitable for the detection of small amplicons, presumably because of stearic hindrance.) PCR products can also be detected by dot hybridization. Alternatively, liquid or oligomer hybridization of the amplicon and probe can be performed before electrophoresis.[44] This relies on the formation of a triple helix by the probe binding in the major groove of the double helix (Hoogsteen bonding). However, it appears to work much better for some probe/target combinations than for others.

There are several other methods of identifying PCR products. Restriction enzyme digestion or RFLP analysis of the amplicon will allow identification of a specific sequence, provided no mutation or misincorporation by *Taq* has occurred at the restriction site. Direct sequencing is useful when it identifies a product amplified with generic primers (e.g., the 5′ NCR of the picornaviruses) as being from a specific virus type. Sequencing is also necessary for comparison of amplicons for epidemiological purposes.[70] Amplicons can also be compared by denaturing gradient gel electrophoresis (DGGE)[71] or single-stranded conformational analysis (SSCP).[72] For both of these methods, a radioactive tracer may be included in the reaction to label the product. Alternatively, fluorescently labeled primers may be used. DGGE allows comparison of double-stranded products that differ by as little as a single base on a denaturing gradient polyacrylamide gel. A population of amplicons from a heterogeneous target sequence can be compared by this technique by melting and then allowing them to reanneal and form heteroduplexes before electrophoresis. SSCP is a method for comparing the melted single strands of an amplicon on a nondenaturing polyacrylamide gel. The separated single strands will migrate depending on their composition and secondary structure, and products with different sequences (including just a single base change) can be differentiated. This is useful for comparing amplicons from different virus isolates of the same species to determine if they are of identical sequence or if they differ.

Other methods for the diagnostic detection of PCR products have been devised involving capture of the amplicons onto a solid phase via a bound specific oligonucleotide probe.[73–77] If the primers contain a biotin tag, the amplicon can be visualized with a standard streptavidin-peroxidase colorimetric system. This is the type of kit-based format for PCR that has been developed by Roche Diagnostic Systems, as part of an effort to make PCR easier to do and interpret.

I. CONTAMINATION

PCR is sensitive to false-positive results from previous amplified products, from target sequences cloned in plasmid vectors, and from other positive specimens.[78,79] The most problematic source of contamination is from previous amplicons. This can be avoided by physical separation of the stages of the PCR, and by physical and chemical sterilization of the amplicon and associated apparatus and reagents.[80–84] Physical separation will also help reduce specimen cross-contamination. The possibility that specimens for PCR diagnosis have been contaminated before they reach the laboratory should be considered and, if possible, negative control specimens should be collected at the time the test specimens are collected.

On arrival at the testing laboratory, specimens should be stored apart from specimens for serological and other tests. At the PHLS Virus Reference Division in London, physical separation is achieved by having separate laboratories for DNA and RNA extraction; reagent preparation (the clean room);

---

* Trademark of FMC Corporation.

amplification, and setting up of nested PCRs; and postamplification analysis (gel electrophoresis, hybridization, etc.).[78] Each laboratory has its own full set of dedicated apparatus and reagents, including positive-displacement pipettes and laboratory coats. Laminar flow cabinets are used in the clean room and in the amplification room (housing the thermal cyclers). Transfer from the primary (outer) reaction to the secondary (inner) reaction is done in a laminar flow cabinet to minimize the danger of contamination. These cabinets have UV lights inside, which are kept on when the cabinets are not in use, to degrade any spilled DNA.[81] Obviously, contamination is monitored by the inclusion of adequate controls, particularly reagent blanks and sterile water (mock specimen) extracted at the same time as the test sample.

Good microbiological practice in handling reagents and specimens will go a long way towards preventing contamination problems. Apparatus (tube racks, pipettes, etc.) can be sterilized with 1 M HCl[79] or with Clorox;[84] by autoclaving; or by exposure to UV light[81] to γ irradiation.[85]

The most commonly used chemical anticontamination measure is the inclusion of dUTP in the reaction mixture in place of TTP.[80] The resultant amplicons will thus contain uracil instead of thymine and be susceptible to degradation by the enzyme uracil N-glycosylase (UNG). This is included when the reaction mixture is set up and allowed to digest any preexisting amplicons (normal target DNA does not naturally contain uracil and will not be digested). It may be necessary to reoptimize the reaction conditions when dUTP and UNG are used. Although UNG is inhibited at the high temperatures used during PCR, it is necessary to process the reaction immediately after thermal cycling has finished as residual enzyme activity can digest the product before it is analyzed. The commercial kits developed by Roche for the detection of pathogens use the dUTP/UNG system.

III. CHARACTERIZATION OF VIRAL GENOMES

The cloning of the genome of the agent of non-A, non-B hepatitis (HCV) was probably the last great heroic act of traditional recombinant DNA technology achieved completely without the benefit of PCR.[86] More than 1 million clones were screened before a positive was found. Since then, cloning of other isolates of HCV, of hepatitis E, and of Norwalk virus have benefitted from the application of PCR.[87,88] If some sequence information is available, PCR can be used to walk along the genome with one specific primer and a primer complementary to either a sequence that is likely to be present on the genome (e.g., a poly-A tract for a positive strand RNA virus), or with a primer complementary to an added sequence (e.g., a poly-C tail, or a ligated synthetic oligonucleotide linker/adapter).[89–91] If no sequence information is available, a linker/adapter can be ligated onto both ends of the RNA/DNA and used as the amplification primer.[92–94] This is known as sequence-independent single-primer amplification (SISPA) and relies on the target sequence being present, if not abundantly, then at least as a major species in the population of nucleic acids. A primer with a random hexamer sequence at its 3' end can be used in a reverse transcriptase reaction to prepare, from a small quantity of RNA, a representative cDNA population that can then be amplified by PCR.[95] While it may be possible to concentrate viral particles from extracellular sources such as serum or feces by physical separation (e.g., centrifugation), it is not easy to do this for viral particles or sequences present in a background of host cellular DNA. This has become possible through a method, developed by Lisitsyn and colleagues, called representational difference analysis (RDA), for preferentially amplifying a sequence present in one population of DNA fragments but not in another.[96] The "tester" population of molecules contains the target sequence, while the "driver" population lacks it. Both of these DNA populations are digested with a restriction enzyme and then ligated with adapters/linkers that serve as primer sequences for PCR. The resultant amplicons are digested with the same enzyme and then new adapters/linkers are religated onto only the tester population. The driver and tester amplicons are denatured and hybridized together with the driver present in great excess. PCR with the new adapter/linker primer results in preferential amplification of the target sequence. This method may prove to be useful for amplifying and cloning viral sequences present in an infected individual, using driver sequences from an uninfected individual.

IV. DEVELOPMENTS OF PCR

The availability of commercial kits for the detection of viral nucleic acids will extend the use of PCR beyond specialized laboratories, although it has yet to be shown that these kits are as easy to use as, for example, ELISA tests. However, the development of PCR kits is still in its infancy.

A technique for fingerprinting complex genomes has been developed using random or arbitrary primers, and is called arbitrary primed PCR (APPCR) or random amplification of polymorphic DNA

(RAPD).[97–99] Most viral genomes are too small for this strategy to be useful. However it could be applied to the large genomes of the herpesviruses or poxviruses. PCR has been extended into the detection of antigen-antibody complexes (immuno-PCR).[100] An antigen is bound to a solid phase, and antibody is added as if the test were an ELISA. A specially made strepavidin-protein A chimera is then added which will bind the antibody via the protein A component. After washing to remove all nonbound strepavidin-protein A, a biotinylated target DNA is added. After further washing, PCR is used to detect any specifically bound target DNA. As few as 580 molecules of antigen are detectable by this technique.[100] If this proves to be robust, reliable, and reproducible, PCR will enhance existing ELISAs. Tissue collections in pathology laboratories are available for the investigation of diseases that were not understood in their time. PCR has already proved useful in examining archival specimens for the presence of viral DNA,[5] for example, in a possible early AIDS case.[101] Even RNA can be amplified from surgical or archival tissue.[102] Perhaps it will be possible with immuno-PCR to examine such specimens for the few antigen or antibody molecules that may survive the fixation process.

Imitation is the sincerest form of flattery and, like most great and beautiful things, PCR has its rivals: the ligase chain reaction,[103] transcription-based systems,[104–106] and Qβ amplification.[107] These are also being commercialized but, at present, PCR is the amplification method of most scientists' choice. Perhaps in the future PCR will be coupled with nucleic acid extraction protocols and sequencing of amplicons to allow manufacture of an automated diagnostic machine, into which the specimen will be fed and the computer will respond with the nucleic acid sequences present. This does not seem beyond the capabilities of technology currently being developed.[108–110]

REFERENCES

1. **Saiki, R. K., Scharf, S., Faloona, F., Mullis, K. B., Horn, G. T., Erlich, H. A., and Arnheim, N.,** Enzymatic amplification of B-globin genomic sequences and restriction site analysis for diagnosis of sickle cell anaemia, *Science,* 230, 1350, 1985.
2. **Mullis, K. B.,** The unusual origins of the polymerase chain reaction, *Sci. Am.,* 262, 56, 1990.
3. **Livesidge, A.,** Interview: Kary Mullis, *Omni,* 14, 69, 1992.
4. **Kleppe, K., Ohtsuka, E., Kleppe, R., Molineux, I., and Khorana, H. G.,** Studies on polynucleotides XCVI: repair replication of short synthetic DNAs as catalyzed by DNA polymerases, *J. Molec. Biol.,* 56, 341, 1971.
5. **Shibata, D. K., Arnheim, N., and Martin, J. W.,** Detection of human papilloma virus in paraffin-embedded tissue using the polymerase chain reaction, *J. Exp. Med.,* 167, 225, 1988.
6. **Erlich, H. A.,** *PCR Technology: Principles and Applications for DNA Amplification,* Stockton, New York, 1989.
7. **Innes, M. A., Gelfand, D. H., Sninsky, J. J., and White, T. J., Eds.,** *PCR Protocols: A Guide to Methods and Applications,* Academic Press, San Diego, CA, 1990.
8. **Erlich, H. A., Gelfand, D., and Sninsky, J. J.,** Recent advances in the polymerase chain reaction, *Science,* 252, 1643, 1991.
9. **Bloch, W.,** A biochemical perspective of the polymerase chain reaction, *Biochemistry,* 30, 2735, 1991.
10. **McPherson, M. J., Quirke, P., and Taylor, G. R.,** *PCR, A Practical Approach,* IRL Press, Oxford, 1991.
11. **Kocher, T. D. and Wilson, A. C.,** DNA amplification by the polymerase chain reaction, in *Essential Molecular Biology: A Practical Approach,* Vol. 2, Brown, T. A., Ed., IRL Press, Oxford, 1991, chap. 7.
12. **Arnheim, N. and Erlich, H.,** Polymerase chain reaction strategy, *Ann. Rev. Biochem.,* 61, 131, 1992.
13. **Erlich, H. A. and Arnheim, N.,** Genetic analysis using the polymerase chain reaction, *Ann. Rev. Genet.,* 26, 479, 1992.
14. **Wright, P. A. and Wynford-Thomas, D.,** The polymerase chain reaction: miracle or mirage? A critical review of its uses and limitations in diagnosis and research, *J. Pathol.,* 162, 99, 1990.
15. **Kitchin, P. A.,** Clinical applications of the polymerase chain reaction, *Int. J. STD & AIDS,* 1, 161, 1990.
16. **Peter, J. B.,** The polymerase chain reaction: amplifying our options, *Rev. Infect. Dis.,* 13, 166, 1991.
17. **Ehrlich, G. D.,** Caveats of PCR, *Clin. Microbiol. Newsl.,* 13, 19, 1991.
18. **Carman, W. F.,** The polymerase chain reaction, *Q. J. Med. New Series,* 78, 195, 1991.
19. **Hayden, J. D., Ho, S. A., Hawkey, P. M., Taylor, G. R., and Quirke, P.,** The promises and pitfalls of PCR, *Rev. Med. Microbiol.,* 2, 129, 1991.

20. **Persing, D. H.,** Polymerase chain reaction: trenches to benches, *J. Clin. Microbiol.,* 29, 1282, 1991.
21. **Williams, S. D. and Kwok, S.,** Polymerase chain reaction: applications for viral detection, in *Laboratory Diagnosis of Infections,* Lennette, E. H., Ed., Dekker, New York, 1991, p. 147.
22. **White, T. J., Madej, R., and Persing, D.,** The polymerase chain reaction: clinical applications, *Adv. Clin. Chem.,* 29, 161, 1992.
23. **Rapley, R., Theophilus, B. D. M., Bevan, I. S., and Walker, M. R.,** Fundamentals of the polymerase chain reaction: a future in clinical diagnosis?, *Med. Lab. Sci.,* 49, 119, 1992.
24. **Skerra, A.,** Phosphorothioate primers improve the amplification of DNA sequences by DNA polymerases with proofreading activity, *Nucl. Acids Res.,* 14, 3551, 1992.
25. **Kwok, S., Kellogg, D. E., McKinney, N., Spasic, D., Goda, L., Levenson, C., and Sninsky, J. J.,** Effects of primer-template mismatches on the polymerase chain reaction: human immunodeficiency type 1 model studies, *Nucl. Acids Res.,* 18, 999, 1990.
26. **Rychlik, W. and Rhoads, R. E.,** A computer program for choosing optimal oligonucleotides for filter hybridization, sequencing and *in vitro* amplification of DNA, *Nucl. Acids Res.,* 17, 8543, 1989.
27. **Rychlik, W., Spencer, W. J., and Rhoads, R. E.,** Optimization of the annealing temperature for DNA amplification *in vitro, Nucl. Acids Res.,* 18, 6409, 1990.
28. **Mullis, K. B.,** The polymerase chain reaction in an anemic mode: how to avoid cold oligodeoxyribonuclear fusion, *PCR Meth. Appl.,* 1, 1, 1991.
29. **Garson, J. A., Ring, C. J. A., and Tuke, P. W.,** Improvement of HCV genome detection with short PCR products, *Lancet,* 338, 1466, 1991.
30. **Clewley, J. P.,** PCR detection of parvovirus B19, in *Diagnostic Molecular Microbiology,* Persing, D. H., Ed., ASM Press, Washington, D.C., 1993, 367.
31. **Balfe, P., Simmonds, P., Ludlam, C. A., Bishop, J. A., and Leigh-Brown, A. J.,** Concurrent evolution of human immunodeficiency virus type 1 in patients infected from the same source: rate of sequence change and low frequency of inactivating mutations, *J. Virol.,* 64, 6221, 1990.
32. **Ruano, G., Fenton, W., and Kidd, K. K.,** Biphasic amplification of very dilute DNA samples via "booster" PCR, *Nucl. Acids Res.,* 17, 5407, 1989.
33. **Candrian, U., Höfelein, C., and Lüthy, J.,** Polymerase chain reaction with additional primers allows identification of amplified DNA and recognition of specific alleles, *Molec. Cell. Probes,* 6, 13, 1992.
34. **Yourno, J.,** A method for nested PCR with single closed reaction tubes, *PCR Meth. Appl.,* 2, 60, 1992.
35. **Ulrich, P. P., Romeo, J. M., Daniel, L. J., and Vyas, G. N.,** An improved method for the detection of hepatitis C virus RNA in plasma utilizing heminested primers and internal control RNA, *PCR Meth. Appl.,* 2, 241, 1993.
36. **Filichkin, S. A. and Stanton, S. B.,** Effect of dimethyl sulfoxide concentration on specificity of primer matching in PCR, *BioTechniques,* 12, 828, 1992.
37. **Sarkar, G., Kapelner, S., and Sommer, S. S.,** Formamide can dramatically improve the specificity of PCR, *Nucl. Acids Res.,* 18, 7465, 1990.
38. **Hung, T., Mak, K., and Fong, K.,** A specificity enhancer for polymerase chain reaction, *Nucl. Acids Res.,* 18, 4953, 1990.
39. **Chou, Q.,** Minimizing deletion mutagenesis artifact during *Taq* DNA polymerase PCR by *E. coli* SSB, *Nucl. Acids Res.,* 20, 4371, 1992.
40. **Oshima, R. G.,** Single-stranded DNA binding protein facilitates amplification of genomic sequences by PCR, *BioTechniques,* 12, 188, 1992.
41. **D'Aquila, R. T., Bechtel, L. J., Videler, J. A., Eron, J. J., Gorczyca, P., and Kaplan, J. C.,** Maximizing sensitivity and specificity of PCR by pre-amplification heating, *Nucl. Acids Res.,* 19, 3749, 1991.
42. **Chou, Q., Russell, M., Birch, D. E., Raymond, J., and Bloch, W.,** Prevention of pre-PCR mis-priming and primer dimerization improves low-copy-number amplifications, *Nucl. Acids Res.,* 20, 1717, 1992.
43. **Don, R. H., Cox, P. T., Wainwright, B. J., Baker, K., and Mattick, J. S.,** Touchdown PCR to circumvent spurious priming during gene amplification, *Nucl. Acids Res.,* 19, 4008, 1991.
44. **Gibson, K. M., McLean, K. A., and Clewley, J. P.,** A simple and rapid method for detecting human immunodeficiency virus by PCR, *J. Virol. Meth.,* 32, 277, 1991.
45. **Frickhofen, N. and Young, N. S.,** A rapid method of sample preparation for detection of DNA viruses in human serum by polymerase chain reaction, *J. Virol. Meth.,* 35, 65, 1991.
46. **Welsh, P. S., Metzger, D. A., and Higuchi, R.,** Chelex 100 as a medium for simple extraction of DNA for PCR-based typing from forensic material, *BioTechniques,* 10, 506, 1991.

47. **Boom, R., Sol, C. J. A., Salimans, M. M. M., Larsen, C. L., Wertheim-van Dillen, P. M. E., and Van der Noorda, J.,** Rapid and simple method for purification of nucleic acids, *J. Clin. Microbiol.,* 28, 495, 1990.

48. **Boom, R., Sol, C. J. A., Salimans, M. M. M., Heijtink, R., Wertheim-van Dillen, P. M. E., and Van der Noorda, J.,** Rapid purification of hepatitis B virus DNA from serum, *J. Clin. Microbiol.,* 29, 1804, 1991.

49. **Pikaart, M. J. and Villeponteau, B.,** Suppression of PCR amplification by high levels of RNA, *BioTechniques,* 14, 24, 1993.

50. **Sharma, J. K., Gopalkrishna, V., and Das, B. C.,** A simple method for elimination of unspecific amplifications in polymerase chain reaction, *Nucl. Acids Res.,* 20, 6117, 1992.

51. **Kureishi, A. and Bryan, L. E.,** Pre-boiling high GC content, mixed primers with 3 complementation allows the successful PCR amplification of *Pseudomonas aeruginosa* DNA, *Nucl. Acids Res.,* 20, 1155, 1992.

52. **Ruano, G., Pagliaro, E. M., Schwartz, T. R., Lamy, K., Messlina, D., Gaensslen, R. E., and Lee, H. C.,** Heat-soaked PCR: an efficient method for DNA amplification with applications to forensic analysis, *BioTechniques,* 13, 266, 1992.

53. **Ferre, F.,** Quantitative or semi-quantitative PCR: reality versus myth, *PCR Meth. Appl.,* 2, 1, 1992.

54. **Gibson, K. M., Mori, J., and Clewley, J. P.,** Detection of HIV-1 in serum, using reverse transcription and the polymerase chain reaction (RT-PCR), *J. Virol. Meth.,* 43, 101, 1993.

55. **Wang, R.-F., Cao, W.-W., and Johnson, M. G.,** A simplified, single tube, single buffer system for RNA-PCR, *BioTechniques,* 12, 702, 1992.

56. **Sellner, L. N., Coelen, R. J., and Mackenzie, J. S.,** Reverse transcriptase inhibits Taq polymerase activity, *Nucl. Acids Res.,* 20, 1487, 1992.

57. **Myers, T. W. and Gelfand, D. H.,** Reverse transcription and DNA amplification by a *Thermus thermophilus* DNA polymerase, *Biochemistry,* 30, 7661, 1991.

58. **Yap, E. P. H. and McGee, J. O'D.,** Short PCR product yields improved by lower denaturation temperatures, *Nucl. Acids Res.,* 19, 1713, 1991.

59. **Mack, D. H. and Sninsky, J. J.,** A sensitive method for the identification of uncharacterized viruses related to known virus groups: Hepadnavirus model system, *Proc. Natl. Acad. Sci. U.S.A.,* 85, 6977, 1988.

60. **Wichman, H. A. and Van Den Bussche, R. A.,** In search of retrotransposons: exploring the potential of PCR, *BioTechniques,* 13, 258, 1992.

61. **Ohler, L. D. and Rose, E. A.,** Optimization of long-distance PCR using a transposon-based model system, *PCR Meth. Appl.,* 2, 51, 1992.

62. **Meyerhans, A., Vartanian, J. P., and Wain-Hobson, S.,** DNA recombination during PCR, *Nucl. Acids Res.,* 18, 1687, 1990.

63. **Bell, D. A. and DeMarini, D. M.,** Excessive cycling converts PCR products to random-length higher molecular weight fragments, *Nucl. Acids Res.,* 19, 5079, 1991.

64. **Sardelli, A. D.,** Cloning PCR products, *Amplifications,* 6, 10, 1991.

65. **Mead, D. A., Pey, N. K., Herrnstadt, C., Marcil, R. A., Smith, L. M.,** A universal method for the direct cloning of PCR amplified nucleic acid, *BioTechnology,* 9, 657, 1991.

66. **Hemsley, A., Arnheim, N., Toney, M. D., Cortopassi, G., and Galas, D. J.,** A simple method for site-directed mutagenesis using the polymerase chain reaction, *Nucl. Acids Res.,* 17, 6545, 1989.

67. **Lohff, C. J. and Cease, K. B.,** PCR using a thermostable polymerase with 3 to 5 exonuclease activity generates blunt end products suitable for direct cloning, *Nucl. Acids Res.,* 20, 144, 1992.

68. **Scharf, S. J., Horn, G. T., and Erlich, H. A.,** Direct cloning and sequence analysis of enzymatically amplified genomic sequences, *Science,* 223, 1076, 1986.

69. **Rashtchian, A., Buchman, G. W., Schuster, D. M., and Berninger, M. S.,** Uracil DNA glycosylase-mediated cloning of polymerase chain reaction-amplified DNA: application to genomic and cDNA cloning, *Anal. Biochem.,* 206, 91, 1992.

70. **Ou, C.-Y., Ciesielski, C. A., Myers, G., Bandea, C. I., Luo, C.-C., Korber, B. T. M., Mullins, J. I., Schochetman, G., Berkelman, R. L., Economou, A. N., Witte, J. J., Furman, L. J., Satten, G. A., MacInnes, K. A., Curran, J. W., and Jaffe, H. W.,** Molecular epidemiology of HIV transmission in a dental practice, *Science,* 256, 1165, 1992.

71. **Ruano, G. and Kidd, K. K.,** Modeling of heteroduplex formation during PCR from mixtures of DNA templates, *PCR Meth. Appl.,* 2, 112, 1992.

72. **Makino, R., Yazyu, H., Kishimoto, K., Sekiya, T., and Hayashi, K.,** F-SSCP: fluorescence-based polymerase chain reaction-single strand conformation polymorphism (PCR-SSCP) analysis, *PCR Meth. Appl.,* 2, 10, 1992.

73. **Keller, G. H., Huang, D.-P., and Manak, M. M.,** A sensitive nonisotopic hybridization assay for HIV-1 DNA, *Anal. Biochem.,* 177, 27, 1989.

74. **Kemp, D. J., Smith, D. B., Foote, S. J., Samaras, N., and Peterson, M. G.,** Colorimetric detection of specific DNA segments amplified by polymerase chain reaction, *Proc. Natl. Sci. U.S.A.,* 86, 2423, 1989.

75. **Inouye, S. and Hondo, R.,** Microplate hybridization of amplified viral DNA segment, *J. Clin. Microbiol.,* 28, 1469, 1990.

76. **Keller, G. H., Huang, D.-P., and Manak, M. M.,** Detection of human immunodeficiency virus type 1 DNA by polymerase chain reaction amplification and capture hybridization in microtitre wells, *J. Clin. Microbiol.,* 29, 638, 1991.

77. **Yolken, R. H., Sierra-Honigmann, A. M., and Viscidi, R. P.,** Solid phase capture methods for the specific amplification of microbial nucleic acids-avoidance of false-positive and false-negative reactions, *Molec. Cell. Probes,* 5, 51, 1991.

78. **Clewley, J. P.,** The polymerase chain reaction, a review of the practical limitations for human immunodeficiency virus diagnosis, *J. Virol. Meth.,* 25, 179, 1989.

79. **Higuchi, R. and Kwok, S.,** Avoiding false positives with PCR, *Nature,* 339, 237, 1989.

80. **Longo, M. C., Berninger, M. S., and Hartley, J. L.,** Use of uracil DNA glycosylase to control carry-over contamination in polymerase chain reactions, *Gene,* 93, 25, 1990.

81. **Sarkar, G. and Sommer, S. E.,** Shedding light on PCR contamination, *Nature,* 343, 27, 1990.

82. **Zhu, Y. S., Isaacs, S. T., Cimino, G. D., and Hearst, J. E.,** The use of exonuclease III for polymerase chain reaction sterilization, *Nucl. Acids Res.,* 19, 2511, 1991.

83. **Cimino, G. D., Metchette, K. C., Tessman, J. W., Hearst, J. E., and Isaacs, S. T.,** Post-PCR sterilization: a method to control carryover contamination for the polymerase chain reaction, *Nucl. Acids Res.,* 19, 99, 1991.

84. **Prince, A. M. and Andrus, L.,** PCR: how to kill unwanted DNA, *BioTechniques,* 12, 358, 1992.

85. **Deragon, J.-M., Sinnett, D., Mitchell, G., Potier, M., and Labuda, D.,** Use of γ irradiation to eliminate DNA contamination for PCR, *Nucl. Acids Res.,* 18, 6149, 1990.

86. **Choo, Q.-L., Kuo, G., Weiner, A., Overby, L. R., Bradley, D. W., and Houghton, M.,** Isolation of a cDNA clone derived from a blood-borne non-A, non-B viral hepatitis genome, *Science,* 244, 359, 1989.

87. **Tam, A. W., Smith, M. M., Guerra, M. E., Huang, C.-C., Bradley, D. W., Fry, K. E., and Reyes, G. R.,** Hepatitis E virus (HEV): molecular cloning and sequencing of the full-length viral genome, *Virology,* 185, 120, 1991.

88. **Matsui, S. M., Kim, J. P., Greenberg, H. B., Su, W., Sun, Q., Johnson, P. C., DuPont, H. L., Oshiro, L. S., and Reyes, G. R.,** The isolation and characterization of a Norwalk virus-specific cDNA, *J. Clin. Invest.,* 87, 1456, 1991.

89. **Parks, C. L., Chang, L.-S., and Shenk, T.,** A polymerase chain reaction mediated by a single primer: cloning of genomic sequences adjacent to a serotonin receptor protein coding region, *Nucl. Acids Res.,* 19, 7155, 1991.

90. **Kriangkum, J., Vainshtein, I., and Elliott, J. F.,** A reliable method for amplifying cDNA using the anchored-polymerase chain reaction (A-PCR), *Nucl. Acids Res.,* 20, 3793, 1992.

91. **Williams, W. V., Sato, A., Rossman, M., Fang, Q., and Weiner, D. B.,** Specific DNA amplification utilizing the polymerase chain reaction and random oligonucleotide primers: application to the analysis of antigen receptor variable regions, *DNA and Cell Biol.,* 11, 707, 1992.

92. **Reyes, G. R. and Kim, J. P.,** Sequence-independent, single-primer amplification (SISPA) of complex DNA populations, *Mol. Cell. Probes,* 5, 473, 1991.

93. **Collasius, M., Puchta, H., Schlenker, S., and Valet, G.,** Analysis of unknown DNA sequences by polymerase chain reaction (PCR) using a single specific primer and a standardized adaptor, *J. Virol. Meth.,* 32, 115, 1991.

94. **Lambden, P. R., Cooke, S. J., Caul, E. O., and Clarke, I. N.,** Cloning of noncultivatable human rotavirus by single primer amplification, *J. Virol.,* 66, 1817, 1992.

95. **Froussard, P.,** rPCR: a powerful tool for random amplification of whole RNA sequences, *PCR Meth. Appl.,* 2, 185, 1993.

96. **Lisitsyn, N., Lisitsyn, N., and Wigler, M.,** Cloning the differences between two complex genomes, *Science,* 259, 946, 1993.

97. **Welsh, J. and McClelland, M.,** Genetic fingerprinting using arbitrary primed PCR and a matrix of pairwise combinations of primers, *Nucl. Acids Res.,* 19, 5275.

98. **Welsh, J., Chada, K., Dalal, S. S., Cheng, R., Ralph, D., and McClelland, M.,** Arbitrary primed PCR fingerprinting of RNA, *Nucl. Acids Res.,* 20, 4965, 1992.

99. **Williams, J. G. K., Kubelik, A. R., Livak, K. J., Rafalski, J. A., and Tingey, S. V.,** DNA polymorphisms amplified by arbitrary primers are useful as genetic markers, *Nucl. Acids Res.,* 18, 6531, 1990.

100. **Sano, T., Smith, C. L., and Cantor, C. R.,** Immuno-PCR: very sensitive antigen detection by means of specific antibody-DNA conjugates, *Science,* 258, 120, 1992.

101. **Corbitt, G., Bailey, A. S., and Williams, G.,** HIV infection in Manchester, 1959, *Lancet,* 336, 51, 1990.

102. **Stanta, G. and Schneider, C.,** RNA extracted from paraffin-embedded human tissues is amenable to analysis by PCR amplification, *BioTechniques,* 11, 304, 1991.

103. **Birkenmeyer, L. G. and Mushahwar, I. K.,** DNA probe amplification methods, *J. Virol. Meth.,* 35, 117, 1991.

104. **Kwoh, D. Y., Davis, G. R., Whitfield, H. L., Chappelle, L., DiMichele, L. J., and Gingeras, T. R.,** Transcription-based amplification system and detection of amplified human immunodeficiency virus type 1 with a bead-based sandwich hybridization format, *Proc. Natl. Acad. Sci. U.S.A.,* 86, 1173, 1989.

105. **Kievets, T., van Gemen, B., van Strijp, D., Schukknk, R., Dircks, M., Adriaanse, H., Malek, L., Sookman, R., and Lens, P.,** NASBA™ isothermal enzymatic *in vitro* nucleic acid amplification optimized for the diagnosis of HIV-1 infection, *J. Virol. Meth.,* 35, 273, 1991.

106. **Walker, G. T., Little, M. C., Nadeau, J. G., and Shank, D. D.,** Isothermal *in vitro* amplification of DNA by a restriction enzyme/DNA polymerase system, *Proc. Natl. Acad. Sci. U.S.A.,* 89, 392, 1992.

107. **Lizardi, P. M., Guerra, C. E., Lomeli, H., Tussie-Luna, I., and Kramer, F. R.,** Exponential amplification of recombinant-RNA hybridisation probes, *BioTechnology,* 6, 1197, 1988.

108. **Blanchard, M. M., Taillon-Miller, P., Nowotny, P., and Nowotny, V.,** PCR buffer optimization with uniform temperature regimen to facilitate automation, *PCR Meth. Appl.,* 2, 234, 1993.

109. **Harrison, D., Baldwin, C., and Procock, D. J.,** Use of an automated workstation to facilitate PCR amplification, loading agarose gels and sequencing of DNA templates, *BioTechniques,* 14, 88, 1993.

110. **Garner, H. R., Armstrong, B., and Lininger, D. M.,** High-throughput PCR, *BioTechniques,* 14, 112, 1993.

Chapter 2

The Current Role of PCR in Diagnostic and Public Health Virology

David W. G. Brown

TABLE OF CONTENTS

I. INTRODUCTION

Diagnostic virology is a rapidly evolving field.[1] For many years the most important component of public health virology was surveillance of infection in the community and within hospitals. This has involved a continuously changing pattern of work. Traditional functions, such as monitoring the antigenic changes of influenza viruses, remain important to enable rational choice of vaccine strains. The development of simple serological techniques has facilitated more precise monitoring of the immune status of the population. Of most significance has been the recognition of "new" virus infections such as human immunodeficiency virus (HIV). This has led to the need to develop novel epidemiological approaches to monitor spread of infection in the community.

The advent of effective antiviral therapy with the discovery of acyclovir in the 1970s has had a dramatic impact on the practice of clinical virology. Several effective antivirals are now in use: zidovudine for HIV, ribavirin for Lassa fever, and acyclovir and gancyclovir for the herpes viruses. Management of the immunocompromised patient, in whom many viruses change their normal pattern of infection and pathogenesis, is now a major focus of work.

These developments have highlighted the need for rapid, accurate viral diagnosis, and this is highly dependent on laboratory investigation. Although a few virus infections, such as measles when infection was common, have a characteristic clinical presentation that makes them reliably recognizable, the majority of viruses produce clinical signs and symptoms that are much less specific and that can be attributable to several different pathogens.

The pace of development of diagnostic virology and the linking of viruses to disease have been largely determined by the techniques available. Each new technical innovation has revealed novel viruses, and in turn each technique has established a place in the clinical diagnostic laboratory. In the 1930s influenza viruses and arboviruses were discovered by inoculation of mice and embryonated eggs. In the 1950s the development of tissue culture techniques led to the description of enteroviruses, adenoviruses, and respiratory syncitial virus. The presence of Norwalk virus, rotaviruses, and astrovirus were revealed by electron microscopy during the 1970s. In recent years molecular techniques have played an increasing role in virus identification. Examples such as the discovery of picobirnaviruses by the visualization of their nucleic acid on polyacrylamide gel electrophoresis (PAGE)[2] and of hepatitis C virus (HCV) by direct cloning of virus nucleic acid from bile illustrate these developments.[3] Indeed HCV has yet to be visualized or cultured, although diagnostic kits are commercially available.

There is no doubt that PCR will play an important role in the discovery and characterization of new viruses, through its use in techniques such as sequence-independent single-primer amplification (SISPA).[4] The diverse role that PCR is playing in research into many aspects of virus infections is fully reviewed in the specific chapters in this book. This contribution will describe some current indications for PCR in diagnosis, and discuss the factors likely to influence future applications of PCR in clinical diagnosis and public health virology.

II. PCR IN VIRAL DIAGNOSIS

A. GENERAL CONSIDERATIONS

Two broad approaches are used to confirm the presence of a virus infection: the detection of an immune response to infection, and direct demonstration of the virus. The most widely used approaches are summarized in Table 2–1. For a new technique, such as PCR, it often takes several years to establish clear indications for its use in the clinical laboratory. Several considerations determine the current and future range of applications for PCR. These include the severity of infection; the availability of effective treatment; and the reliability, sensitivity, specificity, cost, and availability of alternative diagnostic approaches.

B. SENSITIVITY

PCR is of high sensitivity. Detection levels of 1 to 10 genome copies of DNA viruses have been reported using appropriate primer pairs. However, in clinical specimens, the need to extract nucleic acid from complex mixtures (e.g., feces), and the constraints on sensitivity imposed by the need to design specific primers and the use of a reverse transcriptase step for RNA viruses, contribute to a lower sensitivity in practice. Despite this, in many situations PCR is the most sensitive method of virus detection.

To confound the situation the sensitivity of PCR can be a problem, because interpreting the significance of a positive PCR result that cannot be confirmed by another technique is often difficult. This dilemma is well illustrated in studies of genital human papillomavirus infections, where extremely high prevalence rates have been detected by PCR. Because of the sensitivity of PCR it has proved difficult to distinguish between clinically manifest infection, subclinical infection, and latent infection. The development of quantitative PCR[5,6] and detailed longitudinal studies of individual virus infections should help to provide a framework for interpreting PCR results.

Table 2–1. **Current viral diagnosis**

Technique	Time Required to Diagnosis
Virus Detection	
Virus isolation in tissue culture	1–7 Days
Visualization of virus particles by electron microscopy	Hours
Virus antigen detection	Hours
Viral genome detection	1–2 Days
Gel electrophoresis	1–2 Days
Hybridization	1–2 Days
PCR	6 Hours–Days
Detection of viral enzyme activity	Days
Serology	
Virus-specific IgG antibody rise	2 Weeks
Virus-specific IgM	Hours–Days

C. SPECIFICITY

In theory, PCR amplification is highly specific, although this does depend on the primers chosen. However, the extreme sensitivity of the technique has led to specificity problems due to cross contamination of samples. Attempts to contain carryover contamination have focused on the physical separation of the different stages of the reaction and on chemical anticontamination strategies. The importance of designating specific areas for different steps of the test and of including appropriate controls have been previously reviewed.[7] The specificity of a visible PCR product also requires confirmation because with certain primers, and particularly with some specimens (such as tissue samples that contain large quantities of cellular DNA), multiple bands are sometimes seen on gels. Many groups incorporate procedures in their diagnostic PCR tests to confirm the specificity of the amplified product. These include nested PCR and various types of hybridization assays. Except in very unambiguous circumstances, the presence of a visible band on an ethidium bromide stained gel is unlikely to be a satisfactory indicator of the diagnostic presence of viral DNA in a specimen from a patient, as physicians may wish to base their treatment of the patient on the result *(vide infra)*.

III. SPECIFIC APPLICATIONS FOR DIAGNOSIS

Currently, virus diagnosis is based on the detection of virus directly in body fluids or on the detection of a specific immune response (Table 2–1). Potentially, PCR could replace all the methods used for direct virus analysis because it offers a rapid, specific, and sensitive approach to virus detection. However there are at present relatively few clear indications for PCR in the diagnostic laboratory, mainly owing to the cost and complexity of current PCR assays. Nonetheless, its role in investigating neurological disease due to herpes simplex virus (HSV) and JC virus is already established. It has also been widely used to detect HCV. Further, it is likely to find a diagnostic role in severe virus infections for which no culture system exists (hepatitis B virus; HBV) or for which virus culture is difficult (HIV) or hazardous (Lassa fever).

A. HERPES SIMPLEX ENCEPHALITIS (HSE)

HSE is a rare and devastating illness that is amenable to early treatment with acyclovir. Rapid diagnosis is thus desirable. HSE is generally associated with virus reactivation and hence, in most cases, virus cannot be cultured from cerebrospinal fluid samples (CSF). In the past, definitive diagnosis of HSE has required the identification of virus directly in a brain biopsy specimen. This approach to diagnosis has been adopted in the U.S. but has been little used in the U.K. Until the advent of PCR, the detection of intrathecal HSV antibody in CSF samples provided the best hope for early diagnosis.[8] However intrathecal antibody is detected in only a minority of cases during the first 5 days of infection and is not reliably detected until the second week of infection. As a consequence, in many cases, a therapeutic trial of acyclovir is often undertaken without an established diagnosis. Several groups have now used PCR for detecting HSV in CSF specimens. In one study HSV sequences were successfully amplified in CSF samples collected from 42/43 biopsy-proven cases using a nested PCR.[9] Virus was almost always detected in the first CSF sample collected and was reliably amplified for up to 2 weeks after onset of illness, even in the presence of intrathecal antibody. Another study, using a PCR with Southern blot confirmation, was able to confirm HSE infection in all 28 cases studied.[10] Virus was detected for up to 15 days after the onset of illness, which suggested a possible role for PCR in monitoring virus clearance and determining the appropriate length of therapy. PCR is now established as the first-line diagnostic test in suspected HSE. However, the need for extreme care in specimen collection, the need for repeat testing in negative cases, and the theoretical problem of CNS latency, leading to false-positive results, have been highlighted by several authors.

B. PROGRESSIVE MULTIFOCAL LEUKOENCEPHALOPATHY (PML)

PML is a demyelinating disease of the brain caused by JC virus, in which a lytic infection of the oligodendrocytes occurs. The disease has been identified only in the immunocompromised. Common presenting features include dementia, hemiparesis, or hamanopia. Prognosis is very poor, with an average survival time of 9 months. PML is a rare disease, although the incidence is rising as an increasing number of cases are identified in acquired immunodeficiency syndrome (AIDS) patients.[11]

The detection of characteristic histopathological changes, and direct detection of JC virus by immunofluorescence, or by hybridization in a brain biopsy specimen, are still the definitive diagnostic approaches. PCR has been used to detect JC virus directly in brain biopsy material. However its significance is difficult to interpret, since JC virus sequences have also been demonstrated in "healthy" adult brains by PCR.[12,13] JC viral DNA has also been demonstrated by PCR in peripheral blood lymphocytes, and in CSF specimens from PML cases. In a recent retrospective study JC viral sequences were detected by PCR in the CSF samples collected from the majority of PML cases, but not from controls.[14] Prospective studies are required to establish the true sensitivity and specificity of PCR for detecting JC virus in the CSF of PML cases. Despite this caveat, PCR is likely to have an important role in the diagnosis of PML.

C. HEPATITIS C VIRUS (HCV)
Hepatitis C is a newly characterized positive-strand RNA virus that is related to the flaviviruses and pestiviruses. Acute hepatitis C infection is clinically mild compared to hepatitis A or B and may pass unrecognized, particularly as a sporadic infection; 20 to 50% of acute infections progress to chronic liver disease. The diagnosis of acute infection currently depends on the demonstration of seroconversion with the production of anti-HCV IgG, which can be delayed until convalescence. PCR has been used to confirm acute HCV infection in this diagnostic window and to confirm the specificity of the serological assays in both acute and chronic cases.[15]

D. HUMAN IMMUNODEFICIENCY VIRUS (HIV)
PCR has found a particular role in the early detection of HIV infection in newborns and infants. Standard serological diagnosis is compromised in these cases, because the maternal HIV antibody, passively transferred to the infant, is detectable up to 10 months of age. Anti-HIV IgM and IgA antibody detected in infants' sera are specific markers of infection, but these tests are not always reliable. PCR has been evaluated as an alternative to virus culture in these cases, and the diagnosis of HIV can be confirmed in 90 to 100% of infants over 6 months of age.[16,17] In younger infants (less than 3 months) the value of PCR has yet to be established because the low quantity of virus present makes detection unreliable.

E. HUMAN CYTOMEGALOVIRUS (HCMV)
Human cytomegalovirus is a common human infection that frequently causes severe and life-threatening disease in immunocompromised individuals. Effective antiviral therapy is available, but the drugs (gancyclovir and foscarnet) have toxic side effects. Thus rapid and specific diagnosis is essential for the rational use of therapy. The most useful marker of disseminated infection has been the detection of HCMV in peripheral blood leukocytes (PBL).[18] The role of PCR for detecting HCMV in PBLs has been investigated by several groups, but the value of this approach for predicting severe disease is limited by the frequency with which positive PCR signals are detected from PBLs without correlation with HCMV disease.[19,20] A more recent study found a good correlation between HCMV DNA detection in plasma by PCR and active HCMV disease in AIDS patients, suggesting that PCR will become a useful test.[21] Quantitation of virus by PCR may be important in this context.

F. LASSA FEVER
Lassa fever is an endemic infection in several parts of West Africa. It is estimated that there are between 200 and 400,000 cases each year, with an overall mortality rate of 3%, which rises to 16% in more severe hospitalized cases.[22,23] Diagnostic work on this virus is limited to high-containment laboratories in the U.S. and Europe, because it has been responsible for significant hospital outbreaks involving several generations of human-to-human transmission. The early clinical diagnosis of Lassa fever is difficult, since patients present with nonspecific symptoms such as sore throat and fever. Rapid diagnosis is important for public health reasons and because early treatment with ribavirin is effective.[24] Currently, diagnosis is based on detecting specific antibody responses, which are often delayed until the second week of illness. Virus culture is also used, but this is hazardous. A PCR has been developed for Lassa fever and shown to be of similar sensitivity to virus isolation for diagnosis.[25] A major advantage of PCR over virus culture is that the viral nucleic acid extraction procedures used for PCR (such as boiling or using guanidium thiocyanate) inactivate viral infectivity and facilitate diagnostic work outside the high-containment laboratory.

IV. PCR FOR MONITORING ANTIVIRAL THERAPY

Effective antiviral treatment is now available for the herpes viruses, HIV, and hepatitis B and C virus infections. There have been several studies investigating the use of PCR for evaluating antiviral therapy. In non-A, non-B hepatitis, interferon treatment produces an improvement in liver disease in some patients.[26] In a study of interferon treatment of HCV infection using quantitative PCR, a clear correlation was demonstrated between the clinical response to interferon and a decline in circulating HCV RNA titers.[27] In a subsequent study the same group showed that PCR was a sensitive indicator of virological relapse on completion of treatment and suggested that PCR will be a valuable tool for monitoring the effects of drug treatment in HCV infections.[28]

A potential role for PCR in evaluating CMV treatment was suggested by a study of patients following allogenic bone-marrow transplantation. In a group of 15 patients, 11 were found to respond to antiviral treatment, and CMV DNA was not detected by PCR in blood and urine samples collected from all of these 11 cases.[29] In contrast, CMV DNA was detected by PCR in patients who did not respond to therapy, and in those who suffered early relapse or who died. The authors suggested that PCR was a better predictor of the efficacy of antiviral therapy than virus culture. Other studies have produced less clearcut results, but it seems likely that the quantitation of CMV DNA by PCR will find an important role in the management of CMV infection in the immunocompromised.

Treatment with zidovudine has been shown to extend life expectancy and to lower the frequency and severity of opportunistic infections in AIDS patients. The development of zidovudine-resistant HIV strains has been well-documented, but it is technically demanding to study these changes using culture-grown virus. In a recent PCR study, the rate of development of a specific mutation in the reverse transcriptase gene was linked to a reduced sensitivity of HIV isolates to zidovudine. The mutation developed rapidly, and 16/18 isolates made from patients after 2 years of treatment contained the mutation.[30] A more extensive investigation, using PCR-based methods to determine the prevalence of several specific mutations to zidovudine in the reverse transcriptase gene, showed that the mutations tended to appear in a particular order over time, but that there was not a clear correlation between sensitivity of HIV isolates to zidovudine and the presence of these mutations.[31] These research studies indicate how the development of simpler PCR-based methods for monitoring antiviral resistance during treatment holds great potential for patient management.

V. MOLECULAR EPIDEMIOLOGY

The study of epidemiology has contributed immensely to our understanding of the nature and significance of individual viral infections. It has played an important role in defining the broad patterns of infection and also, on a more practical level by demonstrating the important routes of transmission, has enabled the development of specific cross-infection-control measures.

The development of PCR, and of simplified DNA sequencing methods, has facilitated studies of the molecular epidemiology of several virus infections, which were previously difficult to investigate. Many recently published investigations have used a combination of PCR and sequencing to determine the extent of viral genome variation. However, such methods are not currently suitable for large-scale use. The development of simpler rapid methods such as PCR-restriction endonuclease (PCR-RE) analysis[32] and PCR-single-strand-conformation polymorphism (PCR-SSCP)[33] may lead to wider appreciation of the potential of molecular epidemiology for investigating, for example, the routes of transmission of viruses.

There has been particular interest in determining if transmission of HIV occurs between medical staff and patients. The case of the Florida dentist, who was shown to have infected several of his patients, illustrates how sequencing can be used to establish relationships between HIV isolates.[34] However the extreme variability of the HIV genome complicates the analysis of these investigations and also means that they are too expensive to be of widespread use.

Hepatitis B virus (HBV) is a clinically important nosocomial infection. Extensive infection-control measures have been developed to limit its spread within hospitals.[35] HBV has been difficult to type because it does not grow in tissue culture and because only a limited number of antigenic types have been identified. A recent study using PCR sequencing described the sequence polymorphism of HBV strains within several families in Hong Kong, and documented transmission relationships between family members.[36] The description of PCR-RE techniques[32] for HBV should lead to larger-scale studies, which may help to define the routes of transmission between cases in various clinical and epidemiological settings.

PCR may also replace some more traditional typing methods. The epidemiology of group A rotavirus VP7 serotypes has been described using immunoassays based on serotype-specific monoclonal antibodies.[37] This technique requires intact virus particles, and a number of studies have shown that this limits the proportion of strains that can be typed.[38] A PCR technique has been developed using primers based on conserved sequences in the VP7 gene. This gives virus-typing results that correlate well with traditional serotyping.[39] The application of this PCR to clinical specimens has enabled a high proportion of strains to be successfully characterized, and should contribute to a fuller understanding of rotavirus diversity.

The Sabin oral poliovirus vaccine is a live attenuated virus vaccine. It has proved effective in controlling poliomyelitis in many parts of the world. Although it has a very good safety record, rare vaccine-associated cases of poliomyelitis occur. Typing of poliovirus isolates as vaccine-related or wild strains is a critical component of ongoing surveillance programs. At present serological typing of strains using cross-adsorbed polyclonal sera or monoclonal antibodies are the most widely used methods for poliovirus strain characterization.[40] A PCR-RE has been described that can distinguish vaccine strains from circulating wild poliovirus strains,[41] but it remains to be established if this particular PCR approach will be sufficiently reliable to replace conventional serological typing methods. However, it is undoubtedly clear that PCR has much to contribute to the typing of wild-type and vaccine-derived poliovirus strains both from cases of acute flaccid paralysis and in the environment.

VI. DETECTION OF VIRUSES IN BIOLOGICAL PRODUCTS

The transmission of infections such as HIV and the hepatitis viruses by blood products is well established. A number of approaches, including careful donor selection, virological screening of donations, and inactivation of virus in blood concentrates, are used to limit the infectious risks of transfusions. Anti-HCV testing of blood donors has recently been introduced in many areas of the world in order to further control posttransfusion hepatitis. The early serological assays used to screen for anti-HCV antibody were based on a recombinant antigen (C-100), and these gave a significant number of false-positive results. An important role for PCR was established in defining the "infectivity" of individual donations.[42] However, the subsequent development of more specific antibody assays such as immune blotting tests (RIBAs), in which specific antibody detection correlates well with virus detection by PCR, means that PCR may play a less direct diagnostic role in these aspects of HCV control.[43,44] It will, though, continue to be an important reference test.

PCR has also been used to demonstrate both HCV RNA[45] and parvovirus B19 DNA[46] in factor VIII and IX concentrates, confirming these blood products as sources of infection. Although the correlation between PCR positivity and infectivity is not direct, a screening program has been proposed for blood products using parvovirus B19 PCR.[47]

All biological products are potentially hazardous and are required to be checked before use. In the case of live poliovirus vaccine, intraspinal inoculation of cynomologus monkeys is still the definitive test to confirm that a vaccine preparation is fully attenuated. The recent description of a PCR technique that was able to detect a small number of strains carrying a specific mutation within a batch of vaccine[48] points to a way in which PCR could be used for screening poliovirus vaccine preparations. It is easy to see how PCR could have a wide role in direct screening of biological products. An important future role for PCR in the detection of viral nucleic acid directly in food and in the environment is also likely.[49]

VII. CONCLUSIONS

PCR is already established as an important technique in viral diagnosis, although at present it is indicated for diagnosis in relatively few clinical situations. The future scale of use of PCR in clinical diagnostic laboratories is likely to increase and will be influenced by the availability of antiviral therapy and alternative diagnostic techniques. It will depend on the development of simple, quantitative, automated systems for testing specimens. As mentioned, PCR will have an important future role in guiding the use of antiviral therapy for viral infections.

In public health virology PCR will probably become an important technique for direct typing of virus strains. This will facilitate detailed transmission studies of several severe virus infections. In the future, it may well prove to be an important technique for screening biological products, and may be developed to allow the direct detection of viruses in environmental samples.

REFERENCES

1. **Desselberger, U. and Flewett, T. H.,** Clinical and public health virology: a continuous task of changing pattern progress, in *Medical Virology,* Melnick, J. L., Ed., Karger, Basel, 1993, 40, 48–81.

2. **Pereira, H. G., Flewett, T. H., Candeias, J. A., and Barth, O. M.,** A virus with a bisegmented double-stranded RNA genome in rat *(Oryzomys nigripes)* intestines, *J. Gen. Virol.,* 69, 2749–2754, 1988.

3. **Choo, Q. L., Kuo, G., Weiner, A. J., Overby, L. R., Bradley, D. W., and Houghton, M.,** Isolation of a cDNA clone derived from a blood borne non-A non-B viral hepatitis genome, *Science,* 244, 359–362, 1989.

4. **Reyes, G. R. and Kim, J. P.,** Sequence-independent, single primer amplification (SISPA) of complex DNA populations, *Mol. Cell. Probes,* 5, 473–481, 1991.

5. **Gilliland, G., Perrin, S., Blanchard, K., and Franklin Bunn, H.,** Analysis of cytokine mRNA and DNA: detection and quantitation by competitive polymerase chain reaction, *Proc. Natl. Acad. Sci. U.S.A.,* 87, 2725–2729, 1990.

6. **Sykes, P. J., Weoh, S. H., Brisco, M. J., Hughes, E., Condon, J., and Morley, A. A.,** Quantitation of targets by PCR by use of limiting dilution, *BioTechniques,* 13, 444–450, 1992.

7. **Clewley, J. P.,** The polymerase chain reaction: a review of the practical limitations for human immunodeficiency virus diagnosis, *J. Virol. Meth.,* 25, 179–188, 1989.

8. **Klapper, P. E., Laung, I., and Longson, M.,** Rapid non-invasive diagnosis of herpes encephalitis, *Lancet,* 2, 607–609, 1981.

9. **Aurelius, E., Johansson, B., Skoldenberg, B., Staland, R., and Forsgren, M.,** Rapid diagnosis of herpes simplex encephalitis by nested polymerase chain reaction assay of cerebrospinal fluid, *Lancet,* 337, 189–192, 1991.

10. **Rozenberg, F. and Lebon, P.,** Amplification and characterisation of herpesvirus DNA in cerebrospinal fluid from patients with acute encephalitis, *J. Clin. Microbiol.,* 1991, 29, 2412–2417.

11. **Major, E. O., Amemiya, K., Tornatore, C. S., Houff, S. A., and Berger, J. R.,** Pathogenesis and molecular biology of progressive multifocal leukoencephalopathy: the JC virus induced demyelinating disease of the human brain, *Clin. Microbiol. Rev.,* 5, 49–73, 1992.

12. **Elsner, C. and Dorries, K.,** Evidence of human polyomavirus BK and JC infection in normal brain tissue, *Virology,* 191, 72–80, 1992.

13. **White, F. A., Ishaq, M., Stoner, G. L., and Frisque, R. J.,** JC virus DNA is present in many human brain samples from patients without progressive multifocal leukoencephalopathy, *J. Virol.,* 66, 5726–5734, 1992.

14. **Gibson, P. E., Knowles, W. A., Hand, J. F., and Brown, D. W. G.,** Detection of JC virus DNA in the cerebrospinal fluid of patients with progressive multifocal leukoencephalopathy, *J. Med. Virol.,* 39, 278–281, 1993.

15. **Farci, P., Alter, H. J., Wong, D., Miller, R. H., Shih, J. W., Jett, B., and Purcell, R. H.,** A long-term of hepatitis C virus replication in non-A, non-B hepatitis, *N. Engl. J. Med.,* 325, 98–104, 1991.

16. **Yourno, J.,** Direct polymerase chain reaction for detection of human immunodeficiency virus in blood spot residues on filter paper after elution of antibodies: an adjunct to serological surveys for estimating vertical transmission rates among human immunodeficiency virus antibody-positive newborns, *J. Clin. Microbiol.,* 31, 1364–1367, 1993.

17. **Sison, A. V. and Campos, J. M.,** Laboratory methods for early detection of human immunodeficiency virus type 1 in newborns nad infants, *Clin. Microbiol. Rev.,* 5, 238–247, 1992.

18. **van der Bij, W., Schirm, J., Torensma, R., van Jon, W. J., Tegzess, A. M., and The, T. M.,** Comparison between viraemia and antigenemia for detection of cytomegalovirus in blood, *J. Clin. Microbiol.,* 26, 2531–2535, 1989.

19. **Gema, G., Zipeto, D., Parea, M., Revello, M. G., Silini, E., Percivalle, E., Zavattoni, M., Grossi, P., and Milanesi, G.,** Monitoring of human cytomegalovirus infections and gancyclovir treatment in heart transplant recipients by determination of viremia, antigenemia and DNAemia, *J. Infect. Dis.,* 164, 488–498, 1991.

20. **Jiwa, N. M., van Gemert, G. W., Raap, A. K., van de Rijke, F. M., Mulder, A., Leus, P. F., Salimans, M. M., Zwaan, F. E., van Dorp, W., and van der Ploeg, M.,** Rapid detection of human cytomegalovirus DNA in peripheral blood leukocytes of viraemic transplant recipients by the polymerase chain reaction, *Transplantation,* 40, 72–76, 1989.

21. **Spector, S. A., Merrill, R., Wolf, D., and Dankner, W. M.,** Detection of human cytomegalovirus in plasma of AIDS patients during acute visceral disease by DNA amplification, *J. Clin. Microbiol.,* 30, 2359–2365, 1992.

22. **McCormick, J. B., Webb, P. A., Johnson, K. M., and Smith, E. S.,** A prospective study of the epidemiology and ecology of Lassa fever, *J. Infect. Dis.,* 155, 437–444, 1987.

23. **McCormick, J. B., King, J. J., Webb, P. A., Johnson, K. M., O'Sullivan, R., Smith, E. S., Trippel, S., and Tong, T. C.,** A case control study of the clinical diagnosis and course of Lassa fever, *J. Infect. Dis.,* 155, 445–455, 1987.

24. **McCormick, J. B., King, I. J., Webb, P. A., Scribner, C. L., Craven, R. B., Johnson, K. M., Elliot, L. H., and Belmont-Williams, R.,** Lassa fever: effective therapy with Ribavirin, *N. Engl. J. Med.,* 314, 20–26, 1986.

25. **Lunkenheimer, K., Hufert, F. T., and Schmitz, H.,** Detection of Lassa virus RNA in specimens from patients with Lassa fever by using the polymerase chain reaction, *J. Clin. Microbiol.,* 28, 2689–2692, 1990.

26. **Marcellin, P., Giostra, E., Boyer, N., Coriot, M. A., Martinot-Peignonx, M., and Benhamon, J. P.,** Is the response to recombinant interferons related to the presence of antibodies to hepatitis C virus in patients with chronic non A non B hepatitis, *J. Hepatol.,* 11, 77–79, 1990.

27. **Brillianti, S., Garson, J. A., Tuke, P. W., Ring, C., Briggs, M., Masci, C., Miglioli, M., Barbara, J., and Tedder, R. S.,** Effect of interferon therapy on hepatitis C viraemia in community-acquired chronic non-A, non-B hepatitis: a quantitative polymerase chain reaction study, *J. Med. Virol.,* 34, 136–140, 1991.

28. **Garson, J. A., Brillianti, S., Ring, C., Perini, P., Miglioli, M., and Barbara, J.,** Hepatitis C viraemia rebound after successful interferon therapy in patients with chronic non-A, non-B hepatitis, *J. Med. Virol.,* 37, 210–214, 1992.

29. **Einsele, H., Ehninger, G., Steidle, M., Vallbracht, A., Muller, M., Schmidt, H., Saal, J. G., Waller, H. D., and Muller, C. A.,** Polymerase chain reaction to evaluate antiviral therapy for cytomegalovirus disease, *Lancet,* 338, 1991.

30. **Boucher, C. A. B., Tersmette, M., Lange, J. M. A., Kellam, P., De Goede, R. E. Y., Mulder, J. W., Darby, G., Goudsmit, J., and Larder, B. A.,** Zidovudine sensitivity of human immunodeficiency viruses from high risk, symptom free individuals during therapy, *Lancet,* 336, 585–590, 1990.

31. **Boucher, C. A. B., O'Sullivan, E., Mulder, J. W., Ramantarsing, C., Kellam, P., Darby, G., Lange, J. M. A., Goudsmit, J., and Larder, B. A.,** Ordered appearance of zidovudine resistance mutations during treatment of 18 human immunodeficiency virus positive subjects, *J. Infect. Dis.,* 165, 105–110, 1992.

32. **Shih, J. W. K., Cheung, L. C., Alter, H. J., Lee, L. M., and Gu, J. R.,** Strain analysis of hepatitis B virus on the basis of restriction endonuclease analysis of polymerase chain reaction products, *J. Clin. Microbiol.,* 29, 1640–1644, 1991.

33. **Hayashi, K.,** PCR-SSCP, a simple and sensitive method for detection of mutations in the genomic DNA, *PCR Meth. Appl.,* 1, 34–38, 1991.

34. **Ou, C. Y., Ciesielski, C. A., Myers, E., Bandea, C. I., Luo, C., Korber, B. T. M., Mullins, J. I., Schochetman, G., Berkelman, R. L., Economou, A. N., Witte, J. J., Furman, L. J., Satten, G. A., MacInnes, K. A., Curran, J. W., and Jaffe, H. W.,** Molecular epidemiology of HIV transmission in a dental practice, *Science,* 256, 1165–1169, 1992.

35. **Anon.,** Guidance for Clinical Health Workers: Protection against Infection with HIV and Hepatitis Viruses, U.K. Health Departments, Her Majesty's Stationery Office, London, 1990.

36. **Lin, H. J., Lai, C. L., Lauder, J., Wu, P. C., Lau, T. K., and Fong, M. W.,** Application of hepatitis B virus (HBV) DNA sequence polymorphisms to the study of HBV transmission, *J. Infect. Dis.,* 164, 284–288, 1991.

37. **Beards, G. M.,** Serotyping of rotavirus by NADP-enhanced enzyme immunoassay, *J. Virol. Meth.,* 18, 77–85, 1987.

38. **Brown, D. W. G., Mathan, M. M., Martin, R., Beards, G. M., and Mathan, V. I.,** Rotavirus epidemiology in Vellore, South India: group, subgroup, serotype and electropherotype, *J. Clin. Microbiol.,* 26, 2410–2414, 1988.

39. **Gouvea, V., Glass, R. I., Woods, P., Taniguichi, K., Clark, H. F., Forrester, B., and Fang, Z. Y.,** Polymerase chain reaction amplification and typing of rotavirus nucleic acid from stool specimens, *J. Clin. Microbiol.,* 28, 276–282, 1990.

40. **Anon.,** Manual for the Virological Investigation of Poliomyelitis, World Health Organization, Geneva, 1990.

41. **Balanant, J., Guillot, S., Candrea, A., Delpeyroux, F., and Crainic, R.,** The natural genomic variability of poliovirus analysed by a restriction fragment length polymorphism assay, *Virology,* 184, 645–654, 1991.

42. **Garson, J. A., Tedder, R. S., Briggs, M., Tuke, P., Glazebrook, J. A., Truite, A., Barbara, J. A., Contreras, M., and Aloysius, S.,** Detection of hepatitis C viral sequences in blood donations by 'nested' polymerase chain reaction and prediction of infectivity, *Lancet,* 335, 1419–1422, 1990.

43. **Gretch, D., Lee, W., and Corey, L.,** Use of aminotransferase, hepatitis C antibody, and hepatitis C polymerase chain reaction RNA assays to establish the diagnosis of hepatitis C virus infection in a diagnostic virology laboratory, *J. Clin. Microbiol.,* 30, 2145–2149, 1992.

44. **Francois, M., Dubois, F., Brand, D., Bacq, Y., Guerois, C., Mouchet, C., Tichet, J., Goudeau, A., and Barin, F.,** Prevalence and significance of hepatitis C (HCV) viraemia in HCV antibody-positive subjects from various populations, *J. Clin. Microbiol.,* 31, 1189–1193, 1993.

45. **Garson, J. A., Preston, F. E., and Makris, M.,** Detection by PCR of hepatitis C virus in factor VIII concentrates, *Lancet,* 335, 1473, 1990.

46. **Lyon, D. J., Chapman, C. S., Martin, C., Brown, K. E., Clewley, J. P., Flower, A. J., and Mitchell, V. E.,** Symptomatic parvovirus B19 infection and heat treated factor IX concentrate, *Lancet,* i, 1085, 1989.

47. **McOmish, F., Yap, P. L., Jordan, A., Hart, H., Cohen, B. J., and Simmonds, P.,** Detections of parvovirus B19 in donated blood: a model system for screening by polymerase chain reaction, *J. Clin. Microbiol.,* 31, 323–328, 1993.

48. **Chumakov, K. M., Powers, L. B., Noonan, K. E., Roninson, I. B., and Levenbook, I. S.,** Correlation between amount of virus with altered nucleotide sequence and the monkey test for acceptability of oral poliovirus vaccine, *Proc. Natl. Acad. Sci. U.S.A.,* 88, 199–203, 1991.

49. **Alexander, L. M. and Morris, R.,** PCR and environmental monitoring: the way forward, *Water Sci. Technol.,* 24, 291–294, 1991.

Chapter 3

PCR *In Situ:* New Technologies with Single-Cell Resolution for the Detection and Investigation of Viral Latency and Persistence

Katherine A. Staskus, Janet E. Embretson, Ernest F. Retzel, Janet Beneke, and Ashley T. Haase

TABLE OF CONTENTS

I. INTRODUCTION AND BACKGROUND

Many viruses establish latent or slow and persistent infections that may pose diagnostic dilemmas and complicate the monitoring of response to treatment. In contrast to acute viral infections, where the relatively abundant products of replication and the immune response provide the means with which to diagnose disease and to assess therapies, latent or persistent infections may leave no apparent immunological record of recent infection or evidence of viral particles, antigens, or nucleic acids. This will certainly be the case if the mechanisms of host defense involve a repression or downregulation of viral gene expression, well-documented examples of which can be found in the herpesvirus[1] and lentivirus[2,3] families, to cite but two. Nevertheless, even in this complex situation, cells within the host will retain some viral genomes, and possibly a subset of viral transcripts, that in principle will be detected and interpreted as proof of infection. The problem, in practice, is that the infected cells usually constitute a minor fraction of the total population of cells in tissue fluids or tissues and the viral nucleic acids extracted from these specimens are so diluted by normal cellular nucleic acids that they cannot be detected by population-based techniques such as Northern or Southern hybridization.

The development and refinement of sensitive single-cell technology, such as *in situ* hybridization,[4-6] have provided one way to overcome the dilution problem and, perhaps more importantly, have created a window through which events of the virus life cycle can be viewed in the context of the normal cell populations in latent and persistent infections. Not only is it possible to detect the presence of the infectious agent in very few cells within a background of normal cells and tissues, but it is also possible to identify the infected cell type and the nature of the host response to infection. This knowledge is essential for understanding the mechanism of pathogenesis, which will ultimately lead to the design of more effective treatments and preventive measures. However, in the case of true latency, where the cell contains only one or a few copies of transcriptionally silent viral genomes, or where transcriptional activity is minimal, the signal-to-noise ratio may be too low for confident assessment of infected cells, even by *in situ* hybridization, and certainly not in the time frame essential to diagnostic methodology.

The recent development of the polymerase chain reaction (PCR)[7,8] has allowed the amplification of viral sequences represented at extremely low frequencies within extracted nucleic acid pools (one copy of viral DNA per 70,000 cells)[9] to levels detectable by gel electrophoresis and hybridization. Because this technology is very powerful in its ability to specifically amplify a minor population of sequences, it holds great promise for the future of viral diagnostics. By the same token, however, the technology is inherently plagued by potential problems of sample contamination, such as carryover PCR product from previous analyses. In addition, while this technique increases the sensitivity of detection relative to other populational analysis tools, one loses both qualitative and quantitative information regarding the specific cell type infected as well as the spatial organization and numbers of those cells. The logical extension of cellular precision and target expansion involves the coupling of *in situ* technology with the PCR, which provides a new level of sensitivity essential to both diagnosis and the study of latent viral infections. We[10-13] and others[14–20] have developed conditions whereby nucleic acid sequences can now be amplified within the environment of the cell. Each cell, fixed either in suspension or on a solid support, and either as a single cell or in the context of surrounding tissue, functions individually as a reaction chamber for the PCR. With proper fixation and permeabilization conditions, the oligonucleotide primers and other reaction components are able to diffuse into the cells and, upon thermal cycling, are able to amplify available specific target sequences. Product DNA is retained within the source cell and is readily detectable by standard *in situ* hybridization. When this technology is coupled with histological analysis and immunohistochemistry, it is now possible to confidently identify and characterize latent and persistent infections and to determine their pathogenetic context.

II. THE LENTIVIRUSES AS EXPERIMENTAL PROTOTYPE

Our interest in the molecular pathogenesis of the slow diseases caused by lentiviruses, specifically human immunodeficiency virus (HIV) and the ovine agent visna-maedi, prompted us to develop an *in situ*-based amplification protocol for the detection and localization of low-copy viral sequences within tissue. These viruses establish latent and persistent infections in their respective hosts and, following prolonged incubation periods, give rise to slowly progressive degenerative diseases of primarily the pulmonary and central nervous systems (CNS) of sheep (visna-maedi)[21] and cause an acquired immunodeficiency syndrome (AIDS) and CNS disease in humans (HIV).[22] There is considerable evidence that supports restricted viral gene expression as the major mechanism underlying persistence,[23,24] but the limits of sensitivity of conventional *in situ* hybridization have, until recently, prevented identification of cells latently infected with a single copy of viral DNA, and consequently, have precluded reliable estimates of the size of the reservoir of latently infected cells. To address these issues, we undertook the development of amplification methods applicable to individual cells and chose the ovine lentivirus infection of cultured cells as a particularly optimal system in which to develop and test *in situ* PCR.[10]

In this experimental system, we infected monolayers of fibroblasts, derived from explants of the choroid plexus of spring lambs, that support permissive replication of visna-maedi virus. The time course and extent of viral DNA and RNA synthesis in these sheep choroid plexus (SCP) cells is well documented,[25] and as a consequence, the cells could be harvested and fixed at time points postinfection where we could control the approximate copy number of viral DNA and RNA per cell. For the majority of this work, the cells were fixed and collected 1 to 3 h after infection, when they contained one to two copies of viral DNA (the reverse transcription product of incoming viral genomes), or late in the life cycle, when they contained several hundred copies of viral DNA (a consequence of superinfection). With these two extremes as boundaries of the range of detectability, we were able to optimize the conditions of amplification and compare the sensitivity of detection of the product to that achieved with conventional *in situ* hybridization for unamplified DNA. Because the quality of the PCR product may be greatly influenced by slight variations in reaction conditions, we worked to minimize the changes necessary for amplification of nucleic acids within fixed cells relative to amplification of purified nucleic acids. In order to take advantage of the precise thermal control and heat transfer of available thermal cyclers, which accommodate microcentrifuge tubes, the technique was first developed[10] using cells fixed in a suspension.

Using visna-infected SCP cells, we were able to establish conditions of fixation, pretreatment, and amplification that allowed us to detect one or two copies of viral DNA per cell by *in situ* hybridization with a radiolabeled probe (Figure 3–1A).[10] As evidence of specificity for the amplification and detection of visna DNA, we observed no signal over background on slides under the following conditions: (1) with uninfected cells amplified with visna-specific primers and hybridized with visna-specific probe; (2) with infected cells amplified with visna-specific primers and hybridized with a heterologous (HIV) probe (Figure 3–1B) or a visna probe derived from an unamplified region of the genome; (3) with infected cells amplified with heterologous primers and hybridized with visna-specific probe; (4) when an essential component (such as the polymerase) was omitted from the amplification reaction; or (5) when hybridization was carried out for unamplified DNA. The number of copies of amplified product DNA and, consequently, the estimated efficiency of amplification was determined by counting the silver grains over infected cells. The correlation between grain count and nucleic acid copy number was established by quantitative hybridization (solution or blot) for viral nucleic acids extracted from a parallel sample of infected cells.[5,6,25] Statistical analysis of grain counts indicated that we were able to achieve approximately 12% efficiency in the amplification reaction or as much as a 100-fold amplification of the target sequence (a level more than sufficient for reliable detection by *in situ* hybridization). We attributed the observed variability in amplification efficiency from cell to cell within the population to asynchronous infection and reverse transcription processes, as well as to differences in the intracellular microenvironments

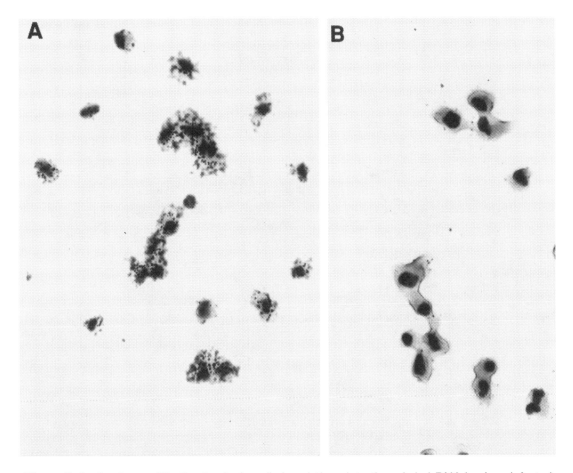

Figure 3–1 *In situ* amplification in single cells in solution: detection of viral DNA in visna-infected SCP cells. At 20 h postinfection, cells were trypsinized, fixed as a suspension in paraformaldehyde, washed with PBS, and resuspended in PCR reaction mixture containing a multiple primer set (MPS) representing 1200 bases of the LTR and *gag* region of the visna genome. Following 50 amplification cycles, the cells were deposited onto Denhardt-coated slides and subjected to *in situ* hybridization with (A) radiolabeled visna-specific or (B) heterologous probes. (C) When amplification was carried out using a single primer pair designed to generate a 600-bp product, hybridization with the visna-specific probe revealed a high extracellular background of silver grains as well as an accumulation of signal at the periphery of the cell. We interpret this artifact as a tendency of small amplimers to diffuse out of the cells under these conditions of amplification, since larger products generated by the MPS or single primer pairs spaced farther apart were more highly localized over the cell. 3-h exposures. (Magnification × 270.)

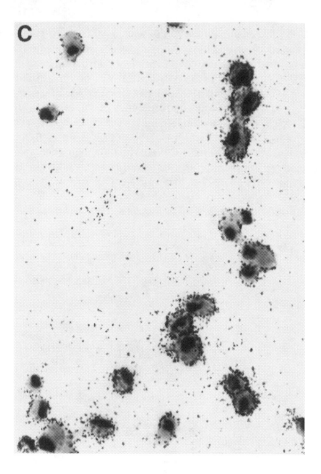

Figure 3–1C

of the DNA targets resulting from the fixation process, which affects the accessibility of reaction components.

Following the establishment of conditions to amplify and detect a single copy of viral DNA in cultured cells, we began the more important refinement of this method for the detection of viral DNA in individual cells in tissue sections.[11] For these experiments, we exploited an *in vivo* model of viral infection of the lungs of sheep.[26] The pulmonary system is a natural route of infection for visna-maedi where viral gene products accumulate over time to levels that elicit and sustain an inflammatory response. In the lung, the massive infiltration of inflammatory cells into the delicate interstitial spaces of the alveoli ultimately leads to tissue destruction, fibrosis, and impaired blood-gas exchange. These events are manifest as shortness of breath, or dyspnea, which is reflected in the Icelandic term *maedi*. In the experimental model, in order to compress these pathological changes in time and space, we introduced a large viral inoculum (1×10^9 plaque-forming units of visna, produced in culture and concentrated by ultracentrifugation) into a defined segment of the lung of a sedated lamb via bronchoscopy. Progress of the infection was followed initially by *in situ* hybridization of a radiolabeled visna-specific probe to pulmonary alveolar macrophages (PAMs) recovered by bronchial lavage, since we knew from previous work[26] that visna-maedi virus infects this cell type. Two weeks after infection, the animal was sacrificed and the lung tissue was removed, embedded in paraffin, sectioned, and prepared for the *in situ* procedures. Using this protocol, we found that, with the exception of the development of fibrosis, within this short period of time we could reproduce the type of histopathology in the lung that normally takes months to years to develop following natural infection (formation of lymphoid follicles, perivascular cuffing, and interstitial inflammation; Figure 3–2A).

In order to amplify target sequences in tissue sections attached to a solid support, we had to modify the use of existing thermal cycling equipment to accommodate glass slides. After numerous experiments employing the heating block design found on many thermal cyclers, we focused on developing the use of a programmable circulating-air oven. While adaptations of the standard block-type thermal cycler have been used successfully,[14–17] a major disadvantage is its lack of capacity to simultaneously process many slides.

We optimized pretreatments and reaction conditions and were able to amplify target viral DNA sequences to levels easily detectable by *in situ* hybridization in individual infected cells within a tissue section (Figure 3–2B).[11] With these tissues, we achieved an approximate 30-fold amplification which, while less efficient than amplification in cells in suspension, was well within levels required for confident detection of single-copy genomes in infected cells. Surprisingly, we found, in addition to PAMs, an expected host cell for visna-maedi virus, a new host cell in the bronchiolar epithelium. A majority of the bronchiolar epithelial cells in regions of heavy inflammation contained viral DNA. *In situ* hybridization for viral RNA in these same regions revealed transcripts not only in the PAMs, again as expected, but also in a very small percentage of the bronchiolar epithelial cells (Figure 3–2D), which corroborated the *in situ* PCR result demonstrating the presence of viral DNA in a cell type that had not previously been described as a host for visna. Consistent with the *in vivo* downregulation of gene expression previously reported for visna,[24,26,27] the number of copies of viral RNA in the transcriptionally active cells was 10- to 100-fold lower than that of permissively infected cells in culture.[25]

III. METHODOLOGY

A. PREPARATION OF CELLS AND TISSUES

Infected and uninfected SCP cells were trypsinized, washed with calcium- and magnesium-free phosphate-buffered saline (PBS-CMF), pelleted (600 *g*, 10 min) and resuspended and fixed for 20 min at

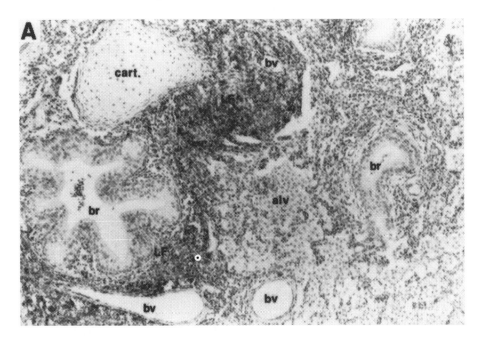

Figure 3–2 *In situ* amplification in tissue sections: detection of viral DNA in visna-infected sheep lung. (A) Experimentally infected tissues exhibited typical visna-induced histopathology characterized by the infiltration of PAMs and inflammatory cells into the alveoli (alv) and the accumulation of lymphocytes and monocytes around the blood vessels (bv) and bronchioles (br) with the formation of lymphoid follicles (LF)[1]; 8-μm sections from formalin-fixed paraffin-embedded tissues were subjected to thermal cycling in the presence of a PCR mixture containing the visna MPS, which targets the LTR and *gag* gene. Subsequent *in situ* hybridization with a radiolabeled visna-specific probe for amplified sequences revealed the presence of viral DNA in a majority of the bronchial epithelial cells (be) in regions of inflammation (B), while epithelium from a region without inflammatory changes shows a background level of silver grains (C). Panels B and C were 24-h exposures. (D) *In situ* hybridization for the presence of viral RNA demonstrates that an occasional cell (→) of the bronchial epithelium, in the region of inflammation, is transcriptionally active (2-day exposure). cart., cartilaginous ring; mc, macrophage. (Magnification: A, × 80; B to D, × 240.) (From Staskus, K. A., Couch, L., Bitterman, P., Retzel, E. F., Zupancic, M., List, J., and Haase, A. T., *Microb. Pathog.*, 11, 67, 1991. With permission.)

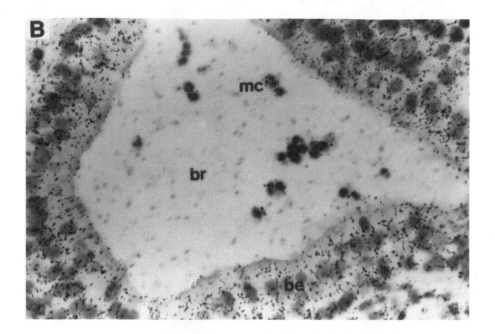

Figure 3–2B

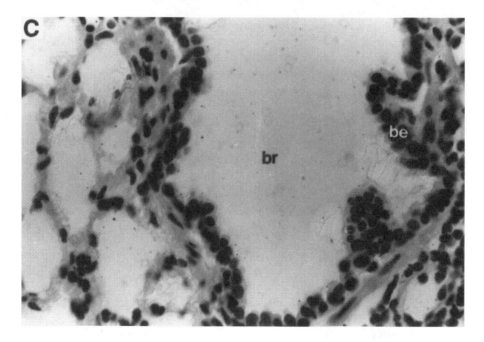

Figure 3–2C

ambient temperature in a freshly prepared solution of 4% (w/v) paraformaldehyde in PBS-CMF.[5] The cells were pelleted from the fixative (1600 *g*, 10 min), washed once in PBS-CMF and then either used immediately for amplification or stored under 70% (v/v) ethanol at 4°C.[10]

Tissues were obtained from the experimentally infected lamb approximately 2 weeks after infection. Following sacrifice and necropsy, the lungs were removed *en bloc,* inflated with a solution of 10% buffered-formalin, and fixed for a period of 72 h. Tissue segments were dissected from the infected and uninfected lobes, dehydrated in 80% (v/v) ethanol for at least 24 h, and using standard techniques, embedded in paraffin for purposes of thin sectioning, histopathological examination, and *in situ* analyses.

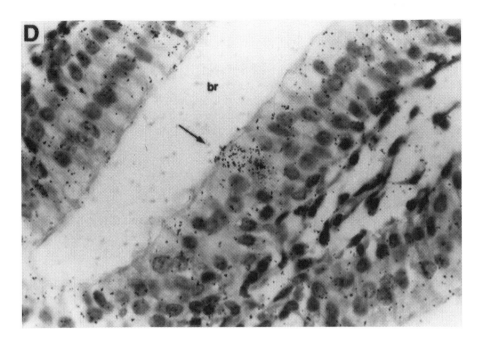

Figure 3–2D

Serial sections (8 μm) of the embedded tissues were attached to Denhardt-coated glass slides[6] with a 3% (v/v) solution of Elmer's white glue in deionized distilled water, dried, and deparaffinized.

Cultured cells may be carried through the amplification process on a solid support in a manner similar to tissue sections, in which case the cells are fixed in formalin or paraformaldehyde, pelleted, embedded in paraffin, and further treated as described for paraffin-embedded tissue.[15,20] Alternatively, trypsinized and washed cells may be resuspended in PBS-CMF and deposited onto Denhardt-coated slides, either by cytocentrifugation or simply by spotting and drying a small volume of the cell suspension (typically, 30,000 to 50,000 cells in 10 μl). These cells are subsequently fixed by immersing the slides in fresh 4% (w/v) paraformaldehyde in PBS-CMF for 20 min at ambient temperature. They are then rinsed in PBS-CMF and dehydrated by passage through a series of graded alcohols (5 min in each of 70%, 80%, and absolute ethanol).

1. The Balance of Fixation and Pretreatment

In situ technologies enable the investigator to localize macromolecules or processes to specific identifiable cells, often within the context of surrounding tissue. To achieve success with these techniques, however, one must empirically determine a set of reagents and conditions that will (1) sufficiently preserve for histological purposes the cellular and tissue architecture of the sample under investigation, and (2) minimize the loss of efficiency to the *in situ* procedure itself. For the purposes of *in situ* PCR, we experimented with a variety of fixatives (both precipitating fixatives and the cross-linking aldehyde fixatives) and fixation conditions, as well as pretreatments that increase the degree of cellular permeability. We found that the aldehyde fixatives are superior in their ability to maintain the structural integrity of the cells and tissues throughout the high-temperature thermal cycling. We[10–13] and others[14–20] have also determined that the degree of fixation of the samples, with respect to the concentration of fixative and length of fixation time, may vary, provided that the fixation is balanced by appropriate pretreatment. When the visna-infected cells and tissues were fixed under the conditions described above, postfixation pretreatments were not required to achieve a good balance between the maintenance of cellular and tissue morphology and a reasonable efficiency of amplification. However, for other types of tissue, for more extensively fixed tissue, or for samples that are fixed under unknown conditions, pretreatment may be necessary to allow components of the amplification reaction access to target nucleic acid template. Generally, with such samples, a 1-h incubation with proteinase K within a standard range of concentrations (5 to 50 μg/ml) and temperatures (37 to 55°C) has been sufficient for this purpose. Other pretreatments employing detergents[18] or denaturants[17,19] have also been used successfully to prepare cells and tissues for *in situ* PCR.

B. PRIMER DESIGN

In preliminary experiments, visna-infected SCP cells were fixed and subjected to amplification in solution with a single primer pair designed to produce a 300-bp product.[10] The primers were so designed as to incorporate, to the extent feasible, the criteria of Kwok[28] and Rychlik.[29] After *in situ* hybridization with a radiolabeled visna-specific probe, we observed large numbers of silver grains located outside of cells or concentrated in a ring-like pattern at the perimeter of the cells, which we interpreted as evidence of the diffusion of amplimers from the cell (Figure 3–1C). When we used single primer pairs that would generate successively larger products (600, 900, or 1200 bp), the high background and ring effect decreased, as did the signal, indicating a decrease in the efficiency of the amplification reaction with increasing size of product. To resolve the opposing requirements for efficient synthesis (short product) and retention (large product), we synthesized a set of nine oligonucleotide primers which were designed to collectively amplify greater than 1200 contiguous base pairs of the LTR and *gag* gene, one of the most highly conserved regions of visna-maedi virus. This multiple primer set (MPS) contained four pairs of primers, whose individual amplification products were approximately 350 bp in length. The boundaries of the products overlap those of adjacent pairs so that upon denaturation, annealing, and extension during subsequent cycles of the PCR, these fragments could combine to form larger product (Figure 3–3A). Using the MPS, we were able to both efficiently synthesize and retain large amounts of DNA in infected cells.[10] Chiu et al.,[17] in determining conditions for amplification of mouse mammary tumor virus (MMTV) sequences in cultured cells, also found that the signal from amplification and detection of short DNA

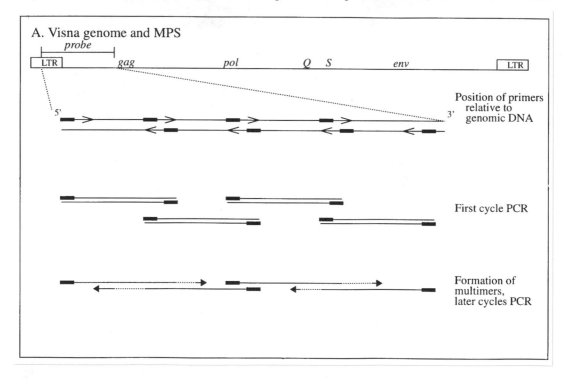

Figure 3–3 Strategies for the concatamerization of *in situ* PCR product. (A) The visna genome and multiple overlapping primer set. Nine primers in the conserved LTR and *gag* regions of the genome were chosen, such that relatively short (~350 bp) fragments would be produced in early rounds of amplification. Plus- and minus-strand primers were spaced approximately 20 bases apart, producing adjacent amplimers with substantial overlap (~60 bp). In later cycles of amplification, these initial products can be further elongated using the adjacent, opposite strand as template. These amplified sequences have the potential for reaching the length of the most distal pair of primers. (B) Amplification and subsequent concatamerization of a single target sequence. Primers defining the bounds of a DNA target (A → B) are engineered to contain unique complementary 5′ tails (D and d). Upon denaturation and annealing, these complementary tails permit the polymerization of a large product from a relatively small original target, providing for both retention and further amplification of the target sequences, *in situ*.

B. Concatamer formation

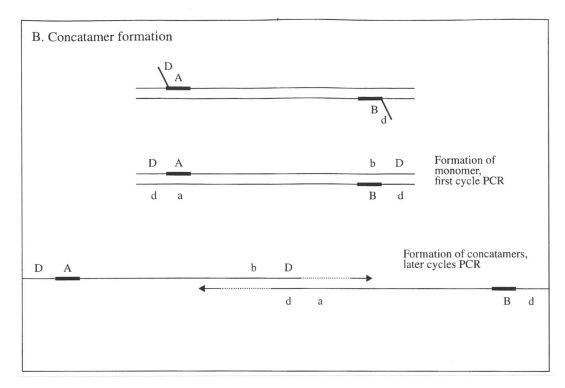

Formation of monomer, first cycle PCR

Formation of concatamers, later cycles PCR

Figure 3–3B

fragments was poorly localized over positive cells and used a similar strategy to increase the size of product (Figure 3–3B). In this situation, a pair of oligonucleotide primers was designed to amplify a short segment of the target DNA (167 bp) and to form large concatamers by annealing of the fragments to each other via complementary sequences engineered onto the 5′ termini of the primers. Advantages of this variation include the requirement for less unique target sequence information and fewer synthetic primers.

C. CONDITIONS FOR AMPLIFICATION
1. Cells in Suspension
In preparation for amplification, the cultured cells that were fixed in suspension and stored under ethanol were pelleted (1600 g, 10 min), resuspended in PBS-CMF, and allowed to rehydrate for at least 20 min at ambient temperature. The cells were then pelleted, resuspended in a few hundred microliters of fresh PBS-CMF, transferred to a siliconized 0.5-ml microcentrifuge tube, pelleted once again and resuspended in PCR reaction mixture (approximately 2×10^6 cells per 100 µl). The reaction mixture was composed of 10 mM Tris-HCl, pH 8.3, 50 mM KCl, 1.5 mM MgCl$_2$, 0.01% (w/v) gelatin, 200 µM deoxynucleoside triphosphates, and 0.1 µM oligonucleotide primers. The cell suspension was overlaid with mineral oil, and the tube was capped and transferred to the heating block of a thermal cycler. Following an initial denaturation at 94°C for 10 min, sample tubes were either quenched to 4°C or maintained at 72°C (when hot-start conditions[30] were used), followed by the addition of 2.5 U of *Thermus aquaticus (Taq)* DNA polymerase under the oil. The cells were subjected to 25 cycles (c) of denaturation (94°C for 2 min), annealing (42°C for 2 min) and polymerization (72°C for 15 min), at which time 5 additional units of *Taq* DNA polymerase were added, and the samples were taken through 25 additional cycles. Following this amplification regimen, the cells were pelleted (16,500 g, 5 min), and the oil and PCR cocktail was aspirated. After washing the cells with PBS-CMF three times, the cells were deposited onto Denhardt-coated slides by spotting or cytocentrifugation (450 rpm, 5 min, Shandon cytocentrifuge).

2. Cells within Tissue Sections
PCR reaction solution (10 mM Tris-HCl, pH 8.3, 50 mM KCl, 1.5 mM MgCl$_2$, 0.01% (w/v) gelatin, 200 µM deoxynucleoside triphosphates, 1 µM oligonucleotide primers, and 5% [v/v] *Taq* DNA polymerase [5 U/µl]) was pipetted onto the slides to fully hydrate the tissue sections (10 to 40 µl, depending on the dimensions of the section). The tissues were then covered with siliconized glass coverslips, excess

solution was blotted to prevent the coverslips from sliding off of the tissue, and the slides were placed in heat-sealable plastic bags (101.6 × 152.4 × 0.1 mm), two slides per bag. Mineral oil (5 to 6 ml) was added to each bag, the air was removed, and the bags were sealed. The oil both facilitates and stabilizes heat transfer to the slide as well as prevents the dehydration of the reaction mixture during thermal cycling. The bags with slides were positioned vertically in a polypropylene 80-pin rack, which was placed inside a thermal cycling circulating-air oven. The thermal sensor, which is mounted on a glass slide, was similarly sealed in a plastic bag with oil and placed in the rack along with the samples. The slides were subjected to 25 c of amplification (94°C for 2 min, 42°C for 2 min, and 72°C for 15 min). They were then removed from the bags, drained, and rinsed in two to three changes of chloroform (5 min each) to remove residual oil film. After the chloroform had evaporated, the coverslips were lifted with forceps for the addition of fresh PCR reaction mixture and *Taq* DNA polymerase. Fresh coverslips were placed over the sections and the slides were once again sealed in oil-filled bags and placed in the oven for another set of 25 c. More recently, sequences in tissue sections have been successfully amplified with a single addition of *Taq* DNA polymerase and 30 to 50 c. A newer model of the thermal cycling oven (BioOven II; BioTherm Corporation, Fairfax, VA) is equipped with a sample rack that holds slides in a horizontal position. Sample preparation for this instrument simply requires a spot of fingernail enamel to anchor the coverslip to the slide over the tissue sample and a thin layer of mineral oil along the edge of the coverslip to prevent dehydration of the reaction mixture underneath. After enzymatic amplification was complete, the slides were removed from the bags and processed through chloroform as described above, and the coverslips were removed under PBS-CMF. The slides were washed for 10 min in PBS-CMF and dehydrated in a series of graded alcohols.

D. DETECTION OF AMPLIFIED SEQUENCES

Amplification product was detected within the cells by standard *in situ* hybridization for DNA.[5,6] Briefly, the slides were treated with ribonucleases, postfixed in 5% (w/v) paraformaldehyde, acetylated, denatured with formamide, and dehydrated prior to the addition of probe. The probe consisted of a fragment of cloned visna DNA, corresponding to the amplified sequence, which we labeled by nick translation[31] to a specific activity of approximately 1×10^9 dpm/μg with either [^{125}I]dCTP or a combination of [^{35}S]dATP and [^{35}S]dCTP. Alternatively, we used PCR to incorporate [^{125}I]dCTP to a specific activity of 4×10^9 dpm/μg.[10,32] Hybridization solution containing the probe (5×10^4 dpm/μl) was applied to the slides (1 to 3×10^6 dpm/section) and, following hybridization for 12 to 16 h, the slides were washed for 12 to 16 h, dehydrated in graded alcohols containing 0.3 *M* ammonium acetate, coated with nuclear track emulsion (Kodak NTB-2), and allowed to expose at 4°C for appropriate periods of time based on test exposures of duplicate slides. The slides were finally developed, stained with hematoxylin and eosin, and viewed under the microscope for the presence of silver grains.

Others have reported the successful use of nonisotopic probes (specifically, biotinylated[14,16] or digoxygenin-labeled[15,19,20]) and the direct incorporation of digoxygenin dUTP[15,19,20] or fluorescence-tagged synthetic oligonucleotides[18] in the *in situ* amplification reaction.

IV. APPLICATION AND RELEVANCE TO HUMAN VIRAL DIAGNOSIS AND RESEARCH OF VIRAL PATHOGENESIS

Within the realm of human viral diagnostics, costly, time-consuming, and often insensitive traditional assays are being replaced by the PCR. It is quickly becoming clear that the gain in speed and sensitivity offered by this technology, even in its relative infancy, far outweigh associated drawbacks, such as the potential for carryover contamination. In many cases, amplification of target sequences from total nucleic acid pools, extracted from blood or tissue specimens, will suffice to provide evidence of viral infection. The latent infections of the lentiviruses, however, present a different challenge, in that the spectrum of infected cells and viral gene expression is beyond the range of mainstream technology.

A case in point is the detection of pediatric AIDS,[33,34] where 13 to 40% of infants born to HIV-positive mothers become infected.[35,36] Pediatric HIV infections progress much more rapidly than infection in adults,[35] and hence, early diagnosis is extremely important for early therapeutic intervention and an improved prognosis. The serological tests for antibodies against HIV are unreliable before 5 to 6 months of age due to contaminating maternal IgG,[37] and both the late time frame of IgA expression[38,39] and assays for serum HIV p24 antigen in the first few months of life have proven to be insensitive.[36,40] Cell culture for virus rescue is costly, hazardous, and time consuming, and it has low sensitivity as well. PCR using extracted nucleic acids from cord blood or peripheral blood lymphocytes (PBLs) is prone to false positives due to carryover or sample contamination, and has been reported to have a sensitivity of only 50% in detecting infected neonates.[40–42] In contrast, the *in situ* PCR applied to fixed cells results in a highly localized hybridization signal over single cells and takes advantage of the power, specificity, and sensitivity of amplification with less concern for either contamination or limits of detection. The cell is both the reaction vessel and the "solid support" medium for the amplification process, and hence, diffusion and degradation of the reaction products are minimized. PCR at the single-cell level will also provide essential information (e.g., the timing of the transmission of infection from mother to infant and the quantitation of viral burden early in infection) that will influence the development, selection, and application of preventive or prophylactic antiviral strategies.

The increased sensitivity and cellular specificity provided by *in situ* amplification will also facilitate the early detection of HIV infections in adults. Previous studies[43,44] suggest that there may be a prolonged period of silent infection prior to seroconversion. The use of *in situ* PCR in prospective studies of seronegative individuals at risk of acquiring HIV-1 will assist not only in bringing early treatment to infected individuals, but also in providing information about the very early events of infection, which will support the process of defining the immunopathogenesis of infection and of developing early preventive and therapeutic interventions. Coupled with fluorescence-activated cell sorting (FACS), amplification of target sequences with fluorescent primers could very quickly and efficiently identify extremely infrequent infected cells within fluid samples such as blood or spinal fluid. This approach has already been successfully used for sorting and analyzing populations of cells in which specific viral nucleic acids.[45] or cellular messenger RNAs have been amplified *in situ*.[18] Indeed, there is enormous potential in the application of *in situ* PCR for the efficient screening of body fluids or tissues for any infectious agent (viral, bacterial, fungal, etc.) for which a unique nucleic acid, DNA or RNA, would be present in cells.

Since the identification of HIV as the etiologic agent of AIDS, investigators have attempted to quantitate infection of the peripheral blood and lymphoid organs and to correlate viral burden with the clinical stages of disease. Early estimates of 10^{-4} to 10^{-5} infected peripheral blood mononuclear cells (PBMCs) were produced by *in situ* hybridization for viral RNA.[46] These values appeared to be too low to be responsible for the dramatic decline in the CD4$^+$ population and ultimate production of immuno-deficiency, and in retrospect, this technique alone was insufficiently sensitive to detect latently infected or severely down-regulated members of the population. An increase in sensitivity (at least tenfold) and accuracy was achieved by using a variety of techniques, from limiting dilution assays[47] to PCR.[48,49] With conventional PCR, however, all information regarding cell type, copy number per cell, and state of latency or expression is lost. By using *in situ* PCR on peripheral blood samples, Bagasra et al.[16] have been able to obviate these problems and, in addition, have demonstrated an approximately tenfold increase in sensitivity over the conventional techniques.

While there is a rapid and profound drop in the amount of virus present in PBLs and the plasma fraction of individuals shortly after infection, it is not clear whether the viral load present in the PBLs necessarily represents a comparable decrease in the virus present in lymphoid organs. Recent *in situ* hybridization studies have discovered high concentrations of HIV-1 RNA in extracellular association with follicular dendritic cells (FDCs) in the germinal centers of lymphoid tissues from adult patients at early stages of infection, and consequently it has been suggested that the lymphoid organs serve as major reservoirs for HIV-1.[50,51] It has been unclear, however, whether or not these cells are actually infected with HIV. By using *in situ* PCR and a double-labeling technique,[52] which combines *in situ* hybridization and immunohistochemistry, we have been able to confirm the association of viral RNA with the FDCs in the germinal centers of lymph nodes and to demonstrate a remarkable intracellular reservoir of HIV DNA in the macrophage and CD4$^+$ lymphocyte population (20 to 30% of cells are positive for viral DNA) residing in and migrating through the lymph node follicles (Figure 3–4).[13] As expected, only a fraction of these infected cells were transcriptionally active and contained viral RNA at levels 10- to 100-fold below those of permissively infected cells. This is consistent with information from *in vivo* studies demonstrating restricted gene expression in lentiviruses.[23,24] By *in situ* PCR, we have also conclusively demonstrated that the FDCs do not contain viral DNA and are therefore not a significant infected cell type in the population. Taking into account electron microscopy data[13] and studies on the tissue distribution of viral antigens, it appears likely that significant numbers of virions are present in immune complexes on the surfaces of the FDCs, and it is this "pool" of virus that may ultimately function as an extracellular reservoir for the continued infection of CD4$^+$ lymphocytes that are trafficked through the lymphoid organs.

We have been working with various tissue specimens from HIV-infected individuals to characterize virus-host interactions in terms of the latently infected cell population, the cellular host range, and level of viral transcriptional activity. Prior to the lymph node studies described above, our initial experiments with *in situ* PCR on human samples were performed on tissue sections from a biopsy of an adenocarcinoma taken from an HIV-seropositive individual. *In situ* amplification revealed that a significant proportion (up to 34%) of the lymphocytes and mononuclear cells (identified by morphological criteria and by immunohistochemistry on adjacent sections with cell-specific antibodies) infiltrating the tumor were positive for viral DNA and, with few exceptions, were transcriptionally silent.[12] Since the CD4+ cells make up about one third of this inflammatory population, it is likely that nearly all the CD4+ cells in this tumor are latently infected and thus represent a potential reservoir for the dissemination of virus over a prolonged period of time. Surprisingly, *in situ* hybridization for RNA (without amplification) showed that the tumor cells, which were epithelial in origin and which lacked detectable CD4 by immunocytochemistry, contained high levels of viral RNA and appeared to be replicating virus permissively while the

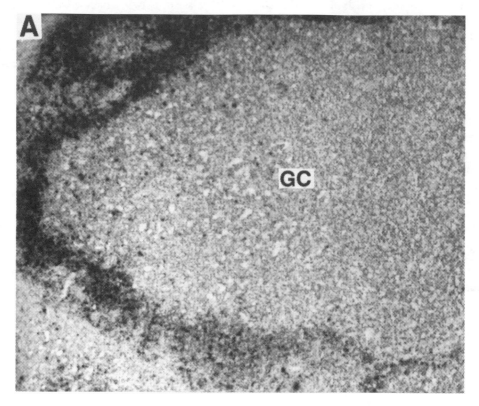

Figure 3–4 *In situ* amplification and detection of HIV DNA in lymph nodes. Lymph node biopsies from HIV-infected individuals were fixed in formalin, embedded in paraffin, sectioned, and analyzed by *in situ* PCR and *in situ* hybridization for the occurrence and distribution of HIV DNA- and RNA-containing cells. A multiple primer set (sequences derived from the HIV HXB2 complete genome) containing six primers (three sense and three antisense) representing approximately 1551 nucleotides of the *gag* gene of HIV-1 was used in the amplification: 5'-GGAACCCACTGCTTAAGCCT-3' (bases 507 to 526), 5'-GCGTCAGTATTAAGCGGGGG-3' (bases 801 to 820), 5'-GTTTTCAGCATTATCAGATG-3' (bases 1301 to 1320), 5'-CATGGTGTTTAAATCTTGTG-3' (bases 1350 to 1331), 5'-CCCTGGCCTTAACCGAATTT-3' (bases 860 to 841), and 5'-TTGGTGTCCTTCCTTTCCAC-3' (bases 2054 to 2035). (A) A combination of long radiographic exposure times (4 to 11 days) and low magnification emphasizes the presence of HIV DNA (black grains) predominantly in cells in the follicular mantle surrounding the germinal center (GC). (B) *In situ* hybridization without amplification reveals HIV RNA primarily over the follicular dendritic cells (FDCs) of the germinal center. The diffuse distribution of silver grains overlying the processes of the FDCs is more clearly seen by the use of epipolarized illumination, which makes the silver grains appear white in panel B. (Magnification: A, × 110; B, × 160.)

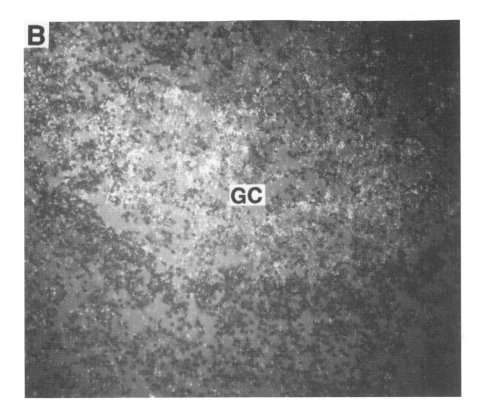

Figure 3–4B

infected lymphocyte population exhibited few cells that were transcriptionally active for HIV. These epithelial cells may illustrate the existence of an alternative cell surface receptor and therefore an expanded host range for HIV. Without the *in situ* PCR these examples of (1) infection of a different cell type and (2) two dramatically different virus life styles (permissive vs. down-regulated) within the same tissue environment would have otherwise gone undetected.

Clinical events such as death and the development of opportunistic infections have been commonly used as endpoints in the early therapeutic trials of HIV-1 infection. The major limitation of this approach has been the length of time (measured in years) that is required for the studies to show clinical significance. Most recently, clinical trials have been incorporating viral measurements, such as changes in viral load and development of resistance, to define study endpoints. By providing a more sensitive and direct estimate of the number of infected blood cells (viral burden) present in HIV-1 infected individuals, the use of *in situ* PCR would be extremely valuable in monitoring changes in viral load secondary to the use of antiretrovirals in monotherapy as well as in combination therapy. Currently, available techniques being evaluated in multicenter clinical trials include quantitative cultures and quantitative PCR of peripheral blood leukocytes and plasma fractions.

A number of recent studies have described the isolation of HIV-1 strains resistant to zidovudine (ZDV),[53,54] didanosine (ddI),[54] zalcitabine (ddC),[55] as well as other non-nucleoside reverse transcriptase inhibitors. Genotypic characterization of these resistant isolates has shown the presence of specific mutations of the *pol* gene.[53–55] These studies have been performed using PCR to amplify the gene region, followed by cloning and sequencing strategies, or secondary PCR reactions using primers specifically designed to detect the mutations of interest. A major limitation of these studies is that they do not provide a clear estimate of the proportion of mutant viruses present in a given individual. Because multiple quasispecies of HIV-1 coexist in the infected individual,[56] the implications of finding a resistant genotype are unclear. With the design of appropriate primers and probes, *in situ* PCR could be applied to quantitate the infected cells containing viruses with specific mutations as well as to define, temporally and spatially, their emergence in the population. The more sensitive single-cell approach, as opposed to a clinical or cultural definition where the mutation has become the predominant variant with concomitant risks of passage of the resistant phenotype, will aide in the fine-tuning of combined therapies already in clinical trials[57] and in the development of new treatment regimens involving alternation of therapeutic agents. ZDV-resistant variants are at a growth disadvantage compared to wild type, and they appear as the primary variant in therapy only because of their ability to replicate under modified nucleoside pressure. A diagnostic tool such as *in situ* PCR would represent a significant advance for the understanding and management of infections caused by resistant HIV. This has not only therapeutic, but also epidemiologic implications, in that transmission of ZDV-resistant virus to adult hosts has been documented.[58] If the emergence of such variants could be suppressed through early and accurate detection and substitution of therapies, one could possibly eliminate or at least control the pandemic consequences of drug-resistant strains.

As *in situ* PCR combines the extreme sensitivity and specificity of nucleic acid amplification with the exquisite localization offered by *in situ* techniques not only to verify the presence of a viral agent, but also to provide information about the context in which it exists, it will be an invaluable tool for research as well as diagnostic purposes. Armed with the ability to reveal a single copy of viral nucleic acid in a single cell, and coupled with histology and immunohistochemistry for the detection of cellular antigens, one is well positioned to determine the cellular host range of the virus, the absolute numbers of cells that are infected, the percentage of infected cells that are transcriptionally active, and therefore the size of the latently infected reservoir. One can also begin to correlate viral burden with disease state and follow the effects of therapies at the cellular level. Relationships between these qualitative and quantitative values can lead to the generation and refinement of mathematical models of viral pathogenesis,[11,56,59,60] the predictions of which must ultimately lead to a clearer understanding of virus-host interactions (including both primary and secondary effects), more accurate diagnosis of infection and staging of disease, and the development of novel and effective therapies and vaccines.

Finally, in addition to its use as a diagnostics and research tool for infections and diseases of known viral etiology, *in situ* PCR will prove extremely useful in the search for links between viruses and diseases of unknown origin (e.g., multiple sclerosis).

ACKNOWLEDGMENTS

We are grateful to Dr. Alejo Erice for insightful suggestions and critical review of this manuscript and Tim Leonard for the preparation of figures. Research support was provided by grants from the National Institutes of Health.

REFERENCES

1. **Fraser, N. W., Spivack, J. G., Wroblewska, Z., Block, T., Deshmane, S. L., Valyi-Nagy, T., Natarajan, R., and Gesser, R. M.,** A review of the molecular mechanism of HSV-1 latency, *Current Eye Research,* Suppl. 10, 1, 1991.
2. **Haase, A. T.,** Pathogenesis of lentivirus infections, *Nature,* 322, 130, 1986.
3. **Bednarik, D. P. and Folks, T. M.,** Mechanisms of HIV-1 latency, *AIDS,* 6, 3, 1992.
4. **Gall, J. G. and Pardue, M. L.,** Formation and detection of RNA-DNA hybrid molecules in cytological preparations, *Proc. Natl. Acad. Sci. U.S.A.,* 63, 378, 1969.
5. **Haase, A. T.,** Analysis of viral infections by *in situ* hybridization, in *In Situ Hybridization— Applications to Neurobiology* (Symp. Monogr.), Valentino, K. L., Eberwine, J. H., and Barchas, J. D., Eds., Oxford University Press, New York, 1987, 197.
6. **Haase, A., Brahic, M., Stowring, L., and Blum, H.,** Detection of viral nucleic acids by *in situ* hybridization, in *Methods in Virology,* Vol. 7, Maramorosch, K. and Kaprowski, H., Eds., Academic Press, New York, 1984, 189.
7. **Saiki, R. K., Scharf, S., Faloona, F., Mullis, K. B., Horn, G. T., Erlich, H. A., and Arnheim, N.,** Enzymatic amplification of β-globin genomic sequences and restriction site analysis for diagnosis of sickle cell anemia, *Science,* 230, 1350, 1985.
8. **Saiki, R. K., Gelfand, D. H., Stoffel, S., Scharf, S. J., Higuchi, R., Horn, G. T., Mullis, K. B., and Erlich, H. A.,** Primer-directed enzymatic amplification of DNA with a thermostable DNA polymerase, *Science,* 239, 487, 1988.
9. **Erlich, H. A., Gelfand, D., and Sninsky, J. J.,** Recent advances in the polymerase chain reaction, *Science,* 252, 1643, 1991.

10. **Haase, A. T., Retzel, E. F., and Staskus, K. A.,** Amplification and detection of lentiviral DNA inside cells, *Proc. Natl. Acad. Sci. U.S.A.,* 87, 4971, 1990.

11. **Staskus, K. A., Couch, L., Bitterman, P., Retzel, E. F., Zupancic, M., List, J., and Haase, A. T.,** *In situ* amplification of visna virus DNA in tissue sections reveals a reservoir of latently infected cells, *Microb. Pathog.,* 11, 67, 1991.

12. **Embretson, J., Zupancic, M., Beneke, J., Till, M., Wolinsky, S., Ribas, J. L., Burke, A., and Haase, A. T.,** Analysis of human immunodeficiency virus-infected tissues by amplification and *in situ* hybridization reveals latent and permissive infections at single-cell resolution, *Proc. Natl. Acad. Sci. U.S.A.,* 90, 357, 1993.

13. **Embretson, J., Zupancic, M., Ribas, J. L., Burke, A., Rácz, P., Tenner-Rácz, K., and Haase, A. T.,** Massive covert infection of helper T lymphocytes and macrophages by human immunodeficiency virus during the incubation period of AIDS, *Nature,* 362, 359, 1993.

14. **Nuovo, G. J., MacConnell, P., Forde, A., and Delvenne, P.,** Detection of human papillomavirus DNA in formalin-fixed tissues by *in situ* hybridization after amplification by polymerase chain reaction, *Am. J. Pathol.,* 139, 847, 1991.

15. **Nuovo, G. J., Gallery, F., MacConnell, P., Becker, J., and Bloch, W.,** An improved technique for the *in situ* detection of DNA after polymerase chain reaction amplification, *Am. J. Pathol.,* 139, 1239, 1991.

16. **Bagasra, O., Hauptman, S. P., Lischner, H. W., Sachs, M., and Pomerantz, R. J.,** Detection of human immunodeficiency virus type 1 provirus in mononuclear cells by *in situ* polymerase chain reaction, *N. Engl. J. Med.,* 326, 1385, 1992.

17. **Chiu, K., Cohen, S. H., Morris, D. W., and Jordan, G. W.,** Intracellular amplification of proviral DNA in tissue sections using the polymerase chain reaction, *J. Histochem. Cytochem.,* 40, 333, 1992.

18. **Embleton, M. J., Gorochov, G., Jones, P. T., and Winter, G.,** In-cell PCR from mRNA: amplifying and linking the rearranged immunoglobulin heavy and light chain V-genes within single cells, *Nucleic Acids Res.,* 20, 3831, 1992.

19. **Komminoth, P., Long, A. A., Ray, R., and Wolfe, H. J.,** *In situ* polymerase chain reaction detection of viral DNA, single-copy genes, and gene rearrangements in cell suspensions and cytospins, *Diagn. Mol. Pathol,* 1, 85, 1992.

20. **Nuovo, G. J., Margiotta, M., MacConnell, P., and Becker, J.,** Rapid *in situ* detection of PCR-amplified HIV-1 DNA, *Diagn. Mol. Pathol.,* 1, 98, 1992.

21. **Pálsson, P. A.,** Maedi and visna in sheep, in *Slow Virus Disease of Animals and Man,* Kimberlin, R. H., Ed., North-Holland, Amsterdam, 1976, 17.

22. **Fauci, A. S.,** The human immunodeficiency virus: infectivity and mechanisms of pathogenesis, *Science,* 239, 617, 1988.

23. **Haase, A. T., Stowring, L., Narayan, O., Griffin, D., and Price, D.,** Slow persistent infection caused by visna virus: role of host restriction, *Science,* 195, 175, 1977.

24. **Brahic, M., Stowring, L., Ventura, P., and Haase, A. T.,** Gene expression in visna virus infection in sheep, *Nature,* 292, 240, 1981.

25. **Haase, A. T., Stowring, L., Harris, J. D., Traynor, B., Ventura, P., Peluso, R., and Brahic, M.,** Visna DNA synthesis and the tempo of infection *in vitro, Virology,* 119, 399, 1982.

26. **Geballe, A. P., Ventura, P., Stowring, L., and Haase, A. T.,** Quantitative analysis of visna virus replication *in vivo, Virology,* 141, 148, 1985.

27. **Peluso, R., Haase, A., Stowring, L., Edwards, M., and Ventura, P.,** A trojan horse mechanism for the spread of visna virus in monocytes, *Virology,* 147, 231, 1985.

28. **Kwok, S., Kellogg, D. E., McKinney, N., Spasic, D., Goda, L., Levenson, C., and Sninsky, J. J.,** Effects of primer-template mismatches on the polymerase chain reaction: human immunodeficiency virus type 1 model studies, *Nucleic Acids Res.,* 18, 999, 1990.

29. **Rychlik, W., Spencer, W. J., and Rhoads, R. E.,** Optimization of the annealing temperature for DNA amplification *in vitro, Nucleic Acids Res.,* 18, 6409, 1990.

30. **Faloona, F., Weiss, S., Ferre, F., and Mullis, K.,** Direct detection of HIV sequences in blood: high gain polymerase chain reaction, abstract 1019 in 6th Int. Conf. on AIDS, San Francisco, 1990.

31. **Sambrook, J., Fritsch, E. F., and Maniatis, T.,** *Molecular Cloning: A Laboratory Manual,* 2nd ed., Cold Spring Harbor Laboratory Press, Cold Spring Harbor, NY, 1989.

32. **Schowalter, D. B. and Sommer, S. S.,** The generation of radiolabeled DNA and RNA probes with polymerase chain reaction, *Anal. Biochem.,* 177, 90, 1989.

33. **Pizzo, P. A.,** Pediatric AIDS: problems within problems, *Journal of Infectious Diseases,* 161, 316, 1990.

34. **Connor, E.,** Advances in early diagnosis of perinatal HIV infection, *JAMA,* 266, 3474, 1991.

35. **Blanche, S., Rouzioux, C., Moscato, M. G., Veber, F., Mayaux, M., Jacomet, C., Tricoire, J., Deville, A., Vial, M., Firtion, G., De Crepy, A., Douard, D., Robin, M., Courpotin, C., Ciraru-Vigneron, N., Le Deist, F., Griscelli, C., and The HIV Infection in Newborns French Collaborative Study Group,** A prospective study of infants born to women seropositive for human immunodeficiency virus type 1, *N. Engl. J. Med.,* 320, 1643, 1989.

36. **European Collaborative Study,** Children born to women with HIV-1 infection: natural history and risk of transmission, *Lancet,* 337, 253, 1991.

37. **Rogers, M. F., Ou, C. Y., Kilbourne, B., and Schochetman, G.,** Advances and problems in the diagnosis of human immunodeficiency virus infection in infants, *Pediatr. Infect. Dis. J.,* 10, 523, 1991.

38. **Quinn, T. C., Kline, R. L., Halsey, N., Hutton, N., Ruff, A., Butz, A., Boulos, R., and Modlin, J. F.,** Early diagnosis of perinatal HIV infection by detection of viral-specific IgA antibodies, *JAMA,* 266, 3439, 1991.

39. **Landesman, S., Weiblen, B., Mendez, H., Willoughby, A., Goedert, J. J., Rubinstein, A., Minkoff, H., Moroso, G., and Hoff, R.,** Clinical utility of HIV-IgA immunoblot assay in the early diagnosis of perinatal HIV infection, *JAMA,* 266, 3443, 1991.

40. **Krivine, A., Yakudima, A., Le May, M., Pena-Cruz, V., Huang, A. S., and McIntosh, K.,** A comparative study of virus isolation, polymerase chain reaction, and antigen detection in children of mothers infected with human immunodeficiency virus, *J. Pediatr.,* 116, 372, 1990.

41. **Rogers, M. F., Ou, C., Rayfield, M., Thomas, P. A., Schoenbaum, E. E., Abrams, E., Krasinski, K., Selwyn, P. A., Moore, J., Kaul, A., Grimm, K. T., Bamji, M., Schochetman, G., and The New York City Collaborative Study of Maternal HIV Transmission and Montefiore Medical Center HIV Perinatal Transmission Study Group,** Use of the polymerase chain reaction for early detection of the proviral sequences of human immunodeficiency virus in infants born to seropositive mothers, *N. Engl. J. Med.,* 320, 1649, 1989.

42. **Husson, R. N., Comeau, A. M., and Hoff, R.,** Diagnosis of human immunodeficiency virus infection in infants and children, *Pediatrics,* 86, 1, 1990.

43. **Imagawa, D. T., Lee, M. H., Wolinsky, S. M., Sano, K., Morales, F., Kwok, S., Sninsky, J. J., Nishanian, P. G., Giorgi, J., Fahey, J. L., Dudley, J., Visscher, B. R., and Detels, R.,** Human immunodeficiency virus type 1 infection in homosexual men who remain seronegative for prolonged periods, *N. Engl. J. Med.,* 320, 1458, 1989.

44. **Wolinsky, S. M., Rinaldo, C. R., Kwok, S., Sninsky, J. J., Gupta, P., Imagawa, D., Farzadegan, H., Jacobson, L. P., Grovit, K. S., Lee, M. H., Chmiel, J. S., Ginzburg, H., Kaslow, R. A., and Phair, J. P.,** Human immunodeficiency virus type 1 (HIV-1) infection a median of 18 months before a diagnostic western blot: evidence from a cohort of homosexual men, *Ann. Intern. Med.,* 111, 961, 1989.

45. **Patterson, B. K., Till, M., Otto, P., Goolsby, C., Furtado, M. R., McBride, L. J., and Wolinsky, S. M.,** Detection of HIV-1 DNA and messenger RNA in individual cells by PCR-driven *in situ* hybridization and flow cytometry, *Science,* 260, 976, 1993.

46. **Harper, M. E., Marselle, L. M., Gallo, R. C., and Wong-Staal, F.,** Detection of lymphocytes expressing human T-lymphotropic virus type III in lymph nodes and peripheral blood from infected individuals by *in situ* hybridization, *Proc. Natl. Acad. Sci. U.S.A.,* 83, 772, 1986.

47. **Ho, D. D., Moudgil, T., and Alam, M.,** Quantitation of human immunodeficiency virus type 1 in the blood of infected persons, *N. Engl. J. Med.,* 321, 1621, 1989.

48. **Schnittman, S. M., Greenhouse, J. J., Psallidopoulos, M. C., Baseler, M., Salzman, N. P., Fauci, A. S., and Lane, H. C.,** Increasing viral burden in CD4+ T cells from patients with human immunodeficiency virus (HIV) infection reflects rapidly progressive immunosuppression and clinical disease, *Ann. Intern. Med.,* 113, 438, 1990.

49. **Hsia, K. and Spector, S. A.,** Human immunodeficiency virus DNA is present in a high percentage of CD4+ lymphocytes of seropositive individuals, *J. Infect. Dis.,* 164, 470, 1991.

50. **Pantaleo, G., Graziosi, C., Butini, L., Pizzo, P. A., Schnittman, S. M., Kotler, D. P., and Fauci, A. S.,** Lymphoid organs function as major reservoirs for human immunodeficiency virus, *Proc. Natl. Acad. Sci. U.S.A.,* 88, 9838, 1991.

51. **Fox, C. H., Tenner-Rácz, K., Rácz, P., Firpo, A., Pizzo, P. A., and Fauci, A. S.,** Lymphoid germinal centers are reservoirs of human immunodeficiency virus type 1 RNA, *J. Infect. Dis.,* 164, 1051, 1991.

52. **Brahic, M. and Haase, A. T.,** Double-label techniques of *in situ* hybridization and immunocytochemistry, *Curr. Top. Microbiol. Immunol.,* 143, 9, 1989.

53. **Larder, B. A. and Kemp, S. D.,** Multiple mutations in HIV-1 reverse transcriptase confer high-level resistance to zidovudine (AZT), *Science,* 246, 1155, 1989.

54. **St. Clair, M. H., Martin, J. L., Tudor-Williams, G., Bach, M. C., Vavro, C. L., King, D. M., Kellam, P., Kemp, S. D., and Larder, B. A.,** Resistance to ddI and sensitivity to AZT induced by a mutation in HIV-1 reverse transcriptase, *Science,* 253, 1557, 1991.

55. **Fitzgibbon, J. E., Howell, R. M., Haberzettl, C. A., Sperber, S. J., Gocke, D. J., and Dubin, D. T.,** Human immunodeficiency virus type 1 *pol* gene mutations which cause decreased susceptibility to 2′,3′-dideoxycytidine, *Antimicrob. Agents Chemother.,* 36, 153, 1992.

56. **Nowak, M. A., May, R. M., and Anderson, R. M.,** The evolutionary dynamics of HIV-1 quasispecies and the development of immunodeficiency disease, *AIDS,* 4, 1095, 1990.

57. **Hirsch, M. S.,** Chemotherapy of human immunodeficiency virus infections: current practice and future prospects, *J. Infect. Dis.,* 161, 845, 1990.

58. **Erice, A., Mayers, D. L., Strike, D. G., Sannerud, K. J., McCutchan, F. E., Henry, K., and Balfour, H. H.,** Primary infection with zidovudine-resistant strain of human immunodeficiency virus type 1 (HIV-1), *N. Engl. J. Med.,* 328, 1163, 1993.

59. **Bremermann, H. J. and Anderson, R. W.,** The HIV cytopathic effect: potential target for therapy?, *J. Acq. Immune Def. Synd.,* 3, 1119, 1990.

60. **McLean, A. R. and Nowak, M. A.,** Competition between zidovudine-sensitive and zidovudine-resistant strains of HIV, *AIDS,* 6, 71, 1992.

Chapter 4

PCR for the Detection and Characterization of Viral Agents of Gastroenteritis

Jonathan Green and John P. Norcott

TABLE OF CONTENTS

I. INTRODUCTION

Worldwide, there are an estimated 3 to 5 billion cases of gastroenteritis and 5 to 10 million deaths per year.[1] In developed countries, gastroenteritis is the second most common illness[2] and, although generally self-limiting, is responsible for considerable misery and time lost from school and work. Also it can be severe or even fatal in the infant, the elderly, or the debilitated patient. In underdeveloped or developing countries, acute gastroenteritis is the leading cause of death in children under the age of 4 years.[3]

A wide range of viruses have been found in the human gastrointestinal tract (Table 4–1), some of which do not cause gastroenteritis (e.g., enteroviruses). Some, the focus of this review, are associated with acute nonbacterial gastroenteritis; others are opportunistic causes of acute gastroenteritis (e.g., herpes simplex virus[4]). Until the early 1970s the causes of acute nonbacterial gastroenteritis were unknown. Norwalk virus, detected in 1972,[5] was the first of a number of agents found by electron microscopy (EM) of stool specimens from patients with gastroenteritis. Several different virus types were subsequently discovered in this way, including rotaviruses,[6] enteric adenoviruses,[7] Norwalk-like viruses,[8-10] caliciviruses,[11] astroviruses,[12] coronaviruses,[13] and more recently, toroviruses.[14] EM remains the only diagnostic method with which all viruses associated with gastroenteritis can be detected. Some of these viruses (rotaviruses, adenoviruses, caliciviruses, astroviruses) have characteristic, readily identifiable surface structures, while others lack obvious identifying features. These generally fall into two categories: the small, 22- to 30-nm particles with a smooth margin (the small, round viruses; SRVs); and the slightly larger, 30- to 38-nm particles with a structured "fuzzy" surface (the small, round, structured viruses; SRSVs[15]).

As a group, gastroenteritis viruses are fastidious and cannot be propagated in routine cell culture or in laboratory animal hosts, although some (group A rotaviruses, enteric adenoviruses, and astroviruses) can be grown in specialized *in vitro* systems. Other detection techniques are available for some viruses, for example, polyacrylamide gel electrophoresis (PAGE) for rotaviruses and immunoassays for rotaviruses, adenoviruses, astroviruses, Norwalk viruses, and the SRSVs. However, for several viruses serological reagents are scarce, and consequently

Table 4–1   Viruses found in the human gastrointestinal tract

Nonenteropathogenic Viruses
Enteroviruses
    Poliovirus
    Coxsackievirus (A and B)
    Echovirus
    Enterovirus (types 68–71)
Adenovirus (type 1–39)
Reovirus hepatitis A

Viruses Associated with Gastroenteritis
Rotaviruses (groups A, B, C, D, and E)
Enteric adenoviruses (types 40 and 41)
Small, round, structured viruses
Caliciviruses
Astroviruses
Fecal parvovirus
Small, round viruses

Opportunistic Gut Viruses
Human immunodeficiency virus
Virus herpes simplex virus
Cytomegalovirus

use of immunoassays for detection and characterization is not widespread. Nucleic acid hybridization procedures are recent additions to the battery of tests available for the better-studied gastroenteritis viruses (rotaviruses and adenoviruses), but again the more fastidious nature of the other agents has limited their application. The polymerase chain reaction (PCR)[16] has opened up new possibilities for the study of these viruses. PCR-assisted cloning techniques (sequence-independent single-primer amplification; SISPA[17,18]) have enabled the cloning of genomic sequences of fastidious viruses that are available only in very small quantities. Subsequent characterization of the virus genome by PCR and sequencing perhaps offers a more valid basis for classification than the possession or lack of virion structures as determined by EM. Also, where serological reagents have previously been lacking, molecular cloning and *in vitro* expression procedures offer a means to obtain well-characterized specific antigens for use in serological tests. Here, we will review recent applications of PCR for the detection and characterization of gastroenteritis viruses and will highlight areas in which PCR is likely to contribute to future work in this area.

II.   VIRUSES ASSOCIATED WITH GASTROENTERITIS: CURRENT PCR STATUS

A. ROTAVIRUSES
1. Historical Aspects
Although viral etiologies for acute gastroenteric illness were long suspected, the role of rotaviruses eluded detection until the last two decades. Rotaviruses ordinarily do not grow in cell cultures used for virus isolation, and consequently they went unrecognized until the specimens from affected animals were examined by other techniques. In Canada, Mebus et al.[19] recognized rotaviruses as the cause of acute diarrhea in calves, and in 1973, Bishop et al.[6] in Australia and Flewett et al.[20] in England described virus particles resembling bovine rotavirus in the stools of diarrheic children. Rotaviruses have subsequently been identified as a common cause of enteric illness in the young of many animal species, and they are now recognized as the single most important cause of acute gastroenteritis in infants and young children throughout the world. There are five recognized groups of rotavirus, groups A to E, each of which possesses a group-specific antigen. Rotaviruses affecting humans were once thought to be limited to one antigenic group, the group A rotaviruses, whereas groups B to E were thought to be strictly zoonotic infections. In 1982, however, an epidemic of group B rotavirus occurred in China, affecting millions of

individuals, including adults.[21] Subsequently, smaller outbreaks of group B rotavirus, the adult diarrhea rotavirus, have occurred in different parts of China, but there has been no observed spread to other countries. Group C rotaviruses, like group B rotaviruses, are primarily pathogens of swine but have been isolated from small outbreaks of human gastroenteritis in Finland,[22] England,[23,24] and from a large outbreak in Japan.[25] Group D rotaviruses have been isolated only from birds and group E rotaviruses only from pigs.

2. Morphology and Genomic Organization

Rotaviruses belong to the Reoviridae family of segmented, double-stranded RNA viruses. The virions have a distinctive appearance on EM, consisting of a characteristic double-shelled capsid, approximately 70 to 75 nm in diameter. The capsid encloses the 11 gene segments, which range from approximately 660 to 3400 bp, and the total genome has a size of approximately 18,600 bp. Each rotavirus genome segment is base paired over its full length, lacks a 3′ terminal polyadenylated sequence, and contains a capped 5′ structure, presumably on the plus strand. Each RNA segment starts with a 5′ guanidine, followed by a region of conserved sequences that are part of the 5′ noncoding region. An open reading frame coding for the protein product follows, and this is flanked by another region of noncoding sequences, which again contain conserved sequences and terminate with a 3′ cytidine residue.[26] Conserved 5′ and 3′ terminal sequences are also found in other virus families with segmented genomes (Orthomyxoviridae, Arenaviridae, and Bunyaviridae) and are thought to contain signals important for genome transcription, replication, and possibly for the assembly of the viral genome segments. The gene assignment of virus-encoded proteins has now been completed for several rotavirus strains (Table 4–2). Three proteins are primarily responsible for the antigenic diversity of rotaviruses. These are a major inner capsid protein, vp6, which specifies

Table 4–2 **Group A rotavirus genome RNA segments and protein products[a]**

Segment	Number of Base Pairs	Protein	Protein Product	Segment Encoding Equivalent Protein Product in	
				Group B Rotavirus[b]	Group C Rotavirus[c]
1	3302	VP1	Subcore protein	NK[d]	1
2	2690	VP2	Core protein	NK[d]	2
3	2591	VP3	Subcore protein	NK[d]	3
4	2362	VP4	Surface spike protein	3	4
5	611	NSP1	Slightly basic	NK[d]	7
6	1365	VP6	Inner capsid protein	6	5
7	104	NSP3	RNA binding protein	NK[d]	6
8	1059	NSP2	Possible role in RNA replication	NK[d]	9
9	1062	VP7	Surface glycoprotein	9	8
10	751	NSP4	Nonstructural, morphogenic role	NK[d]	NK[d]
11	667	NSP5	Nonstructural, possible RNA binding protein	11	10

[a] Genome assignment for Group A/Si/SA11 strain; [b] Strain B/RAT/IDIR; [c] Strain C/Po/Cowden; [d] Not known.

Derived from Estes, M. K. and Cohen, J., *Microbiological Reviews*, 53, 410–449, 1989. With permission.

group and subgroup antigens, and the two outer capsid proteins, vp4 and vp7, which both elicit neutralizing antibody.

3. Classification of Rotaviruses

Serological investigations have revealed the extent of antigenic diversity among rotavirus strains. As implied above, rotaviruses are now subdivided into at least five groups, named A to E, on the basis of group-specific antigenicity (*vide infra*). They also differ considerably in their group-specific terminal RNA sequences.[27] Group A rotaviruses comprise two subgroups (I, II) determined by the major inner capsid protein (vp6) reactivities in neutralization tests. Fourteen serotypes of group A rotaviruses (G types), based on an outer capsid protein (vp7) specificity, have been identified, of which seven serotypes (1 to 4, 8, 9, and 12[28]) occur among the human group A strains. A second outer protein (vp4) also elicits neutralizing antibodies and is a serotype determinant in addition to, but independent of vp7. Antigenic diversity within the vp4 neutralizing antigen is not well understood. Comparative nucleotide sequencing has revealed five genetically distinct gene 4 types among human rotavirus strains on the basis of RNA and predicted amino acid sequence, referred to as P types 1 to 5. Cross-neutralization tests with baculovirus-expressed vp4 polypeptides suggest at least four of the genetic groups represent distinct antigenic groups.[29] It now appears that a complete antigenic characterization of rotaviruses requires a dual serotyping system, in which antigenic specificities are expressed as both P and G types based on vp7 and vp4 reactivities.

4. Detection and Characterization

As most infants with rotavirus infections shed vast numbers of virus particles during the initial period of gastroenteritis, routine diagnosis generally can be readily achieved by the demonstration of virus particles, antigens, or the double-stranded RNA genome in fecal material by means of standard laboratory tests. These tests include immunoassays and immune electron microscopy (IEM) for virus and viral antigen detection, and PAGE of the RNA genome. Serotyping of rotavirus isolates can be determined by *in vitro* neutralization assays. The availability of monoclonal antibodies that specifically bind to the vp7 proteins of human rotavirus strains have facilitated direct serotyping by enzyme immunoassays of stool specimens.[30] These methods constitute the major tools in detection and characterization of rotaviruses in clinical specimens.

5. PCR in the Detection and Characterization of Rotavirus Infection

Several groups have now described PCR assays for the detection and typing of rotaviruses.[31–39] Wilde et al.[33,34,38] designed primers to amplify a conserved region of gene 6 of group A rotaviruses to facilitate detection of a large range of human rotavirus strains with a single primer pair. Gouvea et al.[31] developed a method to amplify the vp7 gene of group A rotaviruses and were able to directly type strains by employing six primer pairs, which gave amplification products of distinct length for serotypes 1, 2, 3, 4, 8, and 9. Subsequently this method was extended to detect group B and C rotaviruses,[35] facilitating simultaneous detection of all three known human groups of rotavirus in a single assay. This group has also developed a nested PCR using primers based on a variable region of gene 4 of group A rotaviruses, which allows typing of the vp4 gene directly from stools.[39]

Given the relative simplicity of virus detection by EM or by RNA PAGE profile, it is perhaps difficult to see a role for PCR in the routine laboratory detection of rotaviruses. However, the sensitivity of rotavirus detection by traditional techniques requires more than 10^7 virions per gram of specimen to be present, and consequently such techniques are unlikely to detect infection in children shedding rotavirus in lower amounts.[40] Failure to identify such individuals, particularly in certain specific environments such as the hospital setting, may allow them to act as reservoirs of infection. Rotavirus detection by PCR is reported to be approximately 1000 times more sensitive, detecting 3×10^3 to 8×10^4 copies,[33,35,36] and can therefore be used to detect low-level shedding. PCR has also been used to highlight the potential role of environmental surfaces and toys in the transmission of rotavirus infection in day care centers.[38] Monitoring such environments using PCR may lead to strategies to diminish disease transmission within hospital wards and nurseries, schools, and similar settings.

The greater sensitivity of PCR is proving particularly useful for the detection and characterization of group B and C rotaviruses, which are more labile than the group As, tend to be shed in small quantities in feces, and cannot be grown in tissue culture. Diagnostic reagents for group B and C rotaviruses are in short supply, and PCR provides a means of characterizing these viruses in terms of genomic relationships and epidemiology.

Current PCR approaches to the study of rotaviruses are being expanded to include analysis of other genes in order to elucidate their natural history. Gene 4 has attracted much recent interest, as vp4 seems to be involved in more than just the straightforward antigenic diversity of the rotaviruses. Besides eliciting neutralizing antibodies,[29] it also appears to be a marker of virulence in that strains isolated from healthy babies ("nursery strains") share a common vp4 gene, which differs from that shared by diarrheogenic strains.[41] The possibilities for the future study of this and other rotavirus genes are much enhanced by the availability of sensitive and specific PCR assays.

B. NORWALK VIRUS AND THE NORWALK-LIKE AGENTS
1. Historical Aspects

Zahorsky first described the characteristics of "epidemic winter vomiting disease" (WVD) in 1929,[42] a clinically distinct entity among the many presentations of nonbacterial gastroenteritis. Investigations of subsequent outbreaks revealed no bacterial pathogen, and a viral etiology was suggested following studies in which illness was successfully transmitted to volunteers orally inoculated with bacteria-free fecal filtrates.[43] In 1969, an explosive outbreak of winter vomiting disease occurred in an elementary school in Norwalk, Ohio; 50% of the pupils and teachers at the school were affected, and secondary cases among family contacts occurred with an incidence of 32%.[44] At the time, no viral or bacterial pathogen was detected. Dolin et al.[45] induced winter vomiting disease in volunteers who ingested a filtrate prepared from specimens taken from a secondary case, and subsequently passaged the agent through volunteers, again inducing WVD. This facilitated some characterization of the agent, which was shown to be less than 66 nm in diameter and resistant to ether, acid, or heating at 60°C for 30 min. Kapikian et al.[5] later examined fecal material from the Norwalk outbreak by IEM with convalescent sera from volunteers and filtrates demonstrated to be capable of transmitting WVD. Clusters of 27-nm virus-like particles were demonstrated, together with rising antibody titers in paired sera from affected individuals. Thus, for the first time, a virus had been detected in association with nonbacterial gastroenteritis. Since the original description of Norwalk virus, a number of viruses with similar physical properties have been found in fecal specimens from adults and children during sporadic infections and outbreaks of gastroenteric disease. These agents have usually been named according to the geographic region of their discovery, hence the descriptions of Hawaii agent (1972[8]), Montgomery County (1972[8]), Otefuke (1979[9]), Snow Mountain (1982[10]), and Taunton (1983[46]) agents.

In the U.K. these viruses are generally referred to as small round structured viruses (SRSVs),[15] while in the U.S. they are known as Norwalk-like viruses. Evidence to suggest that these viruses are members of the Caliciviridae is based on particle size, buoyant density, polypeptide analysis, genome structure, and limited serological data.[47] Attempts to adapt these viruses to growth *in vitro* have been unsuccessful, and this has made it difficult to define the antigenic relationships between SRSVs. Attempts at classification have largely been based on IEM,[48,49] but no widely accepted scheme has emerged. Recently, genomic sequence information obtained from cDNA cloning of members of the SRSV group[50–52] has become available, opening up new possibilities for the detection and characterization of these agents.

2. Clinical Features

Norwalk virus produces a characteristic picture of gastroenteritis in individuals affected in outbreaks or in human volunteers. The onset of diarrhea, which is usually mild to moderate in extent, occurs 24 to 72 h after ingestion and lasts for 18 to 36 h.[45] The diarrhea is usually accompanied by vomiting, which in some cases can occur as often as 20 times a day. Many patients also have headaches, myalgia, and malaise. These symptoms do not usually persist beyond 72 h. The occurrence of chronic infection with Norwalk virus following acute infection has not been reported.

3. Epidemiology

Many of the cases of SRSV-associated gastroenteritis have occurred among individuals living in a closed environment such as schools, recreational camps, and cruise ships. Other outbreaks have occurred in individuals ingesting contaminated shellfish or water. Outbreaks associated with contaminated food-stuffs, e.g., salads or cake icing, have also been reported.[53,54] Outbreaks occur all year round, affecting older children and adults, but not usually infants and young children. As with most enteric infections, person-to-person transmission of SRSVs has been assumed to be by the fecal-oral route, but there is evidence to suggest that transmission may also occur by other routes. It is thought that the vomit of infected individuals may act as a source of infection with projectile vomiting, a feature of Norwalk gastroenteritis, leading to environmental contamination and aerosol production. In an outbreak at a hospital, some affected individuals had merely walked through the area in which vomiting had occurred and had had no contact with staff or patients.[55] This suggests airborne transmission, demonstrates the highly infectious nature of the SRSVs, and emphasizes the difficulties of controlling infection.

4. Currently Available Techniques for the Detection and Characterization of SRSVs

Three approaches for the diagnosis of SRSV infection have been available in recent years: visualization of virus particles by EM; detection of rising antibody titers in paired serum samples; and detection of specific antigen in fecal specimens. Recent developments have pointed toward a new approach: detection of viral RNA by PCR.

In the U.K., EM remains the only method routinely available for detection of SRSVs. The limitations of EM are that the technique is highly labor intensive and that recognition of SRSV particles requires considerable experience and skill on the part of the investigator. During SRSV infections, virus is excreted for only a short period (usually less than 4 days) and in low numbers (10^5 to 10^7 virus particles per gram of feces), and although methods are available for concentrating virus from feces, the sensitivity of detection of SRSVs by EM is low. Solid-phase EM, which involves coating the EM grid with specific antibody in order to capture virus particles present in fecal specimens, has provided both a more sensitive means for the detection of fecal viruses[56] and a means by which antigenic relationships between SRSVs may be investigated. Lewis et al.[48] described a preliminary classification scheme in which SRSV isolates were grouped into five serotypes (Table 4–3).

5. PCR for the Detection of SRSVs

A substantial breakthrough came in 1991, when cDNA cloning of Norwalk virus was achieved and genomic sequence information became available, facilitating application of molecular techniques to the detection and characterization of these agents. Jiang et al.[50] cloned a region of the RNA polymerase gene and subsequently determined the RNA sequence of the entire Norwalk virus genome. Matsui et al.[51] published sequence data from an immunogenic region within the longest open reading frame which contains sequences of the putative 2C polypeptide, the 3C protease, and an RNA-dependent RNA polymerase. Subsequently Lambden and colleagues cloned and sequenced an SRSV, Southampton virus, isolate that is antigenically related to the Snow Mountain agent.[52] Following the publication of Norwalk virus genomic sequences, reverse transcription-polymerase chain reaction (RT-PCR) assays were described for the detection of Norwalk virus in fecal specimens.[57–59]

We have investigated use of PCR with two main objectives:

1. To determine the incidence of Norwalk virus-associated acute gastroenteritis in the UK
2. To investigate the relationship of genomic and antigenic variation among SRSVs

a. The Incidence of Norwalk Virus-Associated Acute Gastroenteritis in the U.K.

By RT-PCR we tested 99 fecal specimens collected in the U.K. from 50 outbreaks and 16 sporadic cases of viral gastroenteritis between 1986 and 1992 and containing small round structured virus (SRSV) particles, as determined by EM. Using primers derived from published Norwalk virus sequences (encoding the putative RNA polymerase gene[50] and an immunogenic region[51]), 15 of 99 specimens were positive by PCR, and all came from three outbreaks. Genomic variation between virus detected in the three outbreaks was investigated by PCR sequencing of amplification products. Sequence comparisons in the two selected regions showed 97% nucleic acid sequence identity among the outbreak strains and 70 to 77% identity between outbreaks strains and published Norwalk sequences. Of the 99 specimens tested by PCR, 21 had been previously antigen typed by solid-phase IEM (SPIEM) and/or ELISA. All PCR-positive strains were typed as SRSV UK2 strains (Table 4–3), suggesting the outbreak strains detected were related at both antigenic and genomic levels.

b. The Relationship of Genomic and Antigenic Variation among SRSVs

Clearly, the primers derived from Norwalk sequences were very specific and detected a narrow range of circulating SRSV strains. For diagnostic purposes, a primer pair that detected a broader range of SRSVs would be required. We have attempted to identify "catch-all" primers based on alignments of the sequences of Norwalk virus,[50] Snow Mountain agent (Dr. Mary Estes, personal communication), and our own data from SRSV UK2 strains.[59] While unable to detect all SRSVs, we have identified primer pairs that detect approximately 38% of strains in fecal specimens collected in the U.K. during 1990 to 1992.[60]

PCR sequencing of a 266-bp region of the RNA polymerase gene amplified from 10 strains has revealed the extent of genomic variation within SRSVs.[60] Antigenic typing of strains by SPIEM has been performed and compared with a phylogenetic analysis of the sequences obtained (Figure 4–1). The primer pair employed amplified members of UK SRSV serotypes 1, 2, 3, and 4 and also some isolates untypeable by SPIEM. Considerable genomic variation within this 266-bp region was demonstrated in the detected strains with as little as 56% identity (isolates ERVL 5 and ERVL 10) being found. The phylogenetic tree shows that SRSV UK2 strains are closely related to each other and distinct from other types. This investigation highlights the difficulties in identifying conserved regions on which to base "catch-all" primers for SRSVs. Accurate determination of antigenic relationships between SRSV strains and, consequently, the identification of group-reactive and type-reactive epitopes, awaits the availability of specific (in vitro expressed), serological reagents.

C. ENTERIC ADENOVIRUSES

Adenovirus types 40 (Ad40) and 41 (Ad41) are considered to be second to only rotaviruses as a cause of gastroenteritis in young children.[61–63] Clinically the disease is usually afebrile, with diarrhea persisting for 24 to 48 h. Vomiting is not usually a syndrome.[7] Morphologically the virion is a nonenveloped isometric particle with icosahedral symmetry. It is between 70 and 90 nm in diameter and consists of 252 capsomeres, each being about 9 nm in diameter. The genome of adenovirus is a single linear molecule of dsDNA of approximately 36 kb (20 to 25×10^6 Da). The GC content ranges from 48 to 61% between strains.[64]

1. Methods for Diagnosis

There are a number of methods for detecting adenovirus virions, antigens, or DNA in stools, including EM, latex agglutination, ELISA, DNA restriction analysis, and dot blot hybridization. Screening for enteric adenoviruses is perhaps most commonly done by EM (or where EM is not available, by adenovirus group-specific ELISA) followed by type-specific ELISA based on monoclonal antibodies. Dot blot hybridization offers an alternative where large numbers of specimens are to be examined.

2. PCR for the Detection and Characterization of Enteric Adenoviruses

Allard et al.[65,66] have described a two-step PCR, the first stage of which detects all six subgenera (A to F) of adenoviruses, using primers derived from the genomic sequences coding for the adenovirus hexon. This is followed by amplification with EAd-specific primers based on Ad40 and Ad41 sequences of the E1B genomic region. The ability to detect a single virus particle was reported using this procedure, and the use of primers based on hexon sequences which are conserved among adenoviruses provides an opportunity to detect previously unrecognized adenoviruses. However, it seems unlikely that PCR will become a routine screening test for enteric adenoviruses in stools, although it may have a role in genomic characterization of isolates for epidemiological studies and in environmental monitoring of sewage and seawater.[67]

D. ASTROVIRUSES

1. Morphology and Genomic Organization

The first report of human astroviruses seen by electron microscope in the diarrheic feces of an infant was by Madeley and Cosgrove in 1975.[12] Astroviruses are round particles 28 to 30 nm in diameter with a smooth edge and a characteristic star configuration with either five or six points. There appear to be at least five distinct serotypes of human astrovirus, distinguished by immunofluorescence and by immune EM.[68]

The astrovirus genome is single-stranded positive-sense RNA of between 7.2 and 7.6 kb with a poly-A tract. It is unclear at present whether astroviruses are a distinct family or whether they belong to either the Picornaviridae or Caliciviridae family. The genomic organization and structural polypeptides of the astroviruses appear to share

Table 4–3 **Serotyping of small, round, structured viruses in the U.K. by SPIEM**

Serotype	Prototype Agents
SRSV UK 1	Taunton
SRSV UK 2	Norwalk
SRSV UK 3	Hawaii
SRSV UK 4	Snow Mountain/Otofuke
SRSV UK 5	—

From Lewis, D. C., Lightfoot, N. R., and Pether, J. V. S., *Journal of Clinical Microbiology,* 26, 938, 1988. With permission.

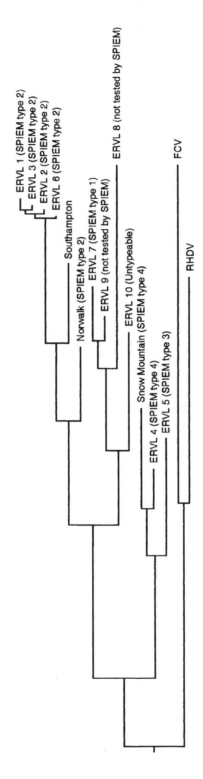

Figure 4–1 Phylogenic tree constructed from a 266-bp region of the RNA polymerase gene of ten SRSV isolates (ERVL 1 to 10) by the method of Jotun Hein. Included are the published sequences of Norwalk virus,[50] Snow Mountain (Dr. M. K. Estes, personal communication), Southampton,[52] feline calcivirus (FCV) (EMBL/GenBank Accession number M86379), and rabbit hemorrhagic disease virus (RHDV) (EMBL/GenBank Accession number M67473). (Modified from Norcott, J. P., Green, J., Lewis, D. C., Estes, M. K., and Brown, D. W. G., *J. Med. Virol.*, 1994, in press. With permission.)

similarities with members of the Picornaviridae. However, astroviruses differ from picornaviruses in their characteristic virion structure compared to the featureless appearance of picornaviruses. Second, astroviruses lack significant amino acid homology within the 3′ terminus region of the genome, which in picornaviruses, encodes the viral RNA polymerase (3D protein). They also lack the Tyr-Gly-Asp-Asp functional motif, believed typical of RNA polymerases, located in the last 150 amino acids of the polyprotein of picornaviruses, suggesting that their genome organization could be different.[69] Third, viral antigens of astrovirus have been shown in the nucleus of infected cells during early stages of replication.[69] No nuclear involvement is seen during picornavirus or calicivirus infection, suggesting that there may be significant differences between replication of these viruses and astroviruses.

Astrovirus has been cultured in HEK, LLC-MK2, and more recently in CaCo-2 cells, a continuous line. As with certain other enteric viruses, the addition of trypsin to the culture medium is a requirement for growth. While not useful in routine diagnosis of astroviruses, the ability to propagate these viruses has facilitated the generation of polyclonal and monoclonal antibodies[75] and their genomic cloning.[69]

2. Astroviruses as a Cause of Human Gastroenteritis

The relative contribution of astroviruses to the total incidence of virus-associated gastroenteritis is not precisely known. They most frequently cause disease in young children, and by the age of 5 years, more than 80% of children show serological evidence of infection. Although similar to rotaviral disease, it is less severe. These astroviruses are generally believed to be a minor cause of gastroenteritis, accounting for 3 to 5% of cases.[70]

3. Methods for Diagnosis

Direct EM is the most useful method for the routine detection of astroviruses. In acute illness, virus particles can be shed in quantities of up to 10^{10} particles per milliliter of feces, and their readily recognizable morphology facilitates identification.[15] However, astroviruses may not always show the typical appearance and may have an indistinct morphology, making them difficult to differentiate from other small round viruses. IEM and immunofluorescence can achieve this differentiation, while monoclonal antibody-based immunoassays have the added advantage of requiring no specimen pretreatment and permit routine testing where EM is not available.[71] A dot blot hybridization assay using an astrovirus-specific RNA probe has been described[72] and was shown to be able to detect high dilutions of tissue-culture-grown virus. However, when compared with immunoassay for routine screening, RNA hybridization did not detect more astrovirus-positive fecal specimens, and immunoassay was the preferred test for routinely screening large numbers of specimens. PCR has been described for astroviruses,[73] specifically serotype 1, but the increasing availability of astrovirus sequence data means that a PCR that detects all current serotypes will probably be available soon. Again, while PCR may not immediately replace other procedures for the routine detection of astroviruses, further characterization of them by PCR may lead to improved serological assays.

II. FUTURE USES OF PCR IN THE STUDY OF NOVEL OR PUTATIVE AGENTS OF GASTROENTERITIS

In approximately 20 to 40% of cases of acute nonbacterial gastroenteritis, no etiological agent is identified.[74] This may be due to virus shedding (at the time of specimen collection) occurring at levels below the limits of detection of the diagnostic procedures being employed. The increased sensitivity of PCR over preexisting techniques may lead to a reduction in the number of virus-negative outbreaks. An alternative explanation for failure to identify a recognized gastroenteric pathogen in these cases is that not all agents of gastroenteritis are known and there may be a number of viral agents whose role in acute gastroenteritis is, as yet, unrecognized. This possibility has led to an ongoing search for new agents of gastroenteritis. Many of the agents associated with human gastroenteritis were originally discovered in animal species. For example, rotaviruses were known to be pathogenic in mice, monkeys, and cows for several years before their identification as a cause of infantile gastroenteritis. Consequently, identification of a new animal enteric pathogen often leads to a search for a human counterpart; described below are examples of candidate pathogens that may or may not have a role in human gastroenteritis.

A. TOROVIRUSES

Toroviruses are enveloped, positive-strand RNA viruses that cause enteric infections in animals. Breda virus was the first torovirus described, isolated from stools of diarrheal calves in 1982[76] and subsequently shown to cause diarrhea in gnotobiotic and conventionally reared calves. Two sera groups of Breda virus have been proposed,[77] and antigenic and morphological relationships have been demonstrated with a second torovirus, Berne virus, which was isolated from a rectal swab taken from a horse in 1973.[78] Unlike Breda virus, Berne virus could be propagated in tissue culture. In 1984, Beards et al.[79] reported the appearance of torovirus-like particles in the stools of patients with diarrhea and a serological cross reaction between these particles and a hyperimmune serum raised against Breda virus II was demonstrated by IEM.[80] In order to determine the prevalence of human torovirus infections, enzyme immunoassays based on Breda virus antigens and antisera were developed, but only 1 of 430 human fecal specimens were positive for Breda virus antigen, and no antibody to Breda or Berne virus was demonstrated in human sera. Hybridization assays with cDNA probes generated from the structural protein and polymerase genes of Berne virus have also been developed. A total of 117 torovirus-like particle-positive human fecal specimens were tested, of which 11 were positive by hybridization.[81] A PCR for toroviruses, using primers derived from 3′ terminal sequences conserved between Breda and Berne viruses, has been described and has been shown to amplify human toroviruses,[82] but no epidemiological studies have as yet been reported. The PCR is likely to prove a useful tool in assessment of the role of toroviruses in human gastroenteritis, as it is more sensitive than EM and antigen ELISA.

B. PESTIVIRUSES

There are three members of the *Pestivirus* genus, bovine viral diarrhea virus (BVDV), hog cholera virus (HCV) of pigs, and Border disease virus (BDV) of sheep. These are single-stranded, positive-polarity RNA viruses and are classified into the Togaviridae family. The prototype virus is BVDV, which causes acute diarrhea and mucus membrane inflammation in young cows. Pigs infected with HCV also have severe gastroenteritis as well as pneumonia and neurological dysfunction. Yolken et al.[83] demonstrated that on an Arizona Indian reservation, 23% of stools from children under 2 years of age with gastroenteritis of unknown etiology contained pestivirus antigen when tested by monoclonal-antibody-based ELISA. A 5′ genomic terminal region conserved among HCV, BVDV, and BDV has been recognized, and a PCR that can detect and differentiate among all three pestiviruses has been described.[84] This may prove to be a useful tool in investigations for a role of pestiviruses in human gastroenteritis.

C. PICOBIRNAVIRUSES

Recently, a new group of viruses has been described in fecal specimens collected from rats (*Oryzomus nigripes*) in Brazil.[85] The name "picobirnaviruses" has been proposed for these viruses, which have a small (30 to 35 nm), featureless virion and a bisegmented dsRNA genome. These viruses have been found in fecal specimens collected from several vertebrate species by PAGE. In humans, approximately 4% of fecal specimens from adults (aged 18 to 90) with diarrhea were positive by PAGE. A similar percentage was found in a control group of hospitalized children without gastroenteritis,[86] suggesting that picobirnavirus infection may be fairly widespread in humans, but may not be causally associated with gastroenteritis. A study by Grohmann et al.,[87] however, demonstrated picobirnaviruses in 9% of fecal specimens from HIV-infected patients with diarrhea but in only 2% of specimens from patients without diarrhea; they concluded that picobirnaviruses may be important etiological agents of diarrhea in these patients. Establishing the true significance of picobirnaviruses in human gastroenteritis awaits the development of more sensitive detection techniques. At present, demonstration of the dsRNA genome in fecal specimens by PAGE is the only available method. As no genomic sequence data are presently available for human picobirnaviruses, molecular methods have yet to be applied.

D. SMALL, ROUND VIRUSES (SRVS)

The SRVs constitute a poorly characterized group of viruses that share a similar morphology by EM of a 22- to 28-nm-diameter particle without detectable ultrastructure. These virions and their DNA most closely resemble parvoviruses,[88] although they have also been described as enterovirus-like.[89,90] Little is known of their role in the etiology of human gastroenteritis, although parvoviruses have been shown to be associated with enteric disease in cattle, cats, dogs, mink, turkeys, and chickens, while entero-like SRVs have been implicated in diarrheal disease in turkeys and chickens (for review of animal SRVs, see Reference 91). There is a clear need for further investigations to determine the true nature of these viruses and their significance as a cause of gastroenteritis in humans.[88] PCR offers great potential as a tool in these investigations.

IV. PCR FOR GASTROENTERITIS VIRUSES: PRACTICAL ASPECTS

A major problem experienced with PCR is that of false-positive results generated by contamination of reagents and specimens with target DNA. This contamination can arise from several sources, including amplified products from previous PCR tests, cloned DNA, culture-grown virus, and positive specimens. The problem of false-positive results and the necessary measures that must be taken to minimize them (e.g., use of dedicated rooms and equipment) are well documented and apply to all PCR assays.[92] For PCR testing of fecal specimens, a particular problem has been that of false-negative results. These are largely due to *Taq* polymerase inhibitors, which co-purify with viral RNA or DNA when extracted by phenol-chloroform and ethanol precipitation-based methods. A variety of extraction procedures have now been described for the detection of gastroenteritis viruses in fecal specimens.[32–35,57–59] We have employed the method of guanidinium isothiocyanate denaturation and binding of nucleic acid to silica, described by Boom et al.,[93] for extraction of RNA for subsequent amplification of sequences from Norwalk-like viruses, toroviruses, and other viruses of the gut, including hepatitis A virus and enteroviruses, and have found it to be a convenient and reliable method. However, no formal comparisons with other methods have been performed.

Environmental applications of PCR also present specific problems in terms of specimen handling.[93] While the sensitivity of PCR makes it particularly suited to environmental specimens such as water samples, problems of inhibitors of *Taq* polymerase still have to be overcome. Filtration procedures have been found to be insensitive and inconsistent,[94] and even methods described for extraction of fecal specimens[32] did not remove inhibitors from surface water samples.[93] However, nucleic acid has been successfully extracted and amplified from 5-l quantities of sewage using a combination of purification steps, but has not been evaluated for seawater or freshwater.[94] Extraction of viral nucleic acids from other environmental specimens (e.g., shellfish) presents particular difficulties and requires specific procedures.[95]

When attempting to PCR RNA viruses, whether from fecal specimens, water, or shellfish, additional care must be taken to prevent degradation of extracted RNA by RNases.[96] The relatively labile nature of RNA compared to DNA may be one factor in explaining the lower PCR sensitivity obtained for RNA viruses compared to DNA viruses. PCRs for rotavirus, for example, generally detect 10^3 to 10^4 copies while sensitivity of single genome copy detection has been reported for adenoviruses.[66] The major factor is most probably the inherent inefficiency of the reverse transcription step,[97,98] although the double-stranded nature of the rotavirus genome may also contribute to the lower sensitivity.[33] However, the sensitivity offered by PCR is still far greater than that offered by more traditional techniques such as EM and immunoassay.

V. CONCLUSIONS

PCR has now been described for a number of viruses associated with gastroenteritis (Table 4–4). The remarkable sensitivity of PCR is well documented. However, its widespread use in the routine screening of fecal specimens for gastroenteritis viruses is perhaps unlikely; compared with EM, PCR cannot simply be applied to detect all viruses associated with gastroenteritis in one test. PCR has facilitated detection of specific viruses in situations where other techniques are not adequately sensitive, for example, detection of group C rotaviruses, which may be shed in small quantities. PCR has also facilitated genomic characterization of noncultivatable viruses, again including group B and C rotaviruses and the SRSVs. It may well be that future detection and characterization of certain gastroenteritis viruses (e.g., the SRSVs) will be by serological assays based on *in vitro* expressed antigens from genomic sequences initially generated by PCR. There may also be a role for PCR in investigations of putative viruses of gastroenteritis, particularly those fastidious viruses (toroviruses and picobirnaviruses) that do not lend themselves to study by more traditional means.

PCR may be useful in outbreak investigations, such as monitoring excretion of Norwalk and Norwalk-like viruses by infected individuals, particularly food handlers,[57] and identifying shedding in asymptomatic individuals; monitoring surfaces within specific environments, including hospitals and day care centers to identify reservoirs of infection;[38] screening of foodstuffs, particularly those commonly associated with outbreaks such as shellfish;[95] and screening water for enteric pathogens.

Thus the impact of PCR on the study of gastroenteritis viruses has been far reaching, and while routine diagnostic roles are presently difficult, PCR seems likely to continue to be a powerful research tool in the study of these viruses.

Table 4–4 **PCR for the detection and characterization of viruses associated with gastroenteritis in humans**

Virus	Strain/Serotype	Amplified Region	Ref.
Rotavirus	Group A	VP7 segment	31,32,37
		Segment 6	33,34,38
	Groups A, B, and C	Segments 9, 8, and 6, respectively	35
	Group B	Segment 11	36
	Group A	Segment 4	39
Norwalk virus		RNA polymerase	57,58
		Immunogenic region	58
Norwalk-like viruses		RNA polymerase	59,60
Enteric adenoviruses	Ad40, Ad41	E1B	65,66
Astroviruses	Serotype 1	3′ terminus	73
Torovirus		3′ terminus	82

REFERENCES

1. **Anon.,** Viral agents of gastroenteritis: public health importance and outbreak management, *Med. Microbiol. Weekly Rep.,* 39, 1–23, 1990.
2. **Kapikian, A. Z., Wyatt, R. G., Greenberg, H. B., Kalica, A. R., Kim, H. W., Brandt, C. D., Rodriguez, W. J., Parrott, R. H., and Chanock, R. M.,** Approaches to immunisation of infants and young children against gastroenteritis due to rotaviruses, *Rev. Infect. Dis.,* 2, 459–469, 1980.
3. **Tolia, V. K. and Dubois, R. S.,** Update of oral rehydration: its place in treatment of acute gastroenteritis, *Paediatica Ann.,* 14, 295–303, 1985.
4. **Farthing, M. J.,** Gut viruses: a role in gastrointestinal disease? in Viruses and the Gut, Proc. 9th BSG.SK&F Int. Workshop, 1988, Farthing, M. J., Ed., SK&F, 1989.
5. **Kapikian, A. Z., Wyatt, R. G., Dolin, R., Thornhill, T. S., Kalika, A. R., and Chanock, R. M.,** Visualisation by immune electron microscopy of a 27-nm particle associated with acute nonbacterial gastroenteritis, *J. Virol.,* 10, 1075, 1972.
6. **Bishop, R. F., Davidson, G. P., Holmes, I. H., and Ruck, B. J.,** Virus particles in epithelial cells of duodenal mucosa from children with acute non-bacterial gastroenteritis, *Lancet,* ii, 1281–1283, 1973.
7. **Flewett, T. H., Bryden, A. S., and Davies, H.,** Epidemic viral enteritis in a long-stay childrens ward, *Lancet,* i, 4, 1975.
8. **Thornhill, T. S., Wyatt, R. G., Kalica, A. R., Dolin, R., Chanock, R. M., and Kapikian, A. Z.,** Detection by immune electron microscopy of 26 to 27-nm virus-like particles associated with two family outbreaks of gastroenteritis, *J. Infect. Dis.,* 135, 20, 1977.
9. **Taniguchi, K., Urasawa, S., and Urasawa, T.,** Virus-like particles, 35 to 40 nm, associated with an institutional outbreak of acute gastroenteritis in adults, *J. Clin. Microbiol.,* 10, 730, 1979.
10. **Dolin, R., Reichman, R. C., Roessner, K. D., Tralk, T. S., Schooley, R. T., Gary, W., and Morens, D.,** Detection by immune electron microscopy of the Snow Mountain agent of acute viral gastroenteritis, *J. Infect. Dis.,* 146, 184, 1982.
11. **Madeley, C. R. and Cosgrove, B. P.,** Caliciviruses in man (letter), *Lancet,* i, 199–200, 1976.
12. **Madeley, C. R. and Cosgrove, B. P.,** 28 nm particles in faeces in infantile gastroenteritis (letter), *Lancet,* ii, 451, 1975.
13. **Caul, E. O., Paver, W. K., and Clarke, S. K. R.,** Coronavirus particles in faeces from patients with gastroenteritis, *Lancet,* i, 1192, 1975.
14. **Beards, G. M., Hall, C., Green, J., Flewett, T. H., Lamouliate, F., and DuPasquier, P.,** An enveloped virus in the stools of children and adults with gastroenteritis that resembles the Breda virus of calves, *Lancet,* ii, 1050, 1984.
15. **Caul, E. O. and Appleton, H.,** The electron microscopical and physical characteristics of small round human faecal viruses; an interim scheme for classification, *J. Med. Virol.,* 9, 257, 1982.

16. **Saiki, R., Scharf, S., Faloona, F., Mullis, K., Horn, G., Erlich, H., and Arnheim, N.,** Enzymatic amplification of beta-globin genomic sequences and restriction site analysis for the diagnosis of sickle cell anaemia, *Science,* 230, 1350–1354, 1985.

17. **Reyes, G. R. and Kim, J. P.,** Sequence-independent, single primer amplification (SISPA) of complex DNA populations, *Mol. Cell. Probes,* 5, 473–481, 1991.

18. **Lambden, P. R., Cooke, S. J., Caul, E. O., and Clarke, I. N.,** Cloning of noncultivatable human rotavirus by single primer amplification, *J. Virol.,* 66, 1817–1822, 1992.

19. **Mebus, C. A., Underdahl, N. R., Rhodes, M. B., and Twiehaus, M. J.,** Calf diarrhoea (scours) reproduced with a virus from a field outbreak, *Univ. Nebraska Res. Bull.,* 233, 1, 1969.

20. **Flewett, T. H., Bryden, A. S., and Davies, H.,** Virus particles in gastroenteritis, *Lancet,* i, 1497, 1973.

21. **Hung, R., Wang, C., Fang, Z., Chou, Z., Chang, X., Liong, X., Chan, G., Yo, H., Chao, T., Ye, W., Den, S., and Chang, W.,** Water borne outbreak of rotavirus diarrhoea in adults in China caused by a novel rotavirus, *Lancet,* i, 1139, 1984.

22. **von Bonsdorff, C. H. and Svensson, L.,** Human serogroup C rotavirus in Finland, Scandinavia, *J. Infect. Dis.,* 20, 475–478, 1988.

23. **Brown, D. W. G., Campbell, L., Tomkins, D. S., and Hambling, M. H.,** School outbreak of gastroenteritis due to atypical rotavirus, *Lancet,* ii, 737–738, 1989.

24. **Caul, E. O., Ashley, C. R., Darville, J. M., and Bridger, J. C.,** Group C rotavirus associated with fatal enteritis in a family outbreak, *J. Med. Virol.,* 30, 201–205, 1990.

25. **Matsumoto, K., Hatana, M., Kobayashi, K., et al.,** An outbreak of gastroenteritis associated with acute rotaviral in school children, *J. Infect. Dis.,* 160, 611–615, 1989.

26. **Theil, K. W.,** Group A Rotaviruses, in *Viral Diarrheas of Man and Animals,* Saif, L. J. and Theil, K. W., Eds., CRC Press, Boca Raton, FL, 1989, chap. 3.

27. **Pedley, S., Bridger, J. C., Brown, J. F., and McCrae, M. A.,** Molecular characterisation of rotaviruses with distinct group antigens, *J. Gen. Virol.,* 64, 2093–2101, 1983.

28. **Woods, P. A., Gentsch, J., Gouvea, V., Mata, L., Santosham, M., Bai, Z.-S., Urasawa, S., and Glass, R. I.,** Distribution of serotypes of human rotavirus in different populations, *J. Clin. Microbiol.,* 30, 781–785, 1992.

29. **Gorziglia, M., Larralde, G., Kapikian, A. Z., and Chanock, R. M.,** Antigenic relationships among human rotaviruses as determined by outer capsid protein vp4, *Proc. Natl. Acad. Sci. U.S.A.,* 83, 7039–7043, 1990.

30. **Taniguchi, K., Urosawa, T., Morita, Y., Greenberg, H. B., and Urasawa, S.,** Direct serotyping of human rotavirus in stools by an enzyme-linked immunosorbent assay using serotype 1-, 2-, 3-, and 4-specific monoclonal antibodies to VP7, *J. Infect. Dis.,* 155, 1159–1166, 1987.

31. **Gouvea, V., Glass, R. I., Woods, P., Taniguchi, K., Clark, H. F., Forrester, B., and Fang, Z.-Y.,** Polymerase chain reaction amplification and typing of rotavirus nucleic acid, *J. Clin. Microbiol.,* 28, 276–282, 1990.

32. **Xu, L., Harbour, D., and McCrae, M. A.,** The application of polymerase chain reaction to the detection of rotaviruses in faeces, *J. Virol. Meth.,* 27, 29–38, 1990.

33. **Wilde, J., Eiden, J., and Yolken, R.,** Removal of inhibitory substances from human faecal specimens for detection of group A rotaviruses by reverse transcriptase and polymerase chain reaction, *J. Clin. Microbiol.,* 28, 1300–1307, 1990.

34. **Wilde, J., Yolken, R., Willoughby, R., and Eiden, J.,** Improved detection of rotavirus shedding by polymerase chain reaction, *Lancet,* 337, 323–326, 1991.

35. **Gouvea, V., Allen, J. R., Glass, R. I., Fang, Z.-Y., Bremont, M., Cohen, J., McCrae, M. A., Saif, L. J., Sinarachatanant, P., and Caul, E. O.,** Detection of group B and C rotaviruses by polymerase chain reaction, *J. Clin. Microbiol.,* 29, 519–523, 1991.

36. **Eiden, J. J., Wilde, J., Firoozmand, F., and Yolken, R.,** Detection of animal and human group B rotaviruses in faecal specimens by polymerase chain reaction, *J. Clin. Microbiol.,* 29, 539–543, 1991.

37. **Nakagomi, O., Oyamada, H., and Nakagomi, T.,** Experience with serotyping rotavirus strains by reverse transcription and two-step polymerase chain reaction with generic and type-specific primers, *Mol. Cell. Probes,* 5, 285–289, 1991.

38. **Wilde, J., Van, R., Pickering, L., Eiden, J., and Yolken, R.,** Detection of rotaviruses in the day care environment by reverse transcriptase polymerase chain reaction, *J. Infect. Dis.,* 166, 507–511, 1992.

39. **Gentsch, J. R., Glass, R. I., Woods, P., Gouvea, V., Gorziglia, M., Flores, J., Das, B. K., Bhan, M. K.,** Identification of group A rotavirus gene 4 types by polymerase chain reaction, *J. Clin. Microbiol.,* 30, 1365–1373, 1992.

40. **Ward, R. L., Bernstein, D. I., Young, E. C., Sherwood, J. R., Knowlton, D. R., and Schiff, G. M.,** Human rotavirus studies in volunteers: determination of infectious dose and serological response to infection, *J. Infect. Dis.,* 154, 871–880, 1986.

41. **Flores, J., Midthun, K., Hoshino, Y., Green, K., Gorziglia, M., Kapikian, A. Z., and Chanock, R. M.,** Conservation of the fourth gene among rotaviruses recovered from asymptomatic newborn infants and its possible role in attenuation, *J. Virol.,* 60, 972–979, 1986.

42. **Zahorsky, J.,** Hyperemesis hiemis or winter vomiting disease, *Arch. Paediatrics,* 46, 391, 1929.

43. **Jordan, W. S., Gordon, I., and Dorrance, W. R.,** A study of illness in a group of Cleveland families. VII. Transmission of acute nonbacterial gastroenteritis to volunteers: evidence for two different etiologic agents, *J. Exp. Med.,* 98, 461, 1953.

44. **Adler, J. L. and Zickl, R.,** Winter vomiting disease, *J. Infect. Dis.,* 119, 668, 1969.

45. **Dolin, R., Blacklow, N. R., DuPont, H., Formal, S. B., Buscho, R. F., Kasel, J. A., Chames, R. P., Hornick, R., and Chanock, R. M.,** Transmission of acute nonbacterial gastroenteritis to volunteers by oral administration of stool filtrates, *J. Infect. Dis.,* 123, 307, 1971.

46. **Pether, J. V. S. and Caul, E. O.,** An outbreak of food-borne gastroenteritis in two hospitals associated with a Norwalk-like virus, *J. Hygiene,* 91, 343, 1983.

47. **Cubitt, W. D., Bradley, D., Carter, M., Chiba, S., Estes, Saif, L., Schaffer, F., Smith, A., Studdert, M., and Thiel, H. J.,** Caliciviridae, in *Classification and Nomenclature of Viruses, Sixth Report of the International Committee on the Taxonomy of Viruses, Archives of Virology,* Suppl. 2, in press.

48. **Lewis, D. C., Lightfoot, N. R., and Pether, J. V. S.,** Solid phase immune electron microscopy with human immunoglobulin M for serotyping of Norwalk-like viruses, *J. Clin. Microbiol.,* 26, 938, 1988.

49. **Okada, S., Sekine, S., Ando, T., Hayashi, Y., Murao, M., Yabuuchi, K., Miki, T., and Ohashi, M.,** Antigenic classification of small round-structured viruses by immune electron microscopy, *J. Clin. Microbiol.,* 28, 1244, 1990.

50. **Jiang, X., Graham, D. Y., Wang, K., and Estes, M. K.,** Norwalk virus genome cloning and characterisation, *Science,* 250, 1580, 1990.

51. **Matsui, S. M., Kim, J. P., Greenberg, H. B., Su, W., Sun, Q., Johnson, P. C., DuPont, H. L., Oshiro, L., and Reyes, G. R.,** The isolation and characterisation of a Norwalk virus-specific cDNA, *J. Clin. Invest.,* 87, 1456, 1991.

52. **Lambden, P. R., Caul, E. O., Ashley, C. R., and Clarke, I. N.,** Sequence and genome organisation of a human small round-structured (Norwalk-like) virus, *Science,* 259, 516, 1993.

53. **Pether, J. V. S. and Caul, E. O.,** An outbreak of food-borne gastroenteritis in two hospitals associated with a Norwalk-like virus, *J. Hygiene,* 91, 343, 1983.

54. **Kuritsky, J. N., Osterholm, M. T., Greenberg, H. B., Korlath, J. A., Godes, J. R., Hedberg, C. W., Forfang, J. C., and Kapikian, A. Z.,** Norwalk gastroenteritis: a community outbreak associated with bakery product consumption, *Ann. Intern. Med.,* 100, 591–521, 1984.

55. **Sawyer, L. A., Murphy, J. J., Kaplan, J. E., Pinsky, P. F., Chacon, D., Walmsley, S., Schonberger, L. B., Phillips, A., Forward, K., Goldman, C., Brunton, J., Fralick, R. A., Carter, A. O., Gary, W. G., Jr., Glass, R. I., and Low, D. E.,** 25- to 30-nm virus particles associated with a hospital outbreak of acute gastroenteritis with evidence for airborne transmission, *Am. J. Epidemiol.,* 127(6), 1261–1271, 1988.

56. **Brandt, C. D., Kim, H. W., Rodriguez, W. J., Thomas, L., Yolken, R. H., Arrobio, J. O., Kapikian, A. Z., Parrott, R. H., and Chanock, R. M.,** Comparison of direct electron microscopy, immune electron microscopy, and rotavirus enzyme-linked immunosorbent assay for detection of gastroenteritis viruses in children, *J. Clin. Microbiol.,* 13, 976, 1981.

57. **Jiang, X., Wang, J., Graham, D. Y., and Estes, M. K.,** Detection of Norwalk virus in stools by polymerase chain reaction, *J. Clin. Microbiol.,* 30(10), 2529–2534, 1992.

58. **De Leon, R., Matsui, S. M., Baric, R. S., Herrman, J. E., Blacklow, N. R., Greenberg, H. B., and Sobsey, M. D.,** Detection of Norwalk virus in stool specimens by reverse transcriptase polymerase chain reaction and nonradioactive oligoprobes, *J. Clin. Microbiol.,* 30(12), 3151–3157, 1992.

59. **Green, J., Norcott, J. P., Lewis, D. C., Arnold, C., and Brown, D. W. G.,** Norwalk-like viruses in the U.K.: demonstration of genomic diversity by PCR sequencing, *J. Clin. Microbiol.,* in press.

60. **Norcott, J. P., Green, J., Lewis, D. C., Estes, M. K., and Brown, D. W. G.,** Genomic diversity among small round structured viruses in the U.K., in preparation.

61. **Brandt, C. D., Rodriguez, W. J., Kim, H. W., Arrobio, J. O., Jeffries, B. C., and Parrott, R. H.,** Rapid presumptive recognition of diarrhea-associated adenoviruses, *J. Clin. Microbiol.,* 20, 1008–1009, 1984.

62. **Kidd, A. H., Rosenblatt, A., Besselaar, T. G., Erasmus, M. J., Tiemessen, C. T., Berkowitz, F. E., and Schoub, B. D.,** Characterization of rotaviruses and subgroup F adenoviruses from acute summer gastroenteritis in South Africa, *J. Med. Virol.,* 18, 159–168, 1986.

63. **Uhnoo, I., Wadell, G., Svensson, L., and Johansson, M.,** Importance of enteric adenoviruses 40 and 41 in acute gastroenteritis in infants and young children, *J. Clin. Microbiol.,* 20, 365–372, 1984.

64. **Matthews, R. E. F.,** Fourth Report of the International Committee on Taxonomy of Viruses. Classification and Nomenclature of Viruses, *Intervirology,* 17, 4–199, 1982.

65. **Allard, A., Girones, R., Juto, P., and Wadell, G.,** Polymerase chain reaction for detection of adenoviruses in stool samples, *J. Clin. Microbiol.,* 28, 2659–2667, 1990.

66. **Allard, A., Albinsson, B., and Wadell, G.,** Detection of adenoviruses in stools from healthy persons and patients with diarrhoea by two-step polymerase chain reaction, *J. Med. Virol.,* 37, 149–157, 1992.

67. **Girones, R., Allard, A., Wadell, G., and Jofre, J.,** Application of PCR to the detection of adenoviruses in polluted waters, *Water Sci. Technol.,* 27, 235–251, 1993.

68. **Kurtz, J. B. and Lee, T. W.,** Human astrovirus serotypes (letter), *Lancet,* ii, 1405, 1984.

69. **Willcocks, M. M., Carter, M. J., and Madeley, C. R.,** Astroviruses, *Rev. Med. Virol.,* 2, 97–106, 1992.

70. **Lew, J. F., Moe, C. L., Monroe, S. S., Allen, J. R., Harrison, B. M., Forrester, B. D., Stine, S. E., Woods, P. A., Hierholzer, J. C., Herrman, J. E., Blacklow, N. R., Bartlett, A. V., and Glass, R. I.,** Astrovirus and adenovirus associated with diarrhoea in children in day care settings, *J. Infect. Dis.,* 164, 673–678, 1991.

71. **Herrman, J. E., Nowak, N. A., Perron-Henry, D. M., Hudson, R. W., Cubbitt, W. D., and Blacklow, N. R.,** Diagnosis of astrovirus gastroenteritis by antigen detection with monoclonal antibodies, *J. Infect. Dis.,* 161, 226–229, 1990.

72. **Moe, C. L., Allen, J. R., Monroe, S. S., Gary, H. E., Jr., Humphrey, C. D., Herrman, J. E., Blacklow, N. R., Carcamo, C., Koch, M., Kim, K.-H., and Glass, R. I.,** Detection of astrovirus in paediatric specimens by immunoassay and RNA probe, *J. Clin. Microbiol.,* 29(11), 2390–2395, 1991.

73. **Major, M. E., Eglin, R. P., and Easton, A. J.,** 3′ Terminal nucleotide sequence of astrovirus type 1 and routine detection of astrovirus nucleic acid and antigen, *J. Virol. Meth.,* 39, 217–225, 1992.

74. **Katloff, K. L., Wasserman, S. S., Steciak, J. Y., et al.,** Acute diarrhoea in Baltimore children attending an outpatient clinic, *Paediatric Infect. Dis.,* 7, 753–759, 1988.

75. **Herrman, J. E., Hudson, R. W., Perron-Henry, D. M., Kurtz, J. B., and Blacklow, N. R.,** Antigenic characterisation of cell-cultivated astrovirus serotypes and development of astrovirus-specific monoclonal antibodies, *J. Infect. Dis.,* 158, 182–185, 1988.

76. **Woode, G. N., Reed, D. E., Runnels, P. L., Herrig, M. A., and Hill, H. T.,** Studies with an unclassified virus isolated from diarrhoeal calves, *Vet. Microbiol.,* 7, 221–240, 1982.

77. **Woode, G. N., Saif, L. J., Quesada, M., Winand, N. J., Pohlenz, J. F., Kelso, Gourley, N.,** Comparative studies on three isolates of Breda virus of calves, *Am. J. Vet. Res.,* 46, 1003–1010, 1985.

78. **Weiss, M., Steck, F., and Horzinek, M. C.,** Purification and partial characterisation of a new enveloped RNA virus (Berne virus), *J. Gen. Virol.,* 64, 1849–1858, 1983.

79. **Beards, G. M., Brown, D. W. G., Green, J., and Flewett, T. H.,** Preliminary characterisation of torovirus-like particles of humans: comparison with Berne virus of horses and Breda virus of calves, *J. Med. Virol.,* 20, 67–68, 1986.

80. **Beards, G. M., Brown, D. W. G., and Flewett, T. H.,** Detection of Breda virus antigen and antibody in humans and animals by enzyme immunoassay, *J. Clin. Microbiol.,* 25, 637–640, 1987.

81. **Koopmans, M., Herrewegh, A., and Horzinek, M. C.,** Diagnosis of torovirus infection, *Lancet,* 337, 859, 1991.

82. **Koopmans, M., Herrewegh, A., and Horzinek, M. C.,** The diagnosis of torovirus infections in animals and humans: recent results, in Koopmans, M., Ed., Diagnosis and Epidemiology of Torovirus Infections in Cattle, Ph.D. thesis, University of Utrecht, The Netherlands, 1990.

83. **Yolken, R., Leister, F., Almeido-Hill, J., Dubovi, E., Reid, R., and Santosham, M.,** Infantile gastroenteritis associated with excretion of pestivirus antigens, *Lancet,* 517–520, 1989.

84. **Katz, J. B., Ridpath, J. F., and Bolin, S. R.,** Presumptive diagnostic differentiation of hog cholera virus from bovine viral diarrhea and border disease viruses by using a cDNA nested-amplification approach, *J. Clin. Microbiol.,* 31, 565–568, 1993.

85. **Pereira, H. G., Flewett, T. H., Candeias, J. A. N., and Barth, O. M.,** A virus with a bisegmented double-stranded RNA genome in rat (*Oryzomys nigripes*) intestines, *J. Gen. Virol.,* 69, 2749–2754, 1988.

86. **Brown, D., Gallimore, C., and Appleton, H.,** Bisegmented ds viruses (Picobirnaviruses) in man: preliminary studies, in 8th Int. Congr. Virology Abstr., Berlin, 334, 1990.

87. **Grohmann, G. S., Glass, R. I., Pereira, H. G., Monroe, S. S., Hightower, A. W., Weber, R., and Bryan, R. T.,** Enteric viruses and diarrhea in HIV-infected patients, *N. Engl. J. Med.,* 329, 14–20, 1993.

88. **Turton, J., Appleton, H., and Clewley, J. P.,** Similarities in nucleotide sequence between serum and faecal human parvovirus DNA, *Epidemiol. Infect.,* 105, 197–201, 1990.

89. **Paver, W. K., Caul, E. O., and Clarke, S. K. R.,** Comparison of a 22 nm virus from human faeces with animal parvovirus, *J. Gen. Virol.,* 22, 447–450, 1974.

90. **Bishop, R. F.,** Other small virus-like particles in humans, in *Virus Infections of the Gastrointestinal Tract,* Tyrell, D. A. J. and Kapikian, A. Z., Eds., Dekker, New York, 1982.

91. **Bridger, J. C.,** Small viruses associated with gastroenteritis in animals, in *Viral Diarrheas of Man and Animals,* Saif, L. J. and Theil, K. W., Eds., CRC Press, Boca Raton, FL, 1990.

92. **Clewley, J. P.,** The polymerase chain reaction: basic procedures and use for virus detection, in *Methods in Gene Technology,* Vol. 1, Dale, J. W. and Sanders, P. G., Eds., JAI Press, London, 1991, 219–238.

93. **Boom, R., Sol, C. J. A., Salimans, M. M. M., Jansen, C. L., Wertheim-van Dillen, P. M. E., and van der Noordaa, J.,** Rapid and simple method for purification of nucleic acids, *J. Clin. Microbiol.,* 28, 495–503, 1990.

94. **Alexander, L. M. and Morris, R.,** PCR and environmental monitoring: the way forward, *Water Sci. Technol.,* 24, 291–294, 1991.

95. **Atmar, R. L., Metcalf, T. G., Neill, F. H., and Estes, M. K.,** Detection of enteric viruses in oysters by using the polymerase chain reaction, *Appl. Environ. Microbiol.,* 59, 631–635, 1993.

96. **Sambrook, J., Fritsch, E. F., and Maniatis, T.,** Extraction, purification and analysis of messenger RNA from eukaryotic cells, in *Molecular Cloning: A Laboratory Manual,* 2nd ed., Cold Spring Harbor Laboratory Press, Cold Spring Harbor, NY, 1989.

97. **Simmonds, P., Zhang, L. Q., Watson, H. G., Rebus, S., Ferguson, E. D., Balfe, P., Leadbetter, G. H., Yap, P. L., Peutherer, J. F., and Ludlam, C. A.,** Hepatitis C quantification and sequencing in blood products, haemophiliacs and drug users, *Lancet,* 336, 1469–1472, 1990.

98. **Henard, D. R., Mehaffey, W. F., and Allain, J. P.,** A sensitive viral capture assay for detection of plasma viremia in HIV infected individuals, *AIDS Res. Human Retroviruses,* 8, 47–52, 1992.

99. **Estes, M. K. and Cohen, J.,** Rotavirus gene structure and function, *Microbiol. Rev.,* 53, 410–449, 1989.

Chapter 5

PCR Detection and Analysis of Hepatitis Viruses

Betty H. Robertson

TABLE OF CONTENTS

I. CLASSIFICATION AND PROPERTIES OF HEPATITIS VIRUSES

The known viral agents that selectively infect the liver causing hepatitis cover a wide spectrum of virus families and include enveloped and nonenveloped DNA and RNA viruses. Currently, there are five defined viral agents causing hepatitis, and their distinguishing properties are listed in Table 5–1. Two of these agents are enterically transmitted icosahedral RNA viruses and cause an acute self-limited disease. Hepatitis A virus (HAV) is a member of the picornavirus family[1,2] while hepatitis E virus (HEV) has calcivirus-like features.[3] The remaining three agents are transmitted by parenteral or sexual routes and can result in chronic infections. Hepatitis C virus (HCV) is an enveloped RNA virus genetically related to flavi- and pestiviruses.[4] Hepatitis B virus (HBV) is an enveloped DNA virus classified as a hepadnavirus[1,2] while hepatitis delta virus (HDV) or the delta agent is a defective RNA viroid-like agent[5] that uses the HBV glycoprotein for transmission of infection.

From an historical perspective, for many years there were two types of transmissible hepatitis identified. One was infectious, or short-incubation, hepatitis, and the other was serum, or long-incubation, hepatitis.[6] With the advent of serological tests to diagnose hepatitis A and B, it became apparent that there were cases of hepatitis that were not caused by HAV or HBV. Thus, for several years, the term non-A,

Table 5–1   **Classification and characteristics of hepatitis viruses**

	HAV	HBV	HCV	HDV	HEV
Family	Picornavirus	Hepadnavirus	Flavivirus	Viroid	Calicivirus
Genus	Hepatovirus				
Physical	Icosahedron	Enveloped	Enveloped	Enveloped	Icosahedral
Size (nm)	27	42	<50	36	30–32
Genome	+ssRNA	c(ds)DNA	+ssRNA	–cc ssRNA	+ssRNA
$S_{20,w}$	157	60	—	>60	183
Density	1.34	1.21	1.08	1.25	1.29
Polypeptides					
Structural	4	4	3	2	1
Nonstructural	6	3	4		4
Cellular site of replication	Cytoplasmic	Nuclear Cytoplasmic	Cytoplasmic	Nuclear	Cytoplasmic

non-B (NANB) hepatitis was used to identify these agents. Three of these agents are now recognized as HCV, HDV, and HEV.

II. PCR OF HEPATITIS A VIRUS

A. BACKGROUND INFORMATION ON HEPATITIS A VIRUS

Hepatitis A virus (HAV), classified as a hepatovirus within the Picornaviradae family,[7] has a positive-sense 7.5-kb RNA genome, as illustrated in Figure 5–1. The single open reading frame (ORF) encodes a large polyprotein that is proteolytically cleaved to generate the final viral products. The genome is functionally delineated into three regions, P1, P2, and P3. P1 encodes the structural proteins that form the viral capsid (VP1, VP2, VP3, and a potential VP4). P2 codes for three potential nonstructural polypeptides, 2A, 2B, and 2C, the function of which is not known while the P3 genome region encodes four protein products, 3A, 3B(VPg), 3C (protease), and 3D (RNA polymerase).

Fecal-oral transmission is the normal route of virus entry into the body. After exposure, an incubation period of 15 to 45 days precedes the development of clinical symptoms. Clinical illness in adults can last for 1 to 3 months, but mortality is limited and no chronic sequelae result from infection. After infection, antibodies generated against the virus persist for life and protect against reinfection. As with many of the other enteroviruses, infants and young children tend to have asymptomatic infections, while older children and adults experience symptomatic disease.

Detection of HAV particles remains largely a research tool, as IgM anti-HAV is used clinically to diagnose acute infection, and most viral excretion occurs before clinical illness. Before the development of PCR, detection of virus particles by immune electron microscopy, cell culture growth, or enzyme immunoassay was used. Identification of wild-type HAV by immune electron microscopy[8] is a technique that is labor intensive and requires experienced personnel and sophisticated equipment. Cell culture growth of wild-type HAV requires long adaptation periods (4 to 8 weeks),[9,10] and some strains are refractive to cell culture growth.[11] The radioimmunofocus assay (RIFA),[12] which detects foci of virus growth based upon binding of iodinated HAV antibody, is technically difficult and time consuming. An enzyme immunoassay for viral antigen has been developed within our laboratory,[13] but sensitivity depends on high-titer antisera. These procedures require special equipment, reagents, or highly skilled personnel when used to detect the presence of virus.

B. APPROACHES FOR PCR OF HAV

Two separate groups[14,15] independently developed a similar approach, called immunocapture or antigen capture, for amplification of HAV genetic sequences. This technique provides antibody selection in addition to primer binding for detection of HAV in heterogeneous mixtures such as stool suspensions, liver homogenates, cell culture lysates, or other aqueous suspensions that contain virus. HAV-specific antibody coated onto a solid phase such as a microfuge tube is used to capture the virus from the aqueous suspension. After proteinase K digestion and nucleic acid extraction[14] or heating,[15,16] positive- and negative-sense primers are added and viral RNA is transcribed to cDNA with reverse transcriptase followed by PCR amplification. Within our laboratory,[16] virus has been detected in serum during the viremic phase by proteinase K digestion, nucleic acid extraction, followed by reverse transcription and PCR using the appropriate primers.

C. PRIMERS USED FOR PCR AMPLIFICATION OF HAV

Two regions have been used for amplification and detection of HAV. Robertson and co-workers amplified a region within the amino terminus of VP1,[14,17] while Jansen and co-workers compared amplification of two genome regions, the carboxy terminus of VP3 and the VP1/2A junction.[15] The relative location of the resulting amplicons is shown by the shaded rectangles in Figure 5–1. The sensitivity of detection following hybridization with internal radiolabeled probes is between 3 and 100 infectious particles.[15,16] For detection purposes, amplification using the more conserved VP3 carboxy terminal primers or the VP1 amino terminal primers are more reliable than the VP1/2A junction primers, and we routinely use the VP1 amino terminal primer pairs for screening unknown samples.[16]

D. USES AND APPLICATION OF PCR-AMPLIFIED HAV SEQUENCES

PCR amplification of HAV sequences to detect or monitor excretion and potential infectivity has been used to augment several epidemiological studies. These have included confirmation of infected neonates who shed infectious virus for up to 6 months after acute infection,[18] and identification of HAV in IgM anti-HAV negative sera, collected during the prodromal phase of infection, in a suspected index case of hepatitis A.[19] PCR amplification of HAV from contaminated shellfish was used in an investigation of a multistate shellfish-associated outbreak of HAV.[20] The studies described above used PCR amplification followed by hybridization analysis to identify the presence of HAV. A more recent investigation of two

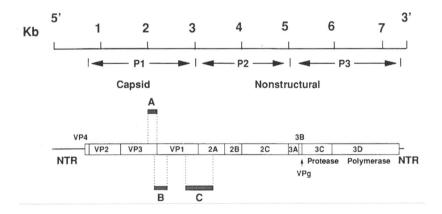

Figure 5–1 Schematic of the HAV genome and PCR-amplified regions. The top line represents the length of the HAV genome in kilobases (kb) from 5′ to 3′. The portions of the genome-encoding structural (capsid) and nonstructural polypeptides are designated between the arrows (P1, P2, and P3). The open rectangle represents the single open reading frame of HAV. Individual viral polypeptides are indicated within or just above the open rectangle, while the lines to the right and left of the open rectangle represent the 5′ and 3′ nontranslated regions (NTR) of the genome. The shaded rectangles labeled A, B and C represent the amplification products (amplicons) of HAV obtained after PCR, and the dotted lines indicate the relative location of these amplicons. Fragment A was amplified using HAV primers +2020 5′ACA GGT ATA CAA AGT CAG 3′ and –2211 5′CTC CAG AAT GAT CTC C 3′; fragment B was amplified using HAV primers +2167 5′GTT TTG CTC CTC TTA TCA TGC TAT G 3′ and –2389 5′GGA AAT GTC TCA GGT ACT TTC TTT G 3′; fragment C was amplified using various primers pairs from the following sets: positive primers –HAV +2799 5′ATT CAG ATT AGA CTG CCT TGG TA 3′; +2891 5′GGT TTC TAT TCA GAT TGC AAA TTA 3′; +2933 5′TTT GTC TTT TAG TTG TTA TTT GTC TGT 3′; +2934 5′TTG TCT TTT AGT TGT TAT TTG TCT G 3′; +2949 5′TAT TTG TCT GTC ACA GAA CAA TCA G 3′; +2984 5′TCC CAG AGC TCC ATT GAA 3′; negative primers –HAV –3192 5′AGG AGG TGG AAG CAC TTC ATT TGA 3′; –3225 5′CAT TTT CCT AGG AGG TGG 3′; –3265 5′CAT TAT TTC ATG CTC CTC AG 3′; –3273 5′CCA AGA AAC CTT CAT TAT TTC ATG 3′; –3285 5′AGT CAC ACC TCT CCA GGA AAA CTT 3′; –3375 5′AGT AAA AAC TCC AGC ATC CAT TTC 3′.

outbreaks of hepatitis A associated with the consumption of frozen strawberries[21] in two separate locations, 3 months apart, were found to be caused by the same virus. Nucleic acid sequencing of PCR-amplified HAV sequences revealed that the virus from these two outbreaks was identical and could be differentiated from more than 150 other strains of HAV found worldwide.

PCR amplification has greatly facilitated identification of new strains of HAV (without the necessity of cloning) and the characterization of other features of the HAV genome. Primers, initially used for the detection studies described above, were valuable for amplifying sequences from simian HAV strains[22,23] and a genetic variant of human HAV,[24] thereby providing preliminary information from which to derive additional primer pairs for sequencing and characterization. Antigen capture, PCR amplification, and nucleic acid sequencing have been used to characterize altered capsid amino acids within neutralization escape mutants.[25,26]

Although HAV is a single serotype,[27-29] genotypes of HAV have been identified by PCR amplification followed by nucleic acid sequencing of portions of the genome.[15-17,30,31] An international collaboration that focused on comparison of the VP1/2A junction from all available wild-type strains and cell-culture-adapted isolates revealed the existence of at least seven discrete genotypes that differ from each other by 15%.[31] The information derived from these investigations has provided a database for distinguishing different strains of HAV and a means of using molecular approaches to identify potentially related outbreaks of HAV.

III. PCR OF HEPATITIS B VIRUS

A. BACKGROUND INFORMATION ON HEPATITIS B VIRUS

Hepatitis B virus (HBV), the prototype of the Hepadnaviradae family,[32] is a 42-nm enveloped virus particle containing a partially double-stranded circular DNA molecule approximately 3.2 kb long (Table 5–1). The genome is enclosed within a nucleocapsid shell composed of core protein (HBc) and then coated with a host-derived lipid envelope containing the surface antigen (HBs) glycoprotein.

The partially double-stranded viral genome is composed of two complementary long and short segments of DNA. The negative-sense long DNA segment (approximately 3200 bases) contains the entire coding information for all the viral polypeptides. A linear representation of the HBV genome is depicted in Figure 5–2A, with the conserved Eco RI site designated as position 0. The position of the four open reading frames (ORF) is shown below this, labeled S, C, X, and P. These ORFs code for seven viral polypeptides by the use of alternative in-frame initiation codons (black boxes). Three in-frame ATG initiation codons within the S region direct the synthesis of three co-linear molecules collectively referred to as HBs in this review. HBs contains a commonly highly antigenic "a" determinant (indicated by the star), and two mutually exclusive antigenic alleles, d/y and w/r, resulting in four major subtypes of HBV (adr, adw, ayr, and ayw). The C genome region uses two in-frame ATG start signals to encode for the HBc and HBe products. HBc is produced by initiation of protein translation at the internal ATG and remains an intracellular protein. Translation of pre-core begins at the initial ATG and is then cleaved by proteases to produce the soluble HBe protein. In addition to these polypeptides, HBV encodes for a DNA polymerase encoded by the P region and a putative transactivator protein derived from the X region. Several recent reviews have excellent detailed descriptions of HBV structure, coding strategy, and replication.[33,34]

Hepatitis B, originally known as serum or long incubation hepatitis, is spread horizontally by blood and body fluids while perinatal transmission from mother to infant accounts for a large number of HBV infections in many parts of the world. After documented exposure, the incubation period ranges from 1 to 3 months before the onset of symptoms. Infection with HBV can result in acute illness, followed by recovery and the development of antibodies against HBs, HBc, and HBe. In other cases, after the acute phase of HBV infection, a chronic carrier state develops with circulating HBs antigen, antibodies against HBc, and variable levels of HBe and anti-HBe. Chronic infection with HBV can result in chronic liver disease (CLD) resulting in cirrhosis or hepatocellular carcinoma (HCC).[35] The development of the chronic carrier state after HBV infection is associated with the age of infection, as infants exposed to HBV from carrier mothers have the highest rate of chronic carriers and the highest rate of development of hepatocellular carcinoma.[36]

B. APPROACHES FOR PCR OF HBV

The titer of HBV particles in serum can approach 10^{11} per milliliter, and in fact, HBV was one of the first viruses subjected to PCR amplification.[37] Classical approaches for extraction of HBV nucleic acid from serum (5 to 25 µl) have included proteinase K digestion or guanidinium isothiocyanate[38] treatment

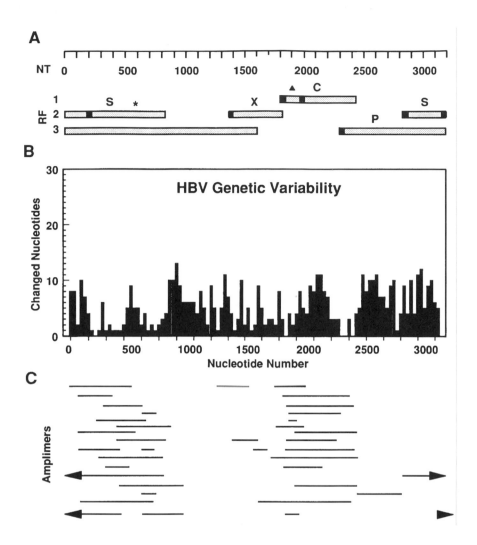

Figure 5–2 HBV open reading frames (A), genetic variability (B) and location of PCR amplicons (C). (A) A linear representation of the HBV genome and the location of the four open reading frames. The scale at the top indicates the approximate nucleotide number (NT) with the conserved Eco RI site designated as position 0. RF indicates the reading frame while the shaded arrows indicate the genome location of the HBV open reading frames (S, X, C, and P). Initiation codons are shown as black boxes within the respective open reading frames. The star indicates the position of the immunodominant "a" determinant of HBs, and the triangle indicates the pre-core stop codon mutation. (B) Analysis of HBV genetic variability. Nine complete HBV sequences[54–61] were aligned from the common Eco RI site. Each bar represents the number of changed nucleotide positions within a 30-nucleotide span, while the approximate genome location is shown on the horizontal axis. (C) Published amplicons of HBV. Each line represents the location and length of products resulting from PCR amplification of HBV. The arrows indicate amplicons that are contiguous within the circular HBV genome.

followed by nucleic acid extraction. Alternative methods to select or disrupt virus in serum have included altering the pH with sodium hydroxide followed by neutralization with HCl,[39] simple adsorption of HBV onto a solid phase,[40] monoclonal antibody capture,[41] microwaving,[42] or heat inactivation.[43]

After amplification, most workers have hybridized a labeled internal oligonucleotide for detection of amplified product at low levels. An alternative method used for detection is the PCR-PCR,[39] or nested, amplification. Nested amplification uses two sets of primers, external and internal, and two rounds of PCR amplification. The external set is used for the first 25 to 30 c of amplification, followed by a small portion of the first PCR reaction being used for amplification with the internal primers for a second round of 25 to 30 c. This approach, commonly used to detect the much-lower-titer HCV viruses, allows the detection of a positive amplification by visualization of an ethidium-bromide-stained band after 50 to 60 c of amplification. With either hybridization or nested PCR detection, stringent control over potential contamination is necessary due to the high titers of HBV that can be present in serum. The sensitivity of detection after hybridization or "nested" PCR is between one and ten viral genome copies,[40,41,44] and studies using a chimpanzee-titered HBV inoculum indicate that PCR detection of HBV is more sensitive than animal infectivity studies.[45] A novel approach for enzyme immunoassay-like detection of amplified sequences uses sandwich hybridization. A capture HBV oligonucleotide, immobilized on microtiter wells, is hybridized to one half of the PCR-amplified HBV sequence. A biotin-labeled HBV probe, homologous to the other half of the PCR-amplified HBV sequence, is added and used to bind peroxidase-conjugated streptavidin. Positive color development with the peroxidase enzyme completes detection of the amplified HBV sequence.[46]

HBV genome sequences have also been amplified from liver tissue and peripheral blood mononuclear cells of chronic carriers. Hepatocellular carcinoma tissue,[47,48] HBV-transformed hepatoma cells,[47] and paraffin-embedded liver sections[47,49–51] have been amplified after appropriate processing.[52] PCR amplification has been used to reveal that peripheral blood mononuclear cells contain and replicate HBV. HBV transcripts were identified by isolation of message-sense RNA (after digestion of HBV DNA and DNA/RNA hybrids with DNase and RNase H), followed by reverse transcription using oligo dT primers and PCR amplification with HBV-specific primers.[53]

C. PRIMERS USED FOR PCR AMPLIFICATION OF HBV

The majority of the primers used for PCR amplification of HBV sequences have been directed toward the C (HBc) and S (HBs) genome regions, as shown by the amplicons illustrated in Figure 5–2C. A schematic representation of the genetic diversity among nine sequenced strains of HBV[54–61] is shown in Figure 5–2B. This analysis reveals that a few regions within the HBV genome appear to be highly conserved (nucleotides 241 to 270, 1891 to 1920, and 2431 to 2460) and that multiple regions contain a single changed nucleotide. This suggests that a single set of primers would reliably amplify HBV. In reality, many investigators have found that multiple primer pairs in different genome regions are necessary to detect HBV. We have experienced similar situations in which amplification with a primer beginning at nucleotide 1860 gave no amplification, yet maintaining the downstream primer (nt 2410) and using a primer beginning at nucleotide 1840 resulted in positive amplification.[62]

D. USES AND APPLICATION OF PCR-AMPLIFIED HBV SEQUENCES

There are limited studies of PCR detection of acute HBV infections. Two groups of investigators, using different primer pairs, arrived at different conclusions. Using S region primers, Zeldis and co-workers[40] found detectable HBs antigen 4 to 8 weeks before PCR-positive reactivity in two human infections. On the other hand, serum from five chimpanzees experimentally infected with HBV and amplified using C region primers revealed detectable HBV DNA 2 to 3 weeks before detection of HBs.[63] These discrepancies might reflect the choice of primers used for amplification or be the result of variable levels of antigenemia vs. viremia in the individual cases.

Numerous studies have verified that HBs-positive chronic carriers who are HBe antigen positive have circulating HBV DNA when tested by PCR.[64–70] However, the common dogma that the presence of HBe antibody indicates the absence of HBV particles and infectivity does not always hold true when evaluated by PCR. The same studies found 80 to 100% of anti-HBe-positive individuals with chronic liver disease (CLD) had PCR-detectable HBV sequences in their serum,[64,65,67–70] and two groups who evaluated asymptomatic, "healthy" anti-HBe-positive individuals found 60 to 70% had HBV DNA in their serum.[67,68] In addition, evaluation of healthy individuals, with anti-HBc antibodies only, revealed 30% had PCR-detectable HBV in their serum.[71]

The contribution of HBV infection to CLD probably varies with the geographic region examined as hyperendemic areas such as China, Southeast Asia, southern Europe, Africa, and parts of South America have a 10 to 30% chronic carrier population. PCR amplification of HBV sequences from CLD patients from these regions, who have other serum or liver markers indicative of past HBV infection, demonstrates that 50 to 100% carry HBV DNA sequences.[49,51,64,65] In addition, the evaluation of CLD or HCC patients who are negative for all markers of HBV infection indicates that 20 to 70% contain HBV DNA sequences within liver biopsy samples.[48,49,51] Unfortunately, none of these studies tested the samples for the presence of HCV antibodies or HCV sequences that also might contribute to CLD and HCC.

The transmission of hepatitis B from individuals who have no markers of HBV infection has been postulated to be caused by a variant of HBV.[72] Two studies evaluated serum from individuals with no markers of HBV infection by PCR amplification and detected HBV sequences in 4 to 10% of the population studied.[73,74] PCR amplification and sequencing of three individual variants after chimpanzee passage revealed no major genetic changes compared with previously characterized HBV strains.[75,76] The lack of detectable markers in such individuals is probably due to low-level viral replication, in addition to a variable host-immune response.

PCR amplification and sequencing have been valuable in correlating genetic mutations within HBV infections with clinical features of HBV infection. The existence of anti-HBe-positive and HBV-DNA-positive chronic carriers[77] prompted investigators to evaluate pre-core sequences within the C genome region.[78,79] In the Far East and Mediterranean regions, individuals with rapidly progressing CLD and fulminant HBV were found to have a common mutation (denoted by the triangle in Figure 5–2A) in the pre-core portion of the C genome region.[80–86] The predominant mutation found is a TGG (tryptophan) to TAG (stop)[78,80,82,84] codon, thereby terminating translation of HBe antigen. The exact relationship between the pre-core stop mutation and aggressive HBV infection is not known, and other mutations have been reported in the C and S genome region in CLD patients[87] and chronic HBV carriers.[88,89]

The predominant antibody response to HBV is directed against the immunodominant "a" determinant of the HBs antigen, indicated by the star in Figure 5–2A. PCR amplification and sequencing have revealed a common glycine-to-arginine mutation at amino acid 145 within the "a" determinant of the HBs antigen in apparent vaccine-induced escape mutants. These mutants were isolated from immunized infants born to chronic carrier mothers[90,91] and from a transplant patient who received anti-HBs monoclonal antibody directed against the "a" determinant.[92] The possibility of generating vaccine-induced escape mutants that can infect despite the presence of antibodies against the "a" determinant has widespread public health implications with the proposed universal immunization of infants.

The major subtypes of HBV (adr, adw, ayr, and ayw) have been further delineated into serologic subgroups termed P1 to P8, and adrq[−93] on the basis of monospecific antibody reactivity. Norder and co-workers[94,95] have attempted to correlate genetic sequence of HBV strains with the adr, adw, ayr, ayw subtypes and defined subgroups of HBV. PCR amplification and nucleic acid sequencing of 680 nucleotides encoded within the S gene from nine representative strains were compared with other published HBV sequences. There was no clear relationship between the genetic sequence and the corresponding subtype or subgroup nor did the sequences from this region cluster into obvious geographic groups.

The use of sequence similarities can be useful in identifying transmission patters. Lin et al.[96] used PCR amplification and sequencing of 100 nucleotides within the P/X region to evaluate HBV genetic variation in 82 children from Hong Kong, most of whom were perinatally infected. Thirty-three sequences were identical, while 31 variants, differing by 1 to 19% compared with the predominant sequence, were found among the remaining 49 children. Within mother-infant pairs there was less than 1% nucleotide difference, and the same variant was found in two generations within the majority of families.

IV. PCR OF HEPATITIS C VIRUS

A. BACKGROUND INFORMATION ON HEPATITIS C VIRUS

Hepatitis C virus (HCV), identified by cloning and characterization of its nucleic acid sequence,[97,98] appears to be genetically related to flavi- and pestiviruses[4] (Table 5–1). Figure 5–3A depicts the 9.4-kb genome that contains a single open reading frame of approximately 3000 amino acids flanked by the 5′ and 3′ noncoding (NC) regions of 341 and 56 nucleotides, respectively. As with HAV, proteolytic processing of the large polyprotein is necessary to generate the functional viral polypeptides. Computer alignments with flavivirus and pestivirus genomes have been used to predict potential viral polypeptides.[99] The putative nucleocapsid or core (C) protein and two potential glycoproteins, E1 and E2/NS1, are at the 5′ end of the genome. The remainder of the genome encodes four possible nonstructural proteins, NS2, NS3, NS4, and NS5. Sequence motifs within the NS3 protein suggest it has helicase and protease activity while the NS5 protein appears to contain sequence similarities with RNA polymerase molecules.[99–101] HCV infections are currently diagnosed based upon antibodies detected using expressed fragments of the core, NS3, and NS4 antigens. These assays detect HCV antibodies in 70 to 80% of NANB hepatitis cases.

Infection with HCV is known to be transmitted by parenteral exposure to blood or to blood products, but there are indications that household and sexual transmission occurs and that a low socioeconomic condition is a risk factor for infection.[102] The time between an identified exposure (such as blood transfusion) and the development of clinical symptoms is between 5 and 15 weeks. Identification of the acute phase of HCV infection is difficult since only 25% of infected individuals develop clinical symptoms while the remaining 75% are asymptomatic.[103] The most insidious feature of HCV infection is the development of chronic infection as at least 50% of individuals infected with HCV will develop some form of chronic liver disease leading to cirrhosis or hepatocellular carcinoma.

B. APPROACHES FOR PCR OF HCV

The titer of HCV within serum is low, 10^2 to 10^4 per milliliter,[104] and therefore one of the primary considerations with HCV PCR is the volume of serum used for amplification. Investigators have used volumes ranging from 10 μl to 6 ml; however, amplification of viral nucleic acid sequences from volumes greater than 200 μl have been performed after high-speed pelleting of virus[105] or polyethylene glycol precipitation[106] to separate virus from serum components that inhibit amplification. All the procedures have used either proteinase K digestion or guanidinium isothiocyanate treatment followed by phenol-chloroform extraction to purify the viral RNA.

Initial studies used Southern blot hybridization with radiolabeled internal probes[105,107,108] to detect amplified sequences; however, subsequent investigators have used "nested" amplification,[106,109,110] as described in the section on HBV to facilitate detection. These studies have shown that between 3 and 0.1 chimpanzee infectious particles can be visualized by hybridization or as an ethidium-bromide-stained product after nested PCR.

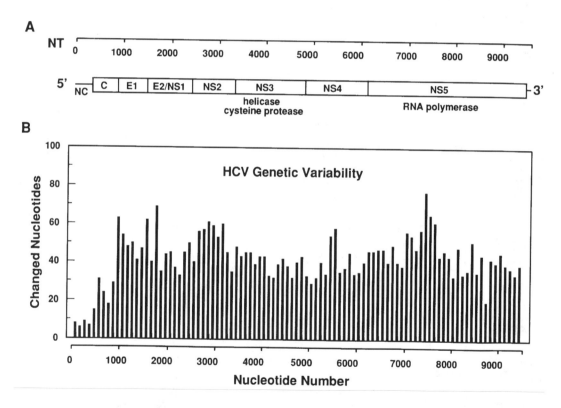

Figure 5–3 Genome organization of HCV (A) and analysis of HCV genetic variability (B). (A) The HCV genome and putative viral proteins. The top scale indicates the nucleotide (NT) position of the 9.4-kb genome. The genome organization of HCV is shown below this and includes the 5′ and 3′ noncoding (NC) regions (solid line) that flank the single continuous open reading frame (rectangle). Within the rectangle, the relative position of the seven putative viral polypeptides that result from proteolytic cleavage are separated by vertical lines. (B) Analysis of HCV genetic variability. Five complete HCV sequences[98,100,101,116,117] were aligned, and the number of changed nucleotide positions within each 100-nucleotide segment was determined. The histogram is a graphic illustration of these nucleotide changes, with each bar representing the number of changed nucleotides for each 100-nucleotide segment. The approximate genome location is shown on the horizontal axis.

The stability of HCV to various environmental conditions has not been explored in depth. Parallel collection of serum, citrated plasma, and heparinized plasma from anti-HCV-positive individuals followed by PCR amplification revealed that heparin inhibited PCR amplification and suggested that −70°C storage minimized virus degradation.[111] Our experience with samples that contain low levels of virus and that have been repeatedly frozen and thawed indicates that virus degradation occurs quite readily.[112]

C. PRIMERS USED FOR PCR AMPLIFICATION OF HCV

Although initial studies used primers from the nonstructural genome region,[105–110] the sequence of the 5′ noncoding (NC) region of HCV has been determined to be remarkably conserved compared with the remainder of the genome.[113–115] This conservation is graphically illustrated by the histogram in Figure 5–3B, which depicts changed nucleotide positions within five published HCV sequences.[98,100,101,116,117] Genetic variability within the coding region averages 20 to 25%, and some regions contain areas with over 50% of the nucleotide positions changed. On the other hand, the 5′ NC region contains limited nucleotide changes. A number of groups have compared amplification using 5′ NC primers to amplification using primers derived from other regions, and found that the 5′ NC primers detect a greater percentage of HCV-positive samples.[118–120] An evaluation of 114 anti-HCV-positive sera using nested primers spanning a highly conserved portion of the 5′ NC region, the entire 5′ NC region, the core region, and the NS3 region found positive amplification of HCV RNA in 74, 68, 52, and 24% of the samples, respectively.[121]

D. USES AND APPLICATION OF PCR-AMPLIFIED HCV SEQUENCES

Serial samples from experimentally infected chimpanzees and individuals who developed transfusion-related HCV have been evaluated for the presence of virus by PCR amplification. These studies have revealed that viremia is first detected about a week after exposure and continues through the peak of liver enzyme elevation.[106,122–127] After the initial phase of the illness, one of two general patterns develop. Some infections apparently resolve, with normalization of liver enzyme levels and loss of detectable virus from the blood. Other infections develop into a chronic infection, with fluctuating liver enzymes and intermittent viremia. Repeated infection with HCV, even in the presence of antibodies against HCV, have been reported by Prince and co-workers.[128] Exposure of previously infected chimpanzees to homologous or heterologous inocula resulted in a second bout of hepatitis and PCR-detectable viremia in some animals. On the other hand, Beach and co-workers[127] found no evidence of hepatitis or HCV viremia by PCR amplification after inoculation of two previously infected chimpanzees with a homologous challenge inoculum, although sera from both animals was infectious when inoculated into naive animals.

In chronic liver disease (CLD) patients who are anti-HCV positive, PCR amplification indicates that 85 to 100% of these individuals have circulating virus,[119,121,125,129–131] and a few studies have reported that 25 to 50% of anti-HCV-negative CLD patients have HCV sequences.[129–131] HCV sequences have also been identified in asymptomatic blood donors, with and without detectable HCV antibodies,[105,108,109,132] some of whom were implicated in virus transmission. PCR amplification used in a study by Alter and co-workers[102] revealed that 100% of individuals who experienced acute HCV infection and recovered with no evidence of chronic liver disease were intermittently HCV RNA positive up to 4 years after infection.

The overall consensus from these investigations suggests that HCV infection results in chronic or intermittent viremia, irrespective of whether clinical symptoms are present. The lack of detection of HCV sequences in "recovered" or asymptomatic individuals is probably due to a low-level viremia, and virus might be detected if larger volumes of sera were processed.

After initial cloning and sequencing of HCV,[97,98,100,101,116] investigators designed oligonucleotide primers to PCR amplify and sequence other strains of HCV. These studies have provided structural information on different HCV isolates that would not have been possible in the absence of PCR. The partial or complete sequence of a number of strains of HCV obtained after PCR amplification has identified genetic species that differ from each other by 10 to 40%,[117,133–147] facilitated characterization of the conserved 5′ end of the genome,[113–115,120] and elucidated two hypervariable regions within the NS1/E2 polypeptide, one located at the extreme amino terminus and the other located approximately 60 amino acids downstream.[148–150] Despite the extreme genetic variability between different strains of HCV, comparison of the virus from a single infected individual isolated after 13 years of chronic infection revealed only 2.5% nucleotide changes compared with virus present during the first year of infection.[151]

Numerous investigators have compared the nucleotide sequences resulting from PCR amplification of selected genome regions and grouped HCV strains based upon genetic similarity.[136,140,142,143,152] Unfortunately, different genome regions (5′ NC, C, NS3, or NS5) have been chosen for analysis, and this has complicated a cohesive picture of the various HCV genotypes. Chan and co-workers[152] performed computer analysis of available sequences in these four regions and found that comparison of core, NS3, or NS5 sequences resulted in delineation of four groups. The average nucleotide variability among the different groups was 15, 37, and 49% within core, NS3, and NS5, respectively. Alternative approaches to identify genetic groups or genotypes based upon restriction fragment length polymorphism (RFLP) analysis of PCR-amplified fragments[153] or the use of genotype-specific primers to generate size-specific bands representative of individual genotypes[154] may be premature, given the wide genetic variability seen within HCV genomes. An agreement on a common genome region for comparative analysis, and a corresponding definition of what percent divergence constitutes a different genotype, will be necessary to facilitate classification and future identification of HCV genotypes.

HCV is transmitted through parenteral exposure to blood or blood products. However, approximately 50% of individuals infected with HCV have no known parenteral exposure, and their only identified risk factors are low socioeconomic status or sexual/household contact with an infected person.[102] Investigators have used PCR amplification to evaluate whether virus present in saliva[155,156] or semen[156] is responsible for transmission in these cases. The results suggest that low levels of HCV can be detected in some samples; however, the possibility of occult blood contamination and the sensitivity of PCR amplification have complicated the reaching of any firm conclusions.

PCR amplification has been used to evaluate interferon treatment as a potential therapy for chronically infected individuals.[157,158] These preliminary studies indicate that PCR-detectable virus disappears in 75 to 81% of treated patients; however, once interferon therapy is terminated, only 30% maintain their PCR-negative status.

V. PCR OF HEPATITIS D VIRUS

A. BACKGROUND INFORMATION ON HEPATITIS D VIRUS

The delta agent, or hepatitis D virus (HDV), is a defective hepatotropic agent that does not fit into any known virus family (Table 5–1). HDV infections occur only in the presence of a hepadnavirus, such as HBV or the related nonhuman hepadnaviruses.[159–161] Physically, it is a 36-nm particle composed of a lipid bilayer containing the HBV glycoprotein (HBs) surrounding a nucleoprotein core composed of delta antigen, HDag.[162–164] The genome of HDV is a 1.7-kb negative-sense covalently closed circular RNA[5,165,166] that forms a base-paired double-stranded rod-like structure. Some of the HDV genome characteristics suggest it is related to viroid-like molecules. These include its high G + C content (60%), the double-stranded rod-like genome structure, regions of nucleotide homology with a viroid central conserved region,[5] and the apparent rolling circle replication intermediates.[167–169] The encoding of its own antigen and the use of HBs as a coat protein suggest that HDV also has features common to satellite RNAs or satellite viruses.[5]

A linear schematic of the HDV genome is shown in Figure 5–4A, with the open reading frame that encodes the delta antigen (HDag). This open reading frame, located on the antigenomic strand, is the only translated open reading frame and results in the expression of two forms of HDag, a large and a small HDag.[170,171] An amber termination codon, UGA, that mutates to UGG (Figure 5–4A, dotted vertical line) during delta replication is responsible for the large and small isoforms of HDag.[172]

Infection with HDV can occur only by co-infection with HBV or superinfection an HBV carrier, as HDV needs the HBs antigen of HBV to maintain transmission. Therefore, transmission patterns of HDV mimic the parenteral or sexual patterns responsible for HBV exposure. HDV superinfection of an HBV carrier usually proceeds to chronic hepatitis and HDV infection. Co-infection with HBV, on the other hand, is more likely to result in self-limited disease and resolution of both HBV and HDV. The acute phase of either type of HDV infection has been correlated with increased probability of fulminant hepatitis, and within certain regions of the world, HDV has been implicated in clusters of fulminant hepatitis.[173,174]

B. APPROACHES FOR PCR OF HDV

Proteinase K digestion or guanidinium isothiocyanate treatment of serum followed by nucleic acid extraction have been used to isolate serum-derived HDV RNA. Liver tissues homogenized in the presence of guanidinium isothiocyanate followed by nucleic acid extraction have also been used as a source of HDV RNA. One of the primary considerations for successful amplification of HDV is adequate transcription of the highly base-paired RNA into cDNA. Extensive denaturation of RNA secondary structure using

Figure 5–4 HDV Genome with delta antigen (HDag) open reading frame and analysis of HCV genetic variability. (A) Linear schematic of HDV genome with the approximate location of the delta antigen (HDag) open reading frame shown by the open rectangle. The position equivalent to the Hind III site of the Makino strain[166] has been designated as position 0. The black bar indicates the location of the amber termination codon that constitutes microheterogeneity of HDV responsible for generating large and small HDag during an HDV infection. (B) Analysis of genetic variability of HDV. Eight complete sequences of HDV[5,165,177,180,181,183,184,211] were aligned using the pileup program within the University of Wisconsin Computer Genetics Group package for DNA analysis.[212] The number of changed nucleotides within each 50-nucleotide span was determined. The bars of this histogram represent these changed nucleotide position within each 50 nucleotides, while the relative genome location is shown on the horizontal axis.

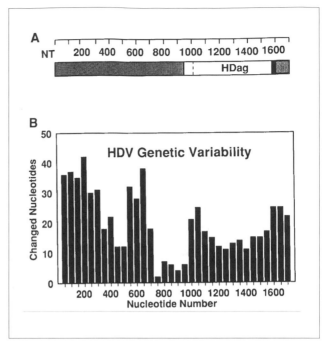

methyl mercury hydroxide during reverse transcription probably provides the most reliable means of overcoming this problem. After synthesis of cDNA with reverse transcriptase, positive- and negative-sense primers are added, and the cDNA is amplified by PCR. Detection of amplified HDV sequences by hybridization with a radiolabeled internal probe indicated that positive amplification could be detected from the equivalent of 8 pl of serum (the infectivity of the serum was not specified).[175] An alternative method for detection of amplified HDV sequences uses streptavidin-coated microtiter plates, followed by binding of a biotinylated HDV oligonucleotide. The amplified PCR solution was heat denatured and then allowed to hybridize to the bound streptavidin-biotin-oligonucleotide. Addition of a monoclonal antibody specific for double-stranded DNA followed by horseradish-peroxidase-labeled rabbit anti-mouse IgG is used for chromogenic detection of the hybridized double-stranded product.[176] The sensitivity of this enzyme immunoassay format was similar to that achieved with parallel samples detected by hybridization.

C. PRIMERS USED FOR PCR AMPLIFICATION OF HDV

Chao and co-workers[177] compared three sequenced strains of HDV (from California,[5] Italy,[166] and Nauru Island in the South Pacific[177]) and proposed three conserved regions (nucleotides 659 to 772, 874 to 966, and 1267 to 1348) for use as primers for detection and diagnosis of HDV using PCR. The histogram shown in Figure 5–4B revises their recommendations by graphically illustrating the genetic conservation between nucleotides 750 and 1000 compared with the remainder of the HDV genome, including the HDag coding region. It is obvious that primers directed toward this conserved genome region have a high probability of binding to all HDV strains. Varying primers from other regions have been used to amplify portions of HDV sequences for cloning and sequence characterization[178–181] or elucidation of structural and functional sequences, depending upon the region of interest for amplification.

D. USES AND APPLICATION OF PCR-AMPLIFIED HDV SEQUENCES

The reports of PCR amplification to diagnose or detect HDV infections have been limited to the demonstration of HDV within serum and liver of chronically infected individuals.[175,182] On the other hand, PCR amplification has been used extensively to amplify and clone or sequence individual HDV strains from different geographic locations.[177,178,180–185] Serial samples from a chronic carrier[181] indicate that the changes within the HDV genome during chronic infection appear to affect 0.5% of the nucleotides. This value is similar to HDV subjected to five serial passages[179] in woodchucks, which had only 0.8% nucleotide divergence from the parental genotype. These values are in sharp contrast with the genetic divergence seen between different strains. The genome region illustrated as nucleotides 1 to 700 and 1600 to 1700 (Figure 5–4B) has numerous insertions and deletions in addition to changed nucleotides that contribute to apparent nucleotide differences of up to 70%.

PCR amplification of selected genome regions of HDV has been used to help elucidate and monitor the transition of the amber termination codon that results in synthesis of the small and large delta antigen,[186,187] and elucidation of the RNA structure necessary for the ribozyme activity of the HDV genome;[188] it has been used to amplify sections of the HDag for characterization of regions involved in the transport of HDag to the nucleus.[189]

VI. PCR OF HEPATITIS E VIRUS

A. BACKGROUND INFORMATION ON HEPATITIS E VIRUS

Hepatitis E virus (HEV), previously referred to in the literature as epidemic, water-borne, or enterically transmitted non-A, non-B hepatitis,[190–193] is a chloroform-resistant 30- to 34-nm icosahedral particle that has many physical properties similar to the caliciviruses and Norwalk viruses[3] (Table 5–1). Actual physical characterization of the virus itself has been difficult due to the labile nature of the virus particle, its lack of growth in cell culture, and the difficulty of obtaining stool samples that contain enough virus for identification by immune electron microscopy.

Cloning and nucleic acid sequencing of the HEV genome revealed that it is a 7.5-kb positive-sense RNA and that it contains three ORFs[194] flanked by 5′ and 3′ noncoding (NC) regions of 28 and 56 nucleotides (Figure 5–5). The longest and most 5′ reading frame encodes polypeptides that have structural motifs consistent with the presence of an RNA polymerase and helicase.[195] The next longest ORF is at the 3′ end of the genome, and is thought to encode the viral capsid polypeptide. Overlapping these two open reading frames is a small open reading frame encoding a cysteine-rich polypeptide of unknown function. Expressed regions from all three of these open reading frames are recognized by HEV antibodies in convalescent sera,[196,197] indicating that protein products from the three reading frames are generated during infection.

Documented outbreaks of hepatitis E in India, Pakistan, Burma, China, Nepal, the former Soviet Union, Africa, Costa Rica, and Mexico[198] have all resulted from contaminated drinking water. Within these outbreaks, the major group to develop clinical illness due to HEV infection were young adults. In addition, fulminant HEV infections were responsible for up to 20% mortality among infected pregnant women. Infection with HEV causes an acute disease, with no chronic sequelae, after an incubation period of approximately 40 days, while the incubation period for HAV is about 30 days. Unlike HAV, there appears to be limited secondary transmission in HEV outbreaks.

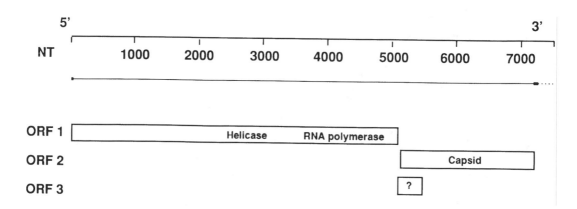

Figure 5–5   Genome organization of HEV. The 7.5-kb genome of HEV is illustrated by the solid line with the 5′ and 3′ noncoding regions, 28 nucleotides (NT) and 56 nucleotides, respectively, depicted by the small black rectangles and the poly-A tail shown as the dotted line. The three overlapping open reading frames (ORFs) are shown by the open rectangles with the approximate location of putative proteins (helicase, RNA polymerase, and capsid).

B. APPROACHES FOR PCR OF HEV

There has been limited use of PCR for detection of HEV compared to HAV, HBV, and HCV. The labile nature of the virus particle may contribute to the difficulty of using PCR amplification for detection and characterization of HEV. The single report related to evaluating clinical samples did not clarify what volume of supernatant from the stool suspensions was processed for RNA extraction and then tested by nested PCR.[199] McCaustland and co-workers[200] found that traditional PCR approaches, including nested amplification, reamplification of the initial product, or increasing the number of cycles, did not result in detection of HEV sequences in samples known to transmit hepatitis E. A comparison of amplification after immunoprecipitation, pelleting, guanidinium extraction, or glass-powder extraction of RNA from virus-containing samples indicated that glass-powder extraction was more sensitive in their hands and detected HEV sequences in all samples known to contain infectious HEV. Other investigators have successfully used nested PCR to detect HEV from stool and bile samples,[199,201,202] although Tsarev and co-workers cautioned that the RNA from serum and stools gave more reliable results if it was extensively purified. Recently, an antigen-capture assay, using IgM from acute-phase serum as the capture antibody coated onto microfuge tubes, has proven to be a reliable selection approach prior to PCR amplification.[203] After coating with IgM antibody, stool suspensions containing HEV are added and allowed to adsorb overnight at 4°C. cDNA and PCR amplification are then carried out as they were done for HAV.[15,16,31]

C. PRIMERS USED FOR PCR AMPLIFICATION OF HEV

The primers used for amplification of HEV have focused on the RNA polymerase region as this was the first sequence published.[194] There are complete sequences for HEV published from Burma,[194] Pakistan,[202] and China.[204] The overall genetic relatedness between these strains is 90 to 97%, and all derive from Southeast Asia. According to Fry and co-workers,[195] the sequence of a strain of HEV from Mexico differs genetically from the Burma strain by about 25%, suggesting that Old and New World groups of these viruses may exist. Until more strains are sequenced from different locations, we have limited information on conserved regions optimal for designing PCR primers or overall genetic variability.

D. USES AND APPLICATION OF PCR-AMPLIFIED HEV SEQUENCES

PCR technology and variations thereof have been used for amplification of viral gene sequences, followed by direct sequencing or cloning. After obtaining a positive HEV clone, Reyes et al.[205] used a modified PCR technique that they termed SISPA (sequence-independent single-primer amplification) to amplify nucleic acid sequences from HEV-infected samples. SISPA involves the ligation of an asymmetric, double-stranded linker-primer onto blunt-end cDNA synthesized by the Gubler and Hoffmann method.[206] Addition of a common primer followed by PCR amplification results in the nonspecific amplification of all cDNAs present. After ligation and cloning, HEV-positive clones were identified by hybridization with a previously cloned HEV fragment. SISPA was also used to construct a library of clones from a strain of HEV obtained from an outbreak in Mexico,[207] and a modification of this approach was used by Tam and co-workers[194] to complete the Burmese HEV sequence. They reverse transcribed HEV RNA to generate cDNA with an HEV-specific primer, tailed the first-strand cDNA with dGTP, and then PCR amplified using oligo(dC) and the initial HEV primer.

Tsarev[202] generated multiple primers based upon the known Burmese HEV sequence, and then proceeded to sequence the entire genome of a Pakistani-derived HEV. A comparison between these two strains indicated that the helicase region contained a region of amino acid variability, although genetically they were greater than 90% homologous. Ray and co-workers[208] have presented additional evidence of unique genetic and antigenic properties of HEV. They identified a 246-nucleotide deletion within the small ORF of viruses involved in two separate outbreaks within India after PCR amplification, cloning, and sequencing.

VII. DISCUSSION AND CONCLUSIONS

The use of PCR technology has greatly expanded the avenues of research on hepatitis viruses. The only source of virus for the majority of these agents has been clinical samples from infected individuals, that results in a finite quantity of virus for characterization. PCR amplification has facilitated cloning and characterization of genetic variants of HAV, HBV, HCV, and HDV that would not have been possible otherwise.

PCR amplification for detection or diagnosis of HAV may have limited use in clinical situations as IgM anti-HAV provides a reliable indicator of acute infection. PCR amplification would be most useful in situations where detection of potentially infectious virus could be used to prevent transmission. The detection of HAV in potentially contaminated water systems and the reliable use of PCR for identification of HAV in shellfish or other contaminated foods are examples of the types of situation where PCR amplification of HAV would be useful.

Mutations within HBV infections will continue to be characterized by PCR amplification and sequencing, and may ultimately provide information on the progression from chronic disease to the development of HCC. The use of PCR amplification and nucleic acid sequencing will help elucidate whether genotypes of HBV can be used to follow transmission and will characterize geographic clusters of HBV. In addition, PCR amplification will be a valuable resource for characterization of the status of HBV carriers and monitoring potential treatment or therapy.

Research on HCV has been, and will be, one of the primary recipients of the exquisite sensitivity of PCR amplification. PCR detection will be invaluable in further defining potential infectivity in recovered and chronically infected individuals. PCR amplification may provide answers regarding the mode of transmission within those cases where no identified exposure can be defined. Carefully controlled studies on the presence of HCV in other body secretions or potential mother-infant transmission[209] will help determine whether transmission can occur in these situations. The other area of HCV research that will benefit from PCR technology is the delineation of genetic groups or genotypes of HCV. Amplification and sequencing of genetically different strains may provide information on unique clinical features or relevant epidemiological characteristics.

The use of PCR for diagnosis of HDV infections would facilitate identification of ongoing HDV replication and provide a potential means of monitoring the status of chronic HDV infection. PCR amplification and sequence characterization of HDV and HBV strains implicated in fulminant disease would clarify whether genetic variants are responsible in these cases.

The routine use of PCR amplification for detection of HEV remains to be adequately explored and yet offers a potentially reliable method to diagnose HEV outbreaks. In addition, PCR could be used to determine whether environmental reservoirs, such as domestic pigs,[210] are sources of HEV within endemic regions.

There remains a small percentage of viral hepatitis that cannot be attributed to HAV, HBV, HCV, HEV, or HDV, based upon serological diagnosis. PCR amplification of acute-phase serum from these cases, with adequate positive and negative controls, might reveal that some cases are the result of one of these five agents, and that a diminished host response is the cause of the apparent negative serologic diagnosis. Once reliable primers are defined for the five viral agents of hepatitis, PCR amplification using a cocktail of hepatitis primers that would amplify a unique size fragment, indicative of the specific viral agent responsible, from virus present in acute-phase serum could be used to diagnose infections.

ACKNOWLEDGMENTS

I would like to acknowledge the patience and understanding of H. L. Robertson and individuals within the Viral Genetics Section during the writing of this review.

REFERENCES

1. **Gust, I. D., Coulepis, A. G., Feinstone, S. M., Locarnini, S. A., Moritsugu, Y., Najera, R., and Siegl, G.,** Taxonomic classification of hepatitis A virus, *Intervirology,* 20, 1, 1983.
2. **Melnick, J. L.,** Classification of hepatitis A virus an enterovirus type 72 and of hepatitis B virus as hepadnavirus type I, *Intervirology,* 18, 105, 1982.
3. **Bradley, D., Andjaparidze, A., Cook, E. H., McCaustland, K., Balayan, M., Stetler, H., Velazquez, O., Robertson, B., Humphrey, C., Kane, M., and Weisfuse, I.,** Aetiological agent of enterically transmitted non-A, non-B hepatitis, *J. Gen. Virol.,* 69, 731, 1988.
4. **Miller, R. H. and Purcell, R. H.,** Hepatitis C virus shares amino acid sequence similarity with pestiviruses and flaviviruses as well as members of two plant virus supergroups, *Proc. Natl. Acad. Sci. U.S.A.,* 87, 2057, 1990.
5. **Wang, K.-S., Choo, Q.-L., Weiner, A. J., Ou, J.-H., Najarian, R. C., Thayer, R. M., Mullenbach, G. T., Denniston, K. J., Gerin, J. L., and Houghton, M.,** Structure, sequence and expression of the hepatitis delta viral genome, *Nature,* 323, 508, 1986.
6. **Krugman, S., Ward, R., and Gilles, W. P.,** The natural history of infectious hepatitis, *Am. J. Med.,* 32, 717, 1962.

7. **Minor, P. D.,** Picornaviridae, in *Classification and Nomenclature of Viruses: The Fifth Report of the International Committee on Taxonomy of Viruses,* Franki, R. I. B., Fauquet, C. M., Knudson, D. L., and Brown, F., Eds., Springer-Verlag, Vienna, 1991, 320.

8. **Feinstone, S. M., Kapikian, A. Z., and Purcell, R. H.,** Hepatitis A: detection by immune electron microscopy of a virus-like antigen association with acute illness, *Science,* 182, 1026, 1973.

9. **Provost, P. J. and Hilleman, M. R.,** Propagation of human hepatitis A virus in cell culture *in vitro, Proc. Soc. Exp. Biol. Med.,* 160, 213, 1979.

10. **Binn, L. N., Lemon, S. M., Marchwicki, R. H., Redfield, R. R., Gates, N. L., and Bancroft, W. H.,** Primary isolation and serial passage of hepatitis A virus strains in primate cell cultures, *J. Clin. Microbiol.,* 20, 28, 1984.

11. **Daemer, R. J., Feinstone, S. M., Gust, I. D., and Purcell, R. H.,** Propagation of human hepatitis A virus in African green monkey kidney cell culture: primary isolation and serial passage, *Infect. Immun.,* 32, 388, 1981.

12. **Lemon, S. M., Binn, L. N., and Marchwicki, R. H.,** Radioimmunofocus assay for quantitation of hepatitis A virus in cell culture, *J. Clin. Microbiol.,* 17, 834, 1983.

13. **Wheeler, C. M., Robertson, B. H., Van Nest, G., Dina, D., Bradley, D. W., and Fields, H. A.,** Structure of hepatitis A virion: peptide mapping of the capsid region, *J. Virol.,* 58, 307, 1986.

14. **Robertson, B. H., Brown, V. K., and Khanna, B.,** Altered hepatitis A VP1 protein resulting from cell culture propagation of virus, *Virus Res.,* 13, 207, 1989.

15. **Jansen, R. W., Siegl, G., and Lemon, S. M.,** Molecular epidemiology of human hepatitis A virus defined by an antigen-capture polymerase chain reaction method, *Proc. Natl. Acad. Sci. U.S.A.,* 87, 2867, 1990.

16. **Robertson, B. H., Khanna, B., Nainan, O. V., and Margolis, H. S.,** Epidemiologic patterns of wild-type hepatitis A virus determined by genetic variation, *J. Infect. Dis.,* 163, 286, 1991.

17. **Robertson, B. H., Khanna, B., Nainan, O. V., and Margolis, H. S.,** Genetic variation of wild-type hepatitis A isolates, in *Viral Hepatitis and Liver Disease,* Hollinger, B. N., Lemon, S. M., and Margolis, H. S., Eds., Williams & Wilkins, Baltimore, 1991, 54.

18. **Rosenblum, L. S., Villarino, M. E., Nainan, O. V., Melish, M. E., Hadler, S. C., Pinsky, P. P., Jarvis, W. R., Ott, C. E., and Margolis, H. S.,** Hepatitis A outbreak in a neonatal intensive care unit: risk factors for transmission and evidence of prolonged viral excretion among preterm infants, *J. Infect. Dis.,* 164, 476, 1991.

19. **Robertson, B. H. and Burkholder, B.,** unpublished observation, 1991.

20. **Desenclos, J. A., Klontx, K. C., Wilder, M. H., Nainan, O. V., Margolis, H. S., and Gunn, R. A.,** A multistate outbreak of hepatitis A caused by the consumption of raw oysters, *Am. J. Public Health,* 81, 1268, 1991.

21. **Niu, M. T., Polish, L. B., Robertson, B. H., Khanna, B., Woodruff, B. A., Shapiro, C. N., Miller, M. A., Smith, J. D., Gedrose, J. K., Alter, M. J., and Margolis, H. S.,** A multistate outbreak of hepatitis A associated with frozen strawberries, *J. Infect. Dis.,* 166, 518, 1992.

22. **Nainan, O. V., Margolis, H. S., Robertson, B. H., Balayan, M., and Brinton, M. A.,** Sequence analysis of a new hepatitis A virus naturally infecting cynomolgus macaques *(Macaca fascicularis), J. Gen. Virol.,* 72, 1685, 1991.

23. **Tsarev, S. A., Emerson, S. U., Balayan, M. S., Ticehurst, J., and Purcell, R. H.,** Simian hepatitis A virus (HAV) strain AGM-27: comparison of genome structure and growth in cell culture with other HAV strains, *J. Gen. Virol.,* 72, 1677, 1991.

24. **Khanna, B., Spelbring, J. E., Innis, B. L., and Robertson, B. H.,** Characterization of a genetic variant of human hepatitis A virus, *J. Med. Virol.,* 36, 118, 1992.

25. **Ping, L.-H. and Lemon, S. M.,** Antigenic structure of human hepatitis A virus defined by analysis of escape mutants selected against murine monoclonal antibodies, *J. Virol.,* 66, 2208, 1992.

26. **Nainan, O. V., Margolis, H. S., and Brinton, M. A.,** Identification of amino acids located in antibody binding sites of human hepatitis A virus, *Virology,* 191, 984–987, 1992.

27. **Lemon, S. M. and Binn, L. N.,** Antigenic relatedness of two strains of hepatitis A virus determined by cross-neutralization, *Infect. Immun.,* 42, 418, 1983.

28. **Provost, P. J., Ittensohn, O. L., Villarejos, V. M., Arguedas, J. A., and Hilleman, M. R.,** Etiologic relationship of marmoset-propagated CR326 hepatitis A to hepatitis in man, *Proc. Soc. Exp. Biol. Med.,* 142, 1257, 1973.

29. **Rakela, J., Fay, O. H., Stevenson, K., Gordon, I., and Mosley, J. W.,** Similarities of two hepatitis A virus strains, *Bull. WHO,* 54, 561, 1976.

30. **Jansen, R. W., Siegl, G., and Lemon, S. M.,** Molecular epidemiology of human hepatitis A virus (HAV), in *Viral Hepatitis and Liver Disease,* Hollinger, B. N., Lemon, S. M., and Margolis, H. S., Eds., Williams & Wilkins, Baltimore, 1991, 58.

31. **Robertson, B. H., Jansen, R. W., Khanna, B., Totsuka, A., Nainan, O. V., Siegl, G., Widell, A., Margolis, H. S., Isomura, S., Ito, K., Ishizu, T., Moritsugu, Y., and Lemon, S. M.,** Genetic relatedness of hepatitis A virus strains recovered from different geographic regions, *J. Gen. Virol.,* 73, 1365, 1992.

32. **Gust, I. D., Burrell, C. J., Coulepis, A. G., Robinson, W. S., and Zuckerman, A. J.,** Taxonomic classification of human hepatitis B virus, *Intervirology,* 25, 14, 1986.

33. **Raney, A. K. and McLachlan, A.,** The biology of hepatitis B virus, in *Molecular Biology of the Hepatitis B Virus,* McLachlan, A., Ed., CRC Press, Boca Raton, FL, 1991, 1.

34. **Foster, G. R., Carman, W. F., and Thomas, H. C.,** Replication of hepatitis B and delta viruses: appearance of viral mutants, *Sem. Liver Dis.,* 11, 121, 1991.

35. **Hoofnagle, J. H., Shafritz, D. A., and Popper, H.,** Chronic type B hepatitis and the healthy HBsAg carrier state, *Hepatology,* 7, 758, 1987.

36. **Beasley, R. P., Hwang, L. Y., Lin, C. C., and Chien, C. S.,** Hepatocellular carcinoma and HBV: a prospective study of 22,707 men in Taiwan, *Lancet,* 2, 1129, 1981.

37. **Larzul, D., Guigue, F., Sninsky, J. J., Mack, D. H., Brechot, C., and Guesdon, J.,** Detection of hepatitis B virus sequences in serum by using in vitro enzymatic amplification, *J. Virol. Meth.,* 20, 227, 1988.

38. **Manzin, A., Salvoni, G., Bagnarelli, P., Menzo, S., Carloni, G., and Clementi, M.,** A single-step DNA extraction procedure for the detection of serum hepatitis B virus sequences by the polymerase chain reaction, *J. Virol. Methods,* 32, 245, 1991.

39. **Kaneko, S., Feinstone, S. M., and Miller, R. H.,** Rapid and sensitive method for the detection of serum hepatitis B virus DNA using the polymerase chain reaction technique, *J. Clin. Microbiol.,* 27, 1930, 1989.

40. **Zeldis, J. B., Lee, J. H., Mamish, K., Finegold, D. J., Sircar, R., Ling, Q., Knudsen, P. J., Kuramoto, I. K., and Mimms, L. T.,** Direct method for detecting small quantities of hepatitis B virus DNA in serum and plasma using the polymerase chain reaction, *J. Clin. Invest.,* 84, 1503, 1989.

41. **Liang, T. J., Isselbacher, K. J., and Wands, J. R.,** Rapid identification of low level hepatitis B-related viral genome in serum, *J. Clin. Invest.,* 84, 1367, 1989.

42. **Cheyrou, A., Guyomarc'h, C., Jasserand, P., and Blouin, P.,** Improved detection of HBV DNA by PCR after microwave treatment of serum, *Nucl. Acids Res.,* 19, 4006, 1991.

43. **Frickhofen, N. and Young, N. S.,** A rapid method of sample preparation for detection of DNA viruses in human serum by polymerase chain reaction, *J. Virol. Meth.,* 35, 65, 1991.

44. **Kaneko, S., Kobayashi, K., and Miller, R. H.,** Detection of hepatitis B virus DNA using the polymerase chain reaction technique, *J. Clin. Lab. Anal.,* 4, 479, 1990.

45. **Ulrich, P. P., Bhat, R. A., Seto, B., Mack, D., Sninsky, J., and Vyas, G. N.,** Enzymatic amplification of hepatitis B virus DNA in serum compared with infectivity testing in chimpanzees, *J. Infect. Dis.,* 160, 37, 1989.

46. **Keller, G. H., Huang, D. P., Shih, J. W., and Manak, M. M.,** Detection of hepatitis B virus DNA in serum by polymerase chain reaction amplification and microtiter sandwich hybridization, *J. Clin. Microbiol.,* 28, 1411, 1990.

47. **Lo, Y. M., Methal, W. Z., and Fleming, K. A.,** *In vitro* amplification of hepatitis B virus sequences from liver tumour DNA and from paraffin wax embedded tissues using the polymerase chain reaction, *J. Clin. Pathol.,* 42, 840, 1989.

48. **Paterlini, P., Gerken, G., Nakajima, E., Terre, S., D'Errico, A., Grigioni, W., Nalpas, B., Franco, D., Wands, J., and Kew, M.,** Polymerase chain reaction to detect hepatitis B virus DNA and RNA sequences in primary liver cancers from patients negative for hepatitis B surface antigen, *N. Engl. J. Med.,* 323, 80, 1990.

49. **Lampertico, P., Malter, J. S., Colombo, M., and Gerber, M. A.,** Detection of hepatitis B virus DNA in formalin-fixed, paraffin-embedded liver tissue by the polymerase chain reaction, *Am. J. Pathol.,* 137, 253, 1990.

50. **Shindo, M., Okuno, T., Arai, K., Matsumoto, M., Takeda, M., Kashima, K., Shimada, M., Fujiwara, Y., and Sokawa, Y.,** Detection of hepatitis B virus DNA in paraffin-embedded liver tissues in chronic hepatitis B or non-A, non-B hepatitis using the polymerase chain reaction, *Hepatology,* 13, 167, 1991.

51. **Diamantis, I. D., McGandy, C., Pult, I., Buhler, H., Schmid, M., Gudat, F., and Bianchi, L.,** Polymerase chain reaction detects hepatitis B virus DNA in paraffin-embedded liver tissue from patients sero- and histo-negative for active hepatitis B, *Virchows Archiv, A Pathological Anatomy and Histopathology (Berlin),* 420, 11, 1992.

52. **Shibata, D. K., Arnheim, N., and Martin, W. J.,** Detection of human papilloma virus in paraffin-embedded tissue using the polymerase chain reaction, *J. Exp. Med.,* 167, 225, 1988.

53. **Baginski, I., Chemin, I., Bouffard, P., Hantz, O., and Trepo, C.,** Detection of polyadenylated RNA in hepatitis B virus-infected peripheral blood mononuclear cells by polymerase chain reaction, *J. Infect. Dis.,* 163, 996, 1991.

54. **Galibert, F., Mandart, E., Fitoussi, F., Tiollais, P., and Charnay, P.,** Nucleotide sequence of the hepatitis B virus genome (subtype ayw) cloned in *E. coli, Nature,* 281, 646, 1979.

55. **Pasek, M., Goto, T., Gilbert, W., Zink, B., Schaller, H., MacKay, P., Leadbetter, G., and Murray, K.,** Hepatitis B virus genes and their expression in *E. coli, Nature,* 282, 575, 1979.

56. **Ono, Y., Onda, H., Sasada, R., Igarashi, K., Sugino, Y., and Niskioka, K.,** The complete nucleotide sequences of the cloned hepatitis B virus DNA: subtype adr and adw, *Nucl. Acids Res.,* 11, 1747, 1983.

57. **Fujiyama, A., Miyanohara, A., Nozaki, C., Yoneyama, T., Ohtomo, N., Matsubara, K.,** Cloning and structural analyses of hepatitis B virus DNAs, subtype adr, *Nucl. Acids Res.,* 11, 4601, 1983.

58. **Kobayashi, M. and Koike, K.,** Complete nucleotide sequence of hepatitis B virus DNA of subtype adr and its conserved gene organization, *Gene,* 30, 227, 1984.

59. **Bichko, V., Pushko, P., Dreilina, D., Pumpen, P., and Gren, E.,** Subtype ayw variant of hepatitis B virus DNA primary structure analysis, *FEBS Lett.,* 185, 208, 1985.

60. **Okamoto, H., Imai, M., Shimozaki, M., Yoshi, Y., Izuka, H., Gotanda, T., Tsuda, F., Miyakawa, Y., and Mayumi, M.,** Nucleotide sequence of a cloned hepatitis B virus genome, subtype ayr: comparison with genomes of the other three subtypes, *J. Gen. Virol.,* 67, 2305, 1986.

61. **Okamoto, H., Tsuda, F., Sakugawa, H., Sastrosoewignjo, R. I., Imai, M., Miyakawa, Y., and Mayumi, M.,** Typing hepatitis B virus by homology in nucleotide sequence: comparison of surface antigen subtypes, *J. Gen. Virol.,* 69, 2527, 1988.

62. **Robertson, B. H. and Kotsopoulou, K.,** unpublished observations, 1992.

63. **Kaneko, S., Miller, R. H., Di Bisceglie, A. M., Feinstone, S. M., Hoofnagle, J. H., and Purcell, R. H.,** Detection of hepatitis B virus DNA in serum by polymerase chain reaction. Application for clinical diagnosis, *Gastroenterology,* 99, 799, 1990.

64. **Kaneko, S., Miller, R. H., Feinstone, S. M., Unoura, M., Kobayashi, K., Hattori, N., and Purcell, R. H.,** Detection of serum hepatitis B virus DNA in patients with chronic hepatitis using the polymerase chain reaction assay, *Proc. Natl. Acad. Sci. U.S.A.,* 86, 312, 1989.

65. **Sumazaki, R., Motz, M., Wolf, H., Heinig, J., Jilg, W., and Deinhardt, F.,** Detection of hepatitis B virus in serum using amplification of viral DNA by means of the polymerase chain reaction, *J. Med. Virol.,* 27, 304, 1989.

66. **Baker, B. L., Di Bisceglie, A. M., Kaneko, S., Miller, R., Feinstone, S. M., Waggoner, J. G., and Hoofnagle, J. H.,** Determination of hepatitis B virus DNA in serum using the polymerase chain reaction: clinical significance and correlation with serological and biochemical markers, *Hepatology,* 13, 632, 1991.

67. **Chemin, I., Baginski, I., Petit, M. A., Zoulim, F., Pichoud, C., Capel, F., Hantz, O., and Trepo, C.,** Correlation between HVB DNA detection by polymerase chain reaction and Pre-S1 antigenemia in symptomatic and asymptomatic hepatitis B virus infections, *J. Med. Virol.,* 33, 51, 1991.

68. **Gerken, G., Paterlini, P., Manns, M., Housset, C., Terre, S., Dienes, H. P., Hess, G., Gerlich, W. H., Berthelot, P., and Meyer zum Buschenfelde, K. H.,** Assay of hepatitis B virus DNA by polymerase chain reaction and its relationship to pre-S- and S-encoded viral surface antigens, *Hepatology,* 13, 158, 1991.

69. **Monjardino, J., Velosa, J., Thomas, H. C., and Carneiro de Moura, M.,** Serum HBV DNA detected by PCR in dot blot negative HBV chronic carriers with active liver disease, *J. Hepatol.,* 13, 44, 1991.

70. **Wirth, S., Mollers, U., Keller, K.-M., and Winterpacht, A.,** Use of the polymerase chain reaction to demonstrate hepatitis B virus DNA in serum of children with chronic hepatitis B, *J. Pediat.,* 120, 438, 1992.

71. **Luo, K.-X., Zhou, R., He, C., Liang, Z.-S., and Jiang, S.,** Hepatitis B virus DNA in sera of virus carriers positive exclusively for antibodies to the hepatitis B core antigen, *J. Med. Virol.,* 35, 55, 1991.

72. **Thiers, V., Kremsdorf, D., Schellekens, H., Goudeau, A., Sninsky, J., Nakajima, E., Mack, D., Driss, F., Wands, J., Tiollais, P., and Brechot, C.,** Transmission of hepatitis B from hepatitis-B-seronegative subjects, *Lancet,* 2, 1273, 1988.

73. **Shih, L. N., Sheu, J. C., Wang, J. T., Huang, G. T., Yang, P. M., Lee, H. S., Sung, J. L., Wang, T. H., and Chen, D. S.,** Serum hepatitis B virus DNA in healthy HBsAg-negative Chinese adults evaluated by polymerase chain reaction, *J. Med. Virol.,* 32, 257, 1990.

74. **Wang, J. T., Wang, T. H., Sheu, J. C., Shih, L. N., Lin, J. T., and Chen, D. S.,** Detection of hepatitis B virus DNA by polymerase chain reaction in plasma of volunteer blood donors negative for hepatitis B surface antigen, *J. Infect. Dis.,* 163, 397, 1991.

75. **Liang, T. J., Blum, H. E., and Wands, J. R.,** Characterization and biological properties of a hepatitis B virus isolated from a patient without hepatitis B virus serologic markers, *Hepatology,* 12, 204, 1990.

76. **Kremsdorf, D., Thiers, V., Garreau, F., Nakajima, E., Chappey, C., Schellenkens, H., Wands, J. R., Sninsky, J., Tiollais, T., and Brechot, C.,** Nucleotide sequence analysis of hepatitis B virus genomes isolated from serologically negative patients, in *Viral Hepatitis and Liver Disease,* Hollinger, F. B., Lemon, S. M., and Margolis, H. S., Eds., Williams & Wilkins, Baltimore, 1991, 222.

77. **Brechot, C., Degos, F., Lugassy, C., Thiers, V., Zafrani, S., Franco, D., Bismuth, H., Trepo, C., Benhamou, J.-P., Wands, J., Isselbacker, K., Tiollais, P., and Berthelot, P.,** Hepatitis B virus DNA in patients with chronic liver disease and negative tests for hepatitis B surface antigen, *N. Engl. J. Med.,* 312, 270, 1985.

78. **Carman, W. F., Jacyna, M. R., Hadziyannis, S., Karayiannis, P., McGarvey, M. J., Makris, A., and Thomas, H. C.,** Mutation preventing formation of hepatitis B e antigen in patients with chronic hepatitis B infection, *Lancet,* 2, 588, 1989.

79. **Okamoto, H., Yotsumoto, S., Akahane, Y., Yamanaka, T., Miyazaki, Y., Sugai, Y., Tsuda, F., Tanaka, T., Miyakawa, Y., and Mayumi, M.,** Hepatitis B viruses with precore region defects prevail in persistently infected hosts along with seroconversion to the antibody against e antigen, *J. Virol.,* 64, 1298, 1990.

80. **Brunetto, M. R., Stemmler, M., Schodel, F., Will, H., Ottobrelli, A., Rizzetto, M., Verme, G., and Bonino, F.,** Identification of HBV variants which cannot produce precore derived HBeAg and may be responsible for severe hepatitis, *Ital. J. Gastroenterol.,* 21, 151, 1989.

81. **Fiordalisi, G., Cariani, E., Mantero, G., Zanetti, A., Tanzi, E., Chiaramonte, M., and Primi, D.,** High genomic variability in the pre-C region of hepatitis B virus in anti-HBe, HBV DNA-positive chronic hepatitis, *J. Med. Virol.,* 31, 297, 1990.

82. **Santantonio, T., Jung, M.-C., Miska, S., Pastore, G., Pape, G. R., and Will, H.,** Prevalence and type of pre-C HBV mutants in anti-HBe positive carriers with chronic liver disease in a highly endemic area, *Virology,* 183, 840, 1991.

83. **Tong, S., Vitvitski, L., and Trepo, C.,** Active hepatitis B virus replication in the presence of anti-HBe is associated with viral variants containing an inactive pre-C region, *Virology,* 176, 596, 1990.

84. **Ulrich, P. P., Bhat, R. A., Kelly, I., Brunetto, M. R., Bonino, F., and Vyas, G. N.,** A precore-defective mutant of hepatitis B virus associated with e antigen-negative chronic liver disease, *J. Virol.,* 32, 109, 1990.

85. **Carman, W. F., Ferrao, M., Lok, A. S. F., Ma, O. C. K., Lai, C. L., and Thomas, H. C.,** Precore sequence variation in Chinese isolates of hepatitis B virus, *J. Infect. Dis.,* 165, 127, 1992.

86. **Hasegawa, K., Huang, J., Wands, J. R., Obata, H., and Liang, T. J.,** Association of hepatitis B viral precore mutations with fulminant hepatitis B in Japan, *Virology,* 185, 460, 1991.

87. **Ehata, T., Omata, M., Yokosuka, O., Hosoda, K., and Ohto, M.,** Variations in codons 84–101 in the core nucleotide sequence correlate with hepatocellular injury in chronic hepatitis B virus infection, *J. Clin. Invest.,* 89, 332, 1992.

88. **Santantonio, T., Jung, M.-C., Schneider, R., Fernholz, D., Milella, M., Monno, L., Pastore, G., Pape, G. R., and Will, H.,** Hepatitis B virus genomes that cannot synthesize pre-S2 proteins occur frequently and as dominant virus populations in chronic carriers in Italy, *Virology,* 188, 948, 1992.

89. **Tran, A., Kremsdorf, D., Capel, F., Housset, C., Dauguet, C., Petit, Marie-A., and Brechot, C.,** Emergence of and takeover by hepatitis B virus (HBV) with rearrangements in the pre-S/S and pre-C/C genes during chronic HBV infection, *J. Virol.,* 65, 3566, 1991.

90. **Carman, W. F., Zanetti, A. R., Karayiannis, P., Waters, J., Manzillo, G., Tanzi, E., Zuckerman, A. J., and Thomas, H. C.,** Vaccine-induced escape mutant of hepatitis B virus, *Lancet,* 336, 325, 1990.

91. **Fujii, H., Moriyama, K., Sakamoto, N., Kondo, T., Yasuda, K., Hiraizumi, Y., Yamazaki, M., Sakaki, Y., Okochi, K., and Nakajima, E.,** Gly145 to Arg substitution in HBs antigen of immune escape mutant of hepatitis B virus, *Biochem. Biophys. Res. Commun.,* 184, 1152, 1992.

92. **McMahon, G., McCarthy, L. A., Dottavio, D., and Ostberg, L.,** Surface antigen and polymerase gene variation in hepatitis B virus isolates from a monoclonal antibody-treated liver transplant patient, in *Viral Hepatitis and Liver Disease,* Hollinger, F. B., Lemon, S. M., and Margolis, H. S., Eds., Williams & Wilkins, Baltimore, 1991, 219.

93. **Courouce-Pauty, A.-M., Plancon, A., and Soulier, J. P.,** Distribution of HBsAg subtypes in the world, *Vox Sanguinis,* 44, 197, 1983.

94. **Norder, H., Hammas, B., and Magnius, L. O.,** Typing of hepatitis B virus genomes by a simplified polymerase chain reaction, *J. Med. Virol.,* 31, 215, 1990.

95. **Norder, H., Hammas, B., Lofdahl, S., Courouce, Anne-M., and Magnius, L. O.,** Comparison of the amino acid sequences of nine different serotypes of hepatitis B surface antigen and genomic classification of the corresponding hepatitis B virus strains, *J. Gen. Virol.,* 73, 1201, 1992.

96. **Lin, H. J., Lai, C., Lauder, I. J., Wu, P., Kau, T. K., and Fong, M.,** Application of hepatitis B virus (HBV) DNA sequence polymorphisms to the study of HBV transmission, *J. Infect. Dis.,* 164, 284, 1991.

97. **Houghton, M., Choo, Q.-L., and Kuo, G.,** Non-A, Non-B Virus Diagnostics and Vaccines, European Patent. 88310922.5, 1988.

98. **Kato, N., Hijikata, M., Ootsuyama, Y., Nakagawa, M., Ohkoshi, S., Sugimura, T., and Shimotohno, K.,** Molecular cloning of the human hepatitis C virus genome from Japanese patients with non-A, non-B hepatitis, *Proc. Natl. Acad. Sci. U.S.A.,* 87, 9524, 1990.

99. **Miller, R. H. and Purcell, R. H.,** Hepatitis C virus shares amino acid sequence similarity with pestiviruses and flaviviruses as well as members of two plant virus supergroups, *Proc. Natl. Acad. Sci. U.S.A.,* 87, 2057, 1990.

100. **Choo, Q.-L., Richman, K. H., Han, J. H., Berger, K., Lee, C., Dong, C., Gallegos, C., Coit, D., Medina-Selby, A., Barr, P. J., Weiner, A. J., Bradley, D. W., Kuo, G., and Houghton, M.,** Genetic organization and diversity of the hepatitis C virus, *Proc. Natl. Acad. Sci. U.S.A.,* 88, 2451, 1991.

101. **Takamizawa, A., Mori, C., Fuke, I., Manabe, S., Murakami, S., Fujita, J., Onishi, E., Andoh, T., Yoshida, I., and Okayama, H.,** Structure and organization of the hepatitis C genome isolated from human carriers, *J. Virol.,* 65, 1105, 1991.

102. **Alter, M. J., Margolis, H. S., Krawczynski, K., Judson, F. N., Mares, A., Alexander, W. J., Hu, P. Y., Miller, J. K., Gerber, M. A., Sampliner, R. E., Meeks, E. L., and Beach, M. J.,** The natural history of community-acquired hepatitis C in the United States, *N. Engl. J. Med.,* 327, 1899–1905, 1992.

103. **Alter, H. J., Purcell, R. H., Holland, P. V., Feinstone, S. M., Morrow, A. G., and Moritsugu, Y.,** Clinical and serological analysis of transfusion associated hepatitis, *Lancet,* 2, 838, 1975.

104. **Bradley, D. W.,** The agents of non-A, non-B hepatitis, *J. Virol. Meth.,* 10, 301, 1985.

105. **Ulrich, P. P., Romeo, J. M., Lane, P. K., Kelly, I., Daniel, L. J., and Vyas, G. N.,** Detection, semiquantitation, and genetic variation in hepatitis C virus sequences amplified from the plasma of blood donors with elevated alanine aminotransferase, *J. Clin. Invest.,* 86, 1609, 1990.

106. **Garson, J. A., Tuke, P. W., Makris, M., Briggs, M., Machin, S. J., Preston, F. E., and Tedder, R. S.,** Demonstration of viraemia patterns in haemophiliacs treated with hepatitis C virus contaminated factor VIII concentrates, *Lancet,* 336, 1022, 1990.

107. **Weiner, A. J., Kuo, G., Bradley, D. W., Bonino, F., Saracco, G., Lee, C., Rosenblatt, J., Choo, Q.-L., and Houghton, M.,** Detection of hepatitis C viral sequences in non-A, non-B hepatitis, *Lancet,* 335, 1, 1990.

108. **Zanetti, A. R., Tanzi, E., Zehender, G., Magni, E., Incarbone, C., Zonaro, A., Primi, D., and Cariani, E.,** Hepatitis C virus RNA in symptomless donors implicated in post-transfusion non-A, non-B hepatitis, *Lancet,* 336, 448, 1990.

109. **Garson, J. A., Tedder, R. S., Briggs, M., Tuke, P., Glazebrook, J. A., Trute, A., Parker, D., Barbara, J. A. J., Contreras, M., and Aloysius, S.,** Detection of hepatitis C viral sequences in blood donations by "nested" polymerase chain reaction and prediction of infectivity, *Lancet,* 335, 1419, 1990.

110. **Simmonds, P., Zhang, L. Q., Watson, H. G., Rebus, S., Ferguson, E. D., Balfe, P., Leadbetter, G. H., Yap, P. L., Peutherer, J. F., and Ludlam, C. A.,** Hepatitis C quantification and sequencing in blood products, haemophiliacs, and drug users, *Lancet,* 336, 1469, 1990.

111. **Wang, J.-T., Wang, T.-H., Sheu, J.-C., Lin, S.-M., Lin, J.-T., and Chen, D.-S.,** Effects of antico-agulants and storage of blood samples on efficacy of the polymerase chain reaction assay for hepatitis C virus, *J. Clin. Microbiol.,* 30, 750, 1992.

112. **Beach, M. J., and Meeks, E. L.,** personal communication, 1992.

113. **Okamoto, H., Okada, S., Sugiyama, Y., Yotsumoto, S., Tanaka, T., Yoshizawa, H., Tsuda, F., Miyakawa, Y., and Mayumi, M.,** The 5′-terminal sequence of the hepatitis C virus genome, *Jpn. J. Exp. Med.,* 60, 167, 1990.

114. **Han, J. H., Shyamala, V., Richman, K. H., Brauer, M. J., Irvine, B., Urdea, M. S., Tekamp-Olson, P., Kuo, G., Choo, L.-L., and Houghton, M.,** Characterization of the terminal regions of hepatitis C viral RNA: identification of conserved sequences in the 5′ untranslated region and poly(A) tails at the 3′ ends, *Proc. Natl. Acad. Sci. U.S.A.,* 88, 1711, 1991.

115. **Bukh, J., Purcell, R. H., and Miller, R. H.,** Sequence analysis of the 5′ noncoding region of hepatitis C virus, *Proc. Natl. Acad. Sci. U.S.A.,* 89, 4942, 1992.

116. **Inchauspe, G., Zebedee, S., Lee, D.-H., Sugitani, M., Nasoff, M., and Prince, A. M.,** Genomic structure of the human prototype strain H of hepatitis C virus: comparison with American and Japanese isolates, *Proc. Natl. Acad. Sci. U.S.A.,* 88, 10292, 1991.

117. **Okamoto, H., Okada, S., Sugiyama, Y., Kurai, K., Iizuka, H., Machida, A., Miyakawa, Y., and Mayumi, M.,** Nucleotide sequence of the genomic RNA of hepatitis C virus isolated from a human carrier: comparison with reported isolates for conserved and divergent regions, *J. Gen. Virol.,* 72, 2697, 1991.

118. **Widell, A., Mansson, A.-S., Sundstrom, G., Hansson, B. G., and Nordenfelt, E.,** Hepatitis C virus RNA in blood donor sera detected by the polymerase chain reaction: comparison with supplementary hepatitis C antibody assays, *J. Med. Virol.,* 35, 253, 1991.

119. **Inchauspe, G., Abe, K., Zebedee, S., Nasoff, M., and Prince, A. M.,** Use of conserved sequences from hepatitis C virus for the detection of viral RNA in infected sera by polymerase chain reaction, *Hepatology,* 14, 595, 1991.

120. **Cha, T.-A., Kolberg, J., Irvine, B., Stempien, M., Beall, E., Yano, M., Choo, Q.-L., Houghton, M., Kuo, G., Han, J. H., and Urdea, M. S.,** Use of a signature nucleotide sequence of hepatitis C virus for detection of viral RNA in human serum and plasma, *J. Clin. Microbiol.,* 29, 2528, 1991.

121. **Bukh, J., Purcell, R. H., and Miller, R. H.,** Importance of primer selection for the detection of hepatitis C virus RNA with the polymerase chain reaction assay, *Proc. Natl. Acad. Sci. U.S.A.,* 89, 187, 1992.

122. **Shimizu, Y. K., Weiner, A. J., Rosenblatt, J., Wong, D. C., Shapiro, M., Popkin, T., Houghton, M., Alter, H. J., and Purcell, R. H.,** Early events in hepatitis C virus infection of chimpanzees, *Proc. Natl. Acad. Sci. U.S.A.,* 87, 6441, 1990.

123. **Farci, P., Alter, H. J., Wong, D., Miller, R. H., Shih, J. W., Jett, B., and Purcell, R. H.,** A long-term study of hepatitis C virus replication in non-A, non-B hepatitis, *N. Engl. J. Med.,* 325, 98, 1991.

124. **Schlauder, G. G., Leverenz, G. J., Amann, C. W., Lesniewski, R. R., and Peterson, D. A.,** Detection of the hepatitis C virus genome in acute and chronic experimental infection in chimpanzees, *J. Clin. Microbiol.,* 29, 2175, 1991.

125. **Schlauder, G. G. and Leverenz, G. J.,** Detection of hepatitis C viral RNA by the polymerase chain reaction in serum of patients with post-transfusion non-A, non-B hepatitis, *J. Virol. Methods,* 37, 189, 1992.

126. **Hilfenhaus, J., Krupka, U., Nowak, T., Cummins, L. B., Fuchs, K., and Roggenforf, M.,** Follow-up of hepatitis C virus infection in chimpanzees: determination of viraemia and specific humoral immune response, *J. Gen. Virol.,* 73, 1015, 1992.

127. **Beach, M. J., Meeks, E. L., Mimms, L. T., Vallari, D., DuCharme, L., Spelbring, J., Taskar, S., Schleicher, J. B., Krawczynski, K., and Bradley, D. W.,** Temporal relationships of hepatitis C virus RNA and antibody responses following experimental infection of chimpanzees, *J. Med. Virol.,* 36, 226, 1992.

128. **Prince, A. M., Brotman, B., Huima, T., Pascual, D., Jaffery, M., and Inchauspe, G.,** Immunity in hepatitis C infection, *J. Infect. Dis.,* 165, 438, 1992.

129. **Kato, N., Yokosuka, O., Omata, M., Hosoda, K., and Ohto, M.,** Detection of hepatitis C virus ribonucleic acid in the serum by amplification with polymerase chain reaction, *J. Clin. Invest.,* 86, 1764, 1990.

130. **Ohkoshi, S., Kato, N., Kinoshita, T., Hijikata, M., Ohtsuyama, Y., Okazaki, N., Ohkura, H., Hirohashi, S., Honma, A., Ozaki, T., Yoshikawa, A., Kojima, H., Asakura, H., and Shimotohno, K.,** Detection of hepatitis C virus RNA in sera and liver tissues of non-A, non-B hepatitis patients using the polymerase chain reaction, *Jpn. J. Cancer Res.,* 81, 862, 1990.

131. **Hagiwara, H., Hayashi, N., Mita, E., Naoki, H., Ueda, K., Takehara, T., Yuki, N., Kasahara, A., Fusamoto, H., and Kamada, T.,** Detection of hepatitis C virus RNA in chronic non-A, non-B liver disease, *Gastroenterology,* 102, 692, 1992.

132. **Villa, E., Ferretti, I., De Palma, M., Melegari, M., Scaglioni, P. P., Trande, P., Vecchi, C., Fratti, N., and Manenti, F.,** HCV RNA in serum of asymptomatic blood donors involved in post-transfusion hepatitis (PTH), *J. Hepatol.,* 13, 256, 1991.

133. **Kubo, Y., Takeuchi, K., Boonmar, S., Katayama, T., Choo, Q.-L., Kuo, G., Weiner, A. J., Bradley, D. W., Houghton, M., Saito, I., and Miyamura, T.,** A cDNA fragment of hepatitis C virus isolated from an implicated donor of post-transfusion non-A, non-B hepatitis in Japan, *Nucl. Acids Res.,* 17, 10367, 1989.

134. **Takeuchi, K., Kubo, Y., Boonmar, S., Watanabe, Y., Katayama, T., Choo, Q.-L., Kuo, G., Houghton, M., Saito, I., and Miyamura, T.,** Nucleotide sequence of core and envelope genes of the hepatitis C virus genome derived directly from human healthy carriers, *Nucl. Acids Res.,* 18, 4626, 1990.

135. **Takeuchi, K., Kubo, Y., Boonmar, S., Watanabe, Y., Katayama, T., Choo, Q.-L., Kuo, G., Houghton, M., Saito, I., and Miyamura, T.,** The putative nucleocapsid and envelope protein genes of hepatitis C virus determined by comparison of the nucleotide sequences of two isolates derived from an experimentally infected chimpanzee and healthy human carriers, *J. Gen. Virol.,* 71, 3027, 1990.

136. **Enomoto, T., Takada, A., Nakao, T., and Date, T.,** There are two major types of hepatitis C virus in Japan, *Biochem. Biophys. Res. Commun.,* 170, 1021, 1990.

137. **Li, J., Tong, S., Vitvitski, L., Lepot, D., and Trepo, C.,** Two French genotypes of hepatitis C virus: homology of the predominant genotype with the prototype American strain, *Gene,* 105, 167, 1991.

138. **Fuchs, K., Motz, M., Schreier, E., Zachoval, R., Deinhardt, F., and Roggendorf, M.,** Characterization of nucleotide sequences from European hepatitis C virus isolates, *Gene,* 103, 163, 1991.

139. **Kremsdorf, D., Porchon, C., Kim, J. P., Reyes, G. R., and Brechot, C.,** Partial nucleotide sequence analysis of a French hepatitis C virus: implications for HCV genetic variability in the E2/NS1 protein, *J. Gen. Virol.,* 72, 2557, 1991.

140. **Kato, N., Ootsuyama, Y., Ohkoshi, S., Nakazawa, T., Mori, S., Hijikata, M., and Shimotohno, K.,** Distribution of plural HCV types in Japan, *Biochem. Biophys. Res. Commun.,* 181, 279, 1991.

141. **Okamoto, H., Kurai, K., Okada, S.-I., Yamamoto, K., Lizuka, H., Tanaka, T., Fukuda, S., Tsuda, F., and Mishiro, S.,** Full-length sequence of a hepatitis C virus genome having poor homology to reported isolates: comparative study of four distinct genotypes, *Virology,* 188, 331, 1992.

142. **Okamoto, H., Kurai, K., Okada, Shun-I., Yamamoto, K., Lizuk, H., Tanaka, T., Fukuda, S., Tsuda, F., and Mishiro, S.,** Full-length sequence of a hepatitis C virus genome having poor homology to reported isolates: comparative study of four distinct genotypes, *Virology,* 188, 331, 1992.

143. **Mori, S., Kato, N., Yagyu, A., Tanaka, T., Ikeda, Y., Petchclai, B., Chiewsilp, P., Kurimura, T., and Shimotohno, K.,** A new type of hepatitis C virus in patients in Thailand, *Biochem. Biophys. Res. Commun.,* 183, 334, 1992.

144. **Chen, P.-J., Lin, M.-H., Tai, K.-F., Liu, P.-C., Lin, C.-J., and Chen, D.-S.,** The Taiwanese hepatitis C virus genome: sequence determination and mapping the 5′ termini of viral genomic and antigenomic RNA, *Virology,* 188, 102, 1992.

145. **Kumar, U. and Cheng, D.,** Cloning and sequencing of the structural region and expression of putative core gene of hepatitis C virus from a British case of chronic sporadic hepatitis, *J. Gen. Virol.,* 73, 1521, 1992.

146. **Fujio, K., Shimomura, H., and Tsuji, T.,** Genetic variation of putative core gene in hepatitis C virus, *Acta Med. (Okayama),* 45, 241, 1991.

147. **Tanaka, T., Kato, N., Nakagawa, M., Ootsuyama, Y., Cho, M.-J., Nakazawa, T., Hijikata, M., Ishimura, Y., and Shimotohno, K.,** Molecular cloning of hepatitis C virus genome from a single Japanese carrier: sequence variation within the same individual and among infected individuals, *Virus Res.,* 23, 39, 1992.

148. **Hijikata, M., Kato, N., Ootsuyama, Y., Nakagawa, M., Ohkoshi, S., and Shimotohno, K.,** Hypervariable regions in the putative glycoprotein of hepatitis C virus, *Biochem. Biophys. Res. Commun.,* 175, 220, 1991.

149. **Weiner, A. J., Brauer, M. J., Rosenblatt, J., Richman, K. H., Tung, J., Crawford, K., Bonino, F., Saracco, G., Choo, Q.-L., Houghton, M., and Han, J. H.,** Variable and hypervariable domains are found in the regions of HCV corresponding to the flavivirus envelope and NS1 proteins and the pestivirus envelope glycoproteins, *Virology,* 180, 842, 1991.

150. **Kato, N., Ootsuyama, Y., Tanake, T., Nakagawa, M., Nakazawa, T., Muraiso, K., Ohkoshi, S., Hijikata, M., and Shimotohno, K.,** Marked sequence diversity in the putative envelope proteins of hepatitis C virus, *Virus Res.,* 22, 107, 1992.

151. **Ogata, N., Alter, H. J., Miller, R. H., and Purcell, R. H.,** Nucleotide sequence and mutation rate of the H strain of hepatitis C virus, *Proc. Natl. Acad. Sci. U.S.A.,* 88, 3392, 1991.

152. **Chan, S. W., McOmish, F., Holmes, E. C., Dow, B., Peutherer, J. F., Follett, E., Yap, P. L., and Simmonds, P.,** Analysis of a new hepatitis C virus type and its phylogenetic relationship to existing variants, *J. Gen. Virol.,* 73, 1131, 1992.

153. **Nakao, T., Enomoto, N., Takada, N., Takada, A., and Date, T.,** Typing of hepatitis C virus genomes by restriction fragment length polymorphism, *J. Gen. Virol.,* 72, 2105, 1991.

154. **Okamoto, H., Sugiyama, Y., Okada, S., Kurai, K., Akahane, Y., Sugai, Y., Tanaka, T., Sato, K., Tsuda, F., Miyakawa, Y., and Mayumi, M.,** Typing hepatitis C virus by polymerase chain reaction with type-specific primers: application to clinical surveys and tracing infectious sources, *J. Gen. Virol.,* 73, 673, 1992.

155. **Wang, J.-T., Wang, T.-H., Sheu, J.-C., Lin, J.-T., and Chen, D.-S.,** Hepatitis C virus RNA in saliva of patients with posttransfusion hepatitis and low efficiency of transmission among spouses, *J. Med. Virol.,* 36, 28, 1992.

156. **Liou, T.-C., Chang, T.-T., Young, K.-C., Lin, X.-Z., Lin, C.-Y., and Wu, H.-L.,** Detection of HCV RNA in saliva, uring seminal fluid, and ascites, *J. Med. Virol.,* 37, 197, 1992.

157. **Chayama, K., Saitoh, S., Arase, Y., Ikeda, K., Matsumoto, T., Sakai, Y., Kobayashi, M., Unakami, M., Morinaga, T., and Kumada, H.,** Effects of interferon administration on serum hepatitis C virus RNA in patients with chronic hepatitis C, *Hepatology,* 13, 1040, 1991.

158. **Hagiwara, H., Hayashi, N., Mita, E., Ueda, K., Takehara, T., Kasahara, A., Fusamoto, H., and Kamada, T.,** Detection of hepatitis C virus RNA in serum of patients with chronic hepatitis C treated with interferon, *Hepatology,* 15, 37, 1992.

159. **Rizzetto, M., Canese, M. G., Arico, S., Crivelli, O., Trepo, C., Bonino, F., and Verme, G.,** Immunofluorescence detection of new antigen-antibody system (delta/anti-delta) associated to hepatitis B virus in liver and serum of HBsAg carriers, *Gut,* 18, 997, 1977.

160. **Rizzetto, M., Hoyer, B., Canese, M. G., Shih, J. W.-K., Purcell, R. H., and Gerin, J. L.,** Delta agent: association of delta agent with hepatitis B surface antigen and RNA in serum of delta-infected chimpanzees, *Proc. Natl. Acad. Sci. U.S.A.,* 77, 6124, 1980.

161. **Ponzetto, A., Cote, P. F., Popper, H., Hoyer, B. H., London, W. T., Ford, E. C., Bonino, F., Purcell, R. H., and Gerin, J. L.,** Transmission of the hepatitis B associated delta agent to the eastern woodchuck, *Proc. Natl. Acad. Sci. U.S.A.,* 81, 2208, 1984.

162. **Bonino, F., Hoyer, B., Ford, E., Shih, J. W.-K., Purcell, R. H., and Gerin, J. L.,** The delta agent: HBsAg particles with delta antigen and RNA in the serum of an HBV carrier, *Hepatology,* 1, 127, 1981.

163. **Bonino, F., Hoyer, B., Shih, W. K., Rizzetto, M., Purcell, R. H., and Gerin, J. L.,** Delta hepatitis agent: structural and antigenic properties of the delta-associated particle, *Infect. Immun.,* 43, 1000, 1984.

164. **Bonino, F., Heermann, K. H., Rizzetto, M., and Gerlich, W. H.,** Hepatitis delta virus: protein composition of delta antigen and its hepatitis B virus-derived envelope, *J. Virol.,* 58, 945, 1986.

165. **Kos, A., Dijkema, R., Arnberg, A. C., van der Meide, P. H., and Schellekens, H.,** The hepatitis delta virus possesses a circular RNA, *Nature,* 323, 558, 1986.

166. **Makino, S., Chang, M. F., Shieh, C. K., Kamahora, T., Vannier, D. M., Govindarajan, S., and Lai, M. M. C.,** Molecular cloning and sequencing of a human hepatitis delta virus RNA, *Nature,* 329, 343, 1987.

167. **Branch, A. D. and Robertson, H. D.,** A replication cycle for viroids and small infectious RNAs, *Science,* 223, 450, 1984.

168. **Kuo, M. Y.-P., Sharmeen, L., Dinter-Gottlieb, G., and Taylor, J.,** Characterization of self-cleaving RNA sequences on the genome and antigenome of human hepatitis delta virus, *J. Virol.,* 62, 4439, 1988.

169. **Wu, H.-N. and Lai, M. M. C.,** Reversible cleavage and ligation of hepatitis delta virus RNA, *Science,* 243, 652, 1988.

170. **Bergmann, K. F. and Gerin, J. L.,** Antigens of hepatitis delta virus in the live and serum of humans and animals, *J. Infect. Dis.,* 154, 702, 1986.

171. **Weiner, A. J., Choo, Q.-L., Wang, K.-S., Govindarajan, S., Redeker, A. G., Gerin, J. L., and Houghton, M.,** A single antigenomic open reading frame of the hepatitis delta virus encodes the epitope(s) of both hepatitis delta antigen polypeptides p24 and p27, *J. Virol.,* 62, 594, 1988.

172. **Luo, G. M., Chao, S.-Y., Hsieh, C., Sureau, C., Nishikura, K., and Taylor, J.,** A specific base transition occurs on replicating hepatitis delta virus RNA, *J. Virol.,* 64, 1021, 1990.

173. **Bonino, F. and Smedile, A.,** Delta agent (Type D) hepatitis, *Semin. Liver Dis.,* 6, 28, 1986.

174. **Rizzetto, M., Ponzetto, A., Bonino, F., and Smedile, A.,** Hepatitis delta virus infection: clinical and epidemiological aspects, in *Viral Hepatitis and Liver Disease,* Zuckerman, A. J., Ed., Alan R. Liss, New York, 1988, 389.

175. **Zignego, A. L., Deny, P., Feray, C., Ponzetto, A., Gentilini, P., Tiollais, P., and Brechot, C.,** Amplification of hepatitis delta virus RNA sequences by polymerase chain reaction: a tool for viral detection and cloning, *Mol. Cell. Probes,* 4, 43, 1990.

176. **Cariani, E., Ravaggi, A., Puoti, M., Mantero, G., Albertini, A., and Primi, D.,** Evaluation of hepatitis delta virus RNA levels during interferon therapy by analysis of polymerase chain reaction products with a nonradioisotopic hybridization assay, *Hepatology,* 15, 685, 1992.

177. **Chao, Y.-C., Chang, M.-F., Gust, I., and Lai, M. M. C.,** Sequence conservation and divergence of hepatitis delta virus RNA, *Virology,* 178, 384, 1990.

178. **Imazeki, F., Omata, M., and Ohto, M.,** Heterogeneity and evolution rates of delta virus RNA sequences, *J. Virol.,* 64, 5594, 1990.

179. **Deny, P., Zignego, A. L., Rascalou, N., Ponzetto, A., Tiollais, P., and Brechot, C.,** Nucleotide sequence analysis of three different hepatitis delta viruses isolated from a woodchuck and humans, *J. Gen. Virol.,* 72, 735, 1991.

180. **Imazeki, F., Omata, M., and Ohto, M.,** Complete nucleotide sequence of hepatitis delta virus RNA in Japan, *Nucl. Acids Res.,* 19, 5439, 1991.

181. **Lee, C. M., Bih, F. Y., Chao, Y. C., Govindarajan, S., and Lai, M. M.,** Evolution of hepatitis delta virus RNA during chronic infection, *Virology,* 188, 265, 1992.

182. **Madejon, A., Castillo, I., Bartolome, J., Melero, M., Campillo, M. L., Porres, J. C., Moreno, A., and Carreno, V.,** Detection of HDV-RNA by PCR in serum of patients with chronic HDV infection, *J. Hepatol.,* 11, 381, 1990.

183. **Saldanha, J. A., Thomas, H. C., and Monjardino, J. P.,** Cloning and sequencing of RNA of hepatitis delta virus isolated from a human serum, *J. Gen. Virol.,* 71, 1603, 1990.

184. **Chao, Y. C., Lee, C. M., Tang, H. S., Govindarajan, S., and Lai, M. M.,** Molecular cloning and characterization of an isolate of hepatitis delta virus from Taiwan, *Hepatology,* 13, 345, 1991.

185. **Oliviero, S., D'Adamio, L., Chiaberge, E., Brunetto, M. R., Smedile, A., De Marchi, M., Negro, F., and Bonino, F.,** Characterization of hepatitis delta antigen gene of a highly pathogenic strain of hepatitis delta virus, *Prog. Clin. Biol. Res.,* 364, 321, 1991.

186. **Luo, G., Chao, M., Hsieh, S.-Y., Sureau, C., Nishikura, K., and Taylor, J.,** A specific base transition occurs on replicating hepatitis delta virus RNA, *J. Virol.,* 64, 1021, 1990.

187. **Zheng, H., Fu, T. B., Lazinski, D., and Taylor, J.,** Editing on the genomic RNA of human hepatitis delta virus, *J. Virol.,* 66, 4693, 1992.

188. **Thill, G., Blumenfeld, M., Lescure, F., and Vasseur, M.,** Self-cleavage of a 71 nucleotide-long ribozyme derived from hepatitis delta virus genomic RNA, *Nucl. Acids Res.,* 19, 6519, 1991.

189. **Xia, Y. P., Yeh, C. T., Ou, J. H., and Lai, M. M.,** Characterization of nuclear targeting signal of hepatitis delta antigen: nuclear transport as a protein complex, *J. Virol.,* 66, 914, 1992.

190. **Khurro, S. M.,** Study of an epidemic of non-A, non-B hepatitis. Possibility of another human hepatitis virus distinct from post-transfusion non-A, non-B type, *Am. J. Med.,* 68, 818, 1980.

191. **Wong, D. C., Purcell, R. H., Sreenivasan, M. A., Prasad, S. R., and Pavri, K. M.,** Epidemic and endemic hepatitis in India: evidence for non-A/non-B hepatitis virus etiology, *Lancet,* 2, 876, 1980.

192. **Balayan, M. S., Andjaparidze, A. G., Savinskaya, S. S., Ketiladze, E. S., Braginsky, D. M., Savinov, A. P., and Poleschuk, V. F.,** Evidence for a virus in non-A/non-B hepatitis transmitted via the fecal-oral route, *Intervirology,* 20, 23, 1983.

193. **Kane, M. A., Bradley, D. W., Shrestha, S. M., Maynard, J. E., Cook, E. H., Mishra, R. P., and Joshi, D. D.,** Epidemic non-A, non-B hepatitis in Nepal, *JAMA,* 252, 3140, 1984.

194. **Tam, A. W., Smith, M. M., Guerra, M. E., Huang, C.-C., Bradley, D. W., Fry, K. E., and Reyes, G. R.,** Hepatitis E virus (HEV): molecular cloning and sequencing of the full-length viral genome, *Virology,* 185, 120, 1991.

195. **Fry, K. E., Tam, A. W., Smith, M. M., Kim, J. P., Luk, K.-C., Young, L. M., Piatak, M., Feldman, R. A., Yun, K. Y., Purdy, M. A., McCaustland, K. A., Bradley, D. W., and Reyes, G. R.,** Hepatitis E virus (HEV): strain variation in the nonstructural gene region encoding consensus motifs for an RNA-dependent RNA polymerase and an ATP/GTP binding site, *Virus Genes,* 6, 173, 1992.

196. **Kaur, M., Hyams, K. C., Purdy, M. A., Krawczynski, K., Ching, W. M., Fry, K. E., Reyes, G. R., Bradley, D. W., and Carl, M.,** Human linear B-cell epitopes encoded by the hepatitis E virus include determinants in the RNA-dependent RNA polymerase, *Proc. Natl. Acad. Sci. U.S.A.,* 89, 3855, 1992.

197. **Favorov, M. O., Fields, H. A., Purdy, M. A., Yashina, T. L., Aleksandrov, A. G., Alter, M. J., Yarasheva, D. M., Bradley, D. W., and Margolis, H. S.,** Serologic identification of hepatitis E virus infections in epidemic and endemic settings, *J. Med. Virol.,* 36, 246, 1992.

198. **Bradley, D. W.,** Enterically-transmitted non-A, non-B hepatitis, *Br. Med. Bull.,* 46, 442, 1990.

199. **Ray, R., Aggarwal, R., Salunke, P. N., Mehrotra, N. N., Talwar, G. P., and Naik, S. R.,** Hepatitis E virus genome in stools of hepatitis patients during large epidemic in north India, *Lancet,* 338, 783, 1991.

200. **McCaustland, K. A., Bi, S., Purdy, M. A., and Bradley, D. W.,** Application of two RNA extraction methods prior to amplification of hepatitis E virus nucleic acid by the polymerase chain reaction, *J. Virol. Methods,* 35, 331, 1991.

201. **Chauhan, A., Dilawari, J. B., Jameel, S., Kaur, U., Chawla, Y. K., Sharma, M. L., and Ganguly, N. K.,** Common aetiological agent for epidemic and sporadic non-A, non-B hepatitis, *Lancet,* 339, 1509, 1992.

202. **Tsarev, S. A., Emerson, S. U., Reyes, G. R., Tsareva, T. S., Legters, L. J., Malik, I. A., Iqbal, M., and Purcell, R. H.,** Characterization of a prototype strain of hepatitis E virus, *Proc. Natl. Acad. Sci. U.S.A.,* 89, 559, 1992.

203. **Bi, S.-L., Purdy, M. A., McCaustland, K. A., Margolis, H. S., and Bradley, D. W.,** The sequence of hepatitis E isolated directly from a single source during an outbreak in China, *Virus Res.,* 28, 233–248, 1992.

204. **Aye, T., Uchida, T., Ma, X. Z., Iida, F., Shikata, T., Zhang, H., and Win, K. M.,** Complete nucleotide sequence of hepatitis E isolated from XinXiang epidemic (1986–1988) of China, *Nucl. Acids Res.,* 20, 3212, 1992.

205. **Reyes, G. R., Purdy, M. A., Kim, J. P., Luk, K.-C., Young, L. M., Fry, K. E., and Bradley, D. W.,** Isolation of a cDNA from the virus responsible for enterically transmitted non-A, non-B hepatitis, *Science,* 247, 1335, 1990.

206. **Gubler, U. and Hoffmann, B. J.,** A simple and very efficient method for generating cDNA libraries, *Gene,* 25, 263, 1983.

207. **Yarbough, P. O., Tam, A. W., Fry, K. E., Krawczynski, K., McCaustland, K. A., Bradley, D. W., and Reyes, G. R.,** Hepatitis E virus: identification of type-common epitopes, *J. Virol.,* 65, 5790, 1991.

208. **Ray, R., Jameel, S., Manivel, V., and Ray, R.,** Indian hepatitis E virus shows a major deletion in the small open reading frame, *Virology,* 189, 359, 1992.

209. **Novati, R., Thiers, V., d'Arminio Monforte, A., Maisonneuve, P., Principi, N., Conti, M., Lazzarin, A., and Brechot, C.,** Mother-to-child transmission of hepatitis C virus detected by nested polymerase chain reaction, *J. Infect. Dis.,* 165, 720, 1992.

210. **Balayan, M. S., Usmanov, R. K., Zamyatina, N. A., Djumalieva, K. I., and Karas, F. R.,** Experimental hepatitis E infection in domestic pigs, *J. Med. Virol.,* 32, 58, 1990.

211. **Kuo, M. Y. P., Goldberg, J., Coates, L., Mason, W., Gerin, J., and Taylor, J.,** Molecular cloning of hepatitis delta virus RNA from an infected woodchuck liver: sequence, structure, and applications, *J. Virol.,* 1988, 1855, 1988.

212. **Devereux, J., Haeberli, P., and Smithies, O.,** A comprehensive set of sequence analysis programs for the VAX, *Nucl. Acids Res.,* 12, 387, 1983.

Chapter 6

Applications of PCR to Influenza, Measles, and Mumps

James S. Robertson and Timothy Forsey

TABLE OF CONTENTS

I. INTRODUCTION

Influenza virus, a member of the Orthomyxoviridae, and measles and mumps virus, both members of the Paramyxoviridae, are structurally similar viruses which possess single, negative strand RNA genomes. The main difference between them is that influenza has a segmented genome while in measles and mumps it is nonsegmented. Virions consist of a nucleocapsid core in which the RNA genome is intimately associated with a nucleoprotein and polymerase protein(s). The cores are surrounded by a shell of matrix protein which is encapsulated within a lipid bilayer membrane derived from the host cell during maturation by budding (for reviews see References 1 and 2).

Protruding from the membrane are viral-encoded glycoproteins. For influenza A and B viruses the surface glycoproteins are the hemagglutinin (HA) and neuraminidase (NA), while influenza C has a single hemagglutinin-esterase-fusion (HEF) glycoprotein. Mumps virus glycoproteins are the hemagglutinin-neuraminidase (HN) and fusion (F) proteins, while measles virus has HA and F proteins and lacks neuraminidase activity. The glycoproteins function in attachment of virions to the host cell and fusion of the viral membrane with host membranes to allow penetration of the nucleocapsid cores into the host cell. They are also instrumental in virus spread by causing cell-to-cell fusion and are implicated in maturation of the virion. The surface glycoproteins are the major antigenic determinants of these viruses.

Influenza viruses are separated serologically into types A, B, and C dependent on their nucleocapsid proteins. Influenza A viruses are further subdivided into serological subtypes dependent on the HA and NA surface antigens. Fourteen HA and nine NA subtypes of influenza A have been described in nature although only three of these have been associated with human disease, the H1N1, the H2N2 (or Asian flu), and the H3N2 (or Hong Kong flu) subtypes. There are four (sub)types of influenza virus currently causing respiratory disease in man: the A(H1N1) and A(H3N2) subtypes and types B and C. Influenza A infections are a source of severe morbidity, with symptoms including fever, sore throat, myalgia, and malaise. Influenza A infection can cause high mortality especially in the very young and the elderly. Influenza B infections are generally less severe while influenza C causes only mild infection and is generally not considered a sufficiently serious human pathogen to warrant consideration for diagnostic or vaccine purposes. Measles and mumps viruses cannot be serotyped by conventional, polyclonal serology, and for many years they were thought to be monotypic viruses. The use of monoclonal antibodies has enabled a more precise examination of the antigenic epitopes of measles to be made, and strain differences have been observed.[3] However, no formal typing scheme for measles or mumps exists. Both these viruses cause acute, febrile illnesses of variable severity. Measles can be very severe, leading to high mortality, and prior to the widespread use of vaccine, very few children escaped infection. Severity of disease is greater when nutrition and general standards of health are poor. Encephalitis occurs in approximately 1 of every 2000 reported cases of measles, and subacute sclerosing panencephalitis is a rare, invariably fatal neurological manifestation which can occur years after a primary episode of measles. Classical measles is characterized by a distinctive maculopapular rash often preceded by an exanthem in the buccal cavity. These raised spots with white centers seen in the mouth, the so-called Kopliks spots, are pathognomonic of measles. However, atypical or mitigated forms of the disease do occur.

Mumps is predominantly a disease of children and young adults, and glandular and nerve tissue are commonly infected. Swelling of the salivary glands, termed parotitis, is a distinctive clinical feature. Meningitis is a common sequela of mumps, and it probably accounts for the majority of reported cases of viral meningitis. However, it is estimated that around one third of mumps infections may be silent and show little or no clinical symptoms. Comprehensive descriptions of the clinical manifestations of measles and mumps infections are given by Carter and ter Meulen[4] and Leinikki.[5] There are effective live attenuated vaccines against measles and mumps which are extensively used during childhood. Influenza vaccines have a lower and variable efficacy and are recommended only for those who are medically at high risk, especially the young and elderly. While there are probably many reasons for the poor efficacy of influenza vaccine, much of the problem is due to the continually changing antigenic nature of the virus, the phenomenon termed antigenic drift.

II. ROUTINE DIAGNOSTIC TECHNIQUES

A. INFLUENZA

Most laboratory diagnosis of influenza is retrospective and based on serological evidence of a recent infection.[6] This is generally based on a high or rising serum IgG level, as measured by hemagglutination inhibition. On a less frequent basis (and usually only for cases of medical concern), clinical specimens derived from the upper respiratory tract are obtained, and this may result in isolation of the causative organism. Although costly and time consuming, virus isolation is essential to provide recent isolates for full antigenic characterization for epidemiological studies and vaccine design. Other, more rapid techniques for diagnosis of influenza infection are available and are based primarily on immunoassays of clinical material. These include nucleoprotein (NP) antigen-detection tests based on immunofluorescence[7] or ELISA,[8] and a rapid culture test based on centrifugation of samples onto cells contained in a 24-well plate (the 24-WPC assay).[9] These assays have the advantage of speed but vary considerably in their sensitivity and specificity of detection. A direct comparison of these methods with conventional virus isolation has been described.[10] The NP detection assays and the 24-WPC virus isolation assay had comparable high sensitivity and specificity. Interestingly, the 24-WPC assay demonstrated fivefold increased sensitivity compared with conventional virus isolation. With its rapidity and capability of providing virus isolates, the 24-WPC method has a promising future in diagnostic laboratories. Speed in diagnosing influenza is useful in epidemiological studies and essential in considering the use of prophylactic drugs such as amantadine and rimantadine in close contacts of infected individuals or health-care personnel.

B. MEASLES AND MUMPS

Measles and mumps infections are often difficult to diagnose solely on clinical grounds. Cases may not always present with the classical features of the disease, and as these infections become less common, for example, following increased use of vaccination, the likelihood of clinicians encountering these diseases becomes less. Therefore laboratory diagnosis of paramyxovirus infection is essential if cases are to be correctly identified and accurate epidemiological data gathered. Although diagnosis of both measles and mumps infections is straightforward by conventional laboratory techniques, the timing of specimen collection is critical as with most viral infections. Infection may be detected by direct demonstration of viral antigens, for example, by fluorescent antibody staining of infected cells; by culturing the agent in a sensitive cell culture system such as monkey kidney cells; or by serology. In the latter technique a rising titer of IgG or the presence of viral specific IgM is diagnostic. A wide range of serological assays are available, including neutralization, hemagglutination-inhibition, hemolysis-in-gel, and ELISA tests. Recent work detecting IgM in saliva looks very promising as a diagnostic technique.[11] Conventional laboratory diagnosis of measles and mumps is comprehensively described by Norrby.[12,13]

III. INFLUENZA AND PCR

A. DIAGNOSIS

A very thorough account of the application of PCR[14] to the detection and identification of influenza virus is given by Zhang and Evans,[15] who describe amplification using primers specific for the various human HA and NA subtypes and for the matrix protein. However, analyses were performed on laboratory-grown virus only, and no clinical specimens were tested. The sensitivity and specificity of PCR as a diagnostic test for human influenza virus in clinical material is described in two papers by Yamada et al.[16,17] By use of a mixture of primers specific for the H1 and H3 genes, it was possible to detect influenza A infections and to determine the virus subtype in a single reaction.[16] In the second paper, the detection of influenza B virus in throat swabs was described.[17] In all cases, the primers were specific for the designated subtype, and no amplification of irrelevant genes was detected. By use of two amplifications of 30 c each, or by the use of nested PCR, as little as one plaque-forming unit (pfu) of virus could be detected in these studies. The efficiency of detection by PCR was compared to traditional isolation on Madin Darby canine kidney (MDCK) cells or in eggs. Each sample from which virus was isolated gave a positive PCR result, and no samples that were negative for virus isolation were positive by PCR. Generally two separate amplifications or nested PCR were required to obtain a positive PCR result from these clinical samples.

B. VACCINE ANALYSIS

PCR has been used to assess the quality and purity of influenza vaccine reference strains.[18] Due to antigenic drift of influenza viruses, it is necessary to continuously assess the nature of virus causing disease so that the most appropriate strains are incorporated into the vaccine. To achieve this, public health laboratories on a worldwide basis isolate influenza viruses for full antigenic characterization. Such data are reviewed annually by the World Health Organization, and often there is a recommendation that at least one of the three components of the vaccine, the A(H1N1), A(H3N2), or B component, is updated. However, since newly isolated virus generally grows poorly in the laboratory, high-growth reassortants (HGRs) containing the surface antigens of the appropriate recommended vaccine strain are prepared for industry by co-infecting eggs with the vaccine strain and a laboratory-adapted high-growth strain. Due to the segmented nature of the influenza genome, genetic reassortment readily occurs in a mixed infection, and a virus with the appropriate surface antigens and high-growth characteristics can be selected by the use of appropriate antiserum. In these circumstances it is important to demonstrate that HGRs for vaccine use have the correct antigenic phenotype and are not contaminated by virus with the surface antigen genes of the laboratory strain. In the past this has been performed by serological techniques that have poor sensitivity in detecting contaminants. PCR is now applied routinely at the National Institute for Biological Standards and Control (Herts, England) (NIBSC) to assess the identity and purity of HGR vaccine strains.[18] Primers specific for H1 and H3 HAs and N1 and N2 NAs used in these analyses are listed in Table 6–1. The primers do not interfere with each other and do not amplify the irrelevant template. We have noted that when two distinct target sequences are amplified in the same reaction, there is often preferential amplification of the smaller of the two targets. This happens especially if the targets are considerably different in size, even to the extent that the larger target apparently fails to be amplified. Also, small amounts of nonspecific products that can arise during PCR of a target present in high concentration may mask the detection of an additional target present at very low levels. Thus detection of contaminating genes within an HGR or within any virus population is best performed in a dedicated reaction and not simultaneously with identification of the major component of the sample. We have shown the level of detection of the influenza HA gene to be in the order of 1 pfu, using either nested PCR primers (see Table 6–1) or Southern blotting after a single PCR.[18] This is comparable to that obtained by Yamada et al. in their studies on diagnosing infection by PCR.[16,17] Using the above approach, an H3N2 HGR developed at the NIBSC and used by vaccine manufacturers has been demonstrated to be free from contamination with the H1 and N1 genes of the original high-growth parent.

Table 6–1 Primers for influenza surface antigen amplification

Gene	Name[a]	Type[b,c]	Sequence (5' → 3')[d]	Size (bp)	Ref.
All	Universal	cDNA synthesis	AGCAAAAGCAGG		36
H1 HA	A/5/1	Outer, forward	AAAGCAGGGGAAAATAAAAACAACC	1113	18,21
	A/1117/2/	Outer, backward	ATCATTCCAGTCCATCCCCCTTCAAT		
	A/303/1	Inner, forward	AAATCATGGTCCTACATTGCAGAAA	513	
	A/815/2	Inner, backward	TGCCTCAAATATTATTGTGTC		
	Bam/A/52/1	Cloning, forward	gcgtatggaTCCTGTTATGTGCATTTACAG	1058	
	Eco/1/1109/2x	Cloning, backward	actgacgaATTCCAGTCCATCCCCTTC		
H3 HA	H3/4/1	Outer, forward	AAAAGCAGGGGATAATTCTATTAA	1074	18
	H3/1077/2/	Outer, backward	ATTTTCTATGAAACCTGCTATTGC		
	H3/339/1	Inner, forward	TTGAACGCAGCAAAGCTTTCAGCA	404	
	H3/742/2	Inner, backward	TTCTACTAGACACAGACCCCTTACCC		
	Eco/H3/11/1	Cloning, forward	GGGGAgAATTCTATTAATCATGAAGAC	1154	
	Bam/H3/1164/2	Cloning, backward	AGATCTGCTGCggaTCCTGTGCCCTC		
N1 NA	N1/454/1[e,f]	Outer, forward	GCTGCCCTGTCGGTGAAGCTCCGT	887	18
	N1/1340/2[e,f]	Outer, backward	CCAATCTACAGTATCACTATTCAC		
	N1/600/1/	Inner, forward	AACGGCATAATAACTGAAACC	439	
	N1/1038/2	Inner, backward	TACCATACCTGTATGAAAACC		
N1 NA	N1/23/1	Outer, forward	GAATCCAAATCAGAAAATAATAAACC	260	
	N1/282/2[e]	Outer, backward	CTTTGCTGTATATAGCCACCCACGG		
	N1/66/1/	Inner, forward	GTAGTCGGACTAATTAGCCTAATATTG	176	
	N1/241/2[e]	Inner, backward	GAATTGCCGGTTAATATCACTGAAG		
N2 NA	N2/498/1	Outer, forward	TGAATGAGTTGGGTGTTCCATTTCA	770	
	N2/1267/2	Outer, backward	GCTTTTGCCCTCAACAGAGAAAT		
	N2/640/1	Inner, forward	CGATGGGAGGCTTGTAGACAG	339	18
	N2/978/2	Inner, backward	CCTGAGCACACATAACTGGAA		
B HA	B/17/1	Forward	TTTCTAATATCCACAAAATGAAGGC	1124	24
	B/1140/2	Backward	ACCAGCAATAGCTCCGAAGAAACC		
	Eco/B/35/1	Cloning, forward	TGAAGGCAAgAATTcTACTACTCATGGT	1058	
	Bam/B/1092/2	Cloning, backward	TCTATATTTGGaTCCATTGGCCAGCTT		

[a] Numerals within slashes refer to the position of the 5' base of each primer within the gene segment cDNA; [b] Outer and inner primer pairs are for use in nested PCR; [c] Forward PCR primers can be used for cDNA synthesis from genomic RNA; [d] Restriction recognition sites in cloning primers are underlined, and mismatched bases are shown in lowercase type; [e] The N1 primers are based on the sequence and numbering of PR8 NA. Due to a 45-base deletion in the PR8 NA segment, the positions of these primers in most other viruses will be 45 bases farther along the segment; [f] N1/454/1 and N1/1340/2 have been found on some occasions to be relatively insensitive, and alternative N1 specific primers (N1/23/1 and N1/282/2) are shown.

Table 6–2 **HA1 codon changes accompanying human influenza egg adaptation**

Virus Subtype	Sample	Codon Positions[a]			No. of Clones with Sequence Shown
Influenza B (B/NIB/15/88)		**196–198**			
	Clinical material	Asn-x-Thr (AAC GAA ACC)			30/30
	MDCK-grown	Asn-x-Thr (AAC GAA ACC)			17/17
	Egg-adapted	Asp-x-Thr (GAC GAA ACC)			13/20
		Asn-x-Ile (AAC GAA ATC)			5/20
		Ser-x-Thr (AGC GAA ACC)			1/20
		Asn-x-Pro (AAC GAA CCC)			1/20
Influenza A(H1N1) (A/NIB/23/89)		**163**	**187**	**190**	
	Clinical material	Asn (AAT)	Asn (AAC)	Asp (GAC)	30/30
	MDCK-grown	Asn (AAT)	Asn (AAC)	Asp (GAC)	28/30
		Ser (AGT)	Asn (AAC)	Asp (GAC)	1/30
		Asn (AAT)	Asn (AAC)	–(AC[b])	1/30
	Egg-adapted	Asn (AAT)	Asn (AAC)	Asn (AAC)	27/30
		Asn (AAT)	Ser (AGC)	Asn (AAC)	1/30
		Asn (AAT)	Asn (AAC)	Asp (GAC)	2/30

Information on Influenza B from Robertson, J. S., Bootman, J. S., Nicolson, C., Major, D., Robertson, E. W., and Wood, J. M., *Virology*, 179, 35, 1990. With permission. Information on Influenza A from Robertson, J. S., Nicolson, C., Bootman, J. S., Major, D., Robertson, E. W., and Wood, J. M., *Journal of General Virology*, 72, 2671, 1991. With permission.

[a] Substitutions in HA1 codons associated with egg adaptation are shown; [b] Base deletion (presumed to be an amplification/cloning artifact).

C. ANALYSIS OF CLINICAL MATERIAL

Isolation of influenza virus in diagnostic laboratories is now generally performed in mammalian cell culture, usually monkey kidney or MDCK cells. However, prior to full antigenic characterization, cell-isolated virus is usually propagated in embryonated eggs to provide sufficient virus for analysis. There is now a considerable body of evidence that genetically distinct influenza viruses are derived in the laboratory from human clinical material, depending on the substrate used for isolation and propagation. We have reported that virus derived in eggs differs from equivalent tissue culture grown virus by one or two amino acid residues in the HA.[19] These substitutions cluster around the receptor binding site of the HA molecule and suggest that the basis for selection of variants is the specificity of this site for different cell substrates. These variants are often antigenically distinct,[19] and animal studies have indicated that a difference in their HA of only a few amino acid residues can have measurable differences in their immune response and on their protective efficacy.[20] Thus for epidemiological studies and vaccine design, it is important to determine which variant(s) is present in original clinical material. Prior to the development of PCR there was insufficient virus in a clinical sample to perform the required analyses.

Use of PCR, however, has enabled the nucleotide sequence of the HA genes of virus present in clinical material to be determined and compared with the sequence from virus grown in the laboratory in cells or in eggs. Such studies have now been performed on all currently circulating human influenza virus types, A(H1),[21,22] A(H3),[23] and B[24] viruses. In our studies on influenza A(H1) and B,[21,24] the PCR-amplified HA gene was cloned prior to sequencing, and many clones were sequenced in order to examine the extent of heterogeneity of these variants within clinical material or laboratory-derived virus. In each case, the HA of naturally occurring human influenza virus was relatively homogeneous and identical to that of virus isolated on tissue culture (Table 6–2). In contrast, analysis of the HA of egg-adapted virus confirmed that variants were selected by passage in the allantoic cavity of embryonated eggs. Furthermore, the progeny from a single egg often contained more than one variant (Table 6–2). Because of this latter finding, we have examined a variety of vaccine reference strains, all of which are egg grown, following the same procedure of PCR amplification of the HA1 region, cloning, and sequence analysis of multiple clones. To date we have found that the influenza B recommended strain, B/Yamagata/16/89, consists of two distinct variants that differ in their HA by three amino acid residues.[44]

IV. MEASLES AND MUMPS AND PCR

Currently, measles and mumps infections are not routinely diagnosed by PCR or other techniques employing molecular biology. Due to the availability of sensitive and rapid conventional diagnostic assays, it is difficult to envisage a need for such techniques. However, PCR is being used in other, nondiagnostic studies of these viruses such as the detection and characterization of viral proteins[25] and transcription studies.[26] As described for influenza, PCR facilitates the generation of sequence data from negative-strand RNA viruses and has been particularly useful in comparing different isolates of paramyxoviruses. For example, Schulz et al.[27] used PCR to compare a measles isolate from a child with Kawasaki disease with vaccine and contemporaneous wild viruses. A more extensive use of PCR has been made in molecular characterization studies of mumps viruses associated with adverse reactions following vaccination.

A. ANALYSIS OF VACCINE-ASSOCIATED MUMPS

Following the introduction of routine mumps vaccination in the U.K. in 1988, cases of postvaccination meningitis occurred in which mumps virus was isolated from cerebrospinal fluid (CSF).[28] It was important to determine whether these infections were due to the vaccine or coincidental infection with wild mumps virus. No means of differentiating vaccine from wild virus was available at that time, and we therefore attempted to develop such a system based on sequencing data. We needed a rapid method suitable for dealing with multiple isolates, and this precluded growing each virus isolate in bulk in order to generate enough RNA to sequence directly. Also we were concerned that the multiple passages needed to produce high-titer virus might introduce genetic changes. PCR provided a solution, and by amplifying and sequencing a small region of the F protein gene, we were able to characterize and differentiate the mumps isolates.[29] The technical details of this procedure are given below. In our hands one round of amplification was sufficient to produce adequate DNA for sequencing from a typical mumps isolate. The sensitivity of the technique was sufficient for around 100 pfu of tissue culture virus per sample. We have also been able to amplify mumps nucleic acid directly from clinical material. We have extended this work to compare the sequence of the putative small hydrophobic (SH) protein gene and flanking regions of mumps vaccine virus isolated from vaccinees and grown on different substrates.[30] This region of the genome contains hypervariable areas and thus shows a greater degree of nucleotide difference between strains. Using this technique we have detected two distinct populations of virus present in the widely used Jeryl Lynn mumps vaccine.[45] Other workers have used similar techniques to characterize vaccine and wild mumps isolates. Yamada and colleagues[31] examined the P gene, while Brown et al.[32] compared the sequence of mumps strains over a region of the HN gene. Different lineages of measles viruses[33] and mumps viruses[34] have also been identified, using PCR and sequencing. All these studies use very similar methodologies, although some workers prefer to clone their PCR products before sequencing rather than sequence the material directly.

V. PRACTICAL ASPECTS OF PCR

One of the prime considerations in applying PCR to orthomyxoviruses and paramyxoviruses is that they possess RNA genomes. Great care must therefore be taken against ribonuclease contamination by using gloves, sterile reagents, and absolute cleanliness. However, we consider good laboratory practice to be sufficient and do not take special precautions such as the use of ribonuclease inhibitors. Another consequence of working with these viruses is that a reverse transcription step is necessary to convert RNA to cDNA prior to amplification. We have found the standard procedure of proteinase K/SDS treatment of virus samples followed by phenol/chloroform extraction and ethanol precipitation[35] to be effective in extracting RNA from influenza, measles, and mumps. In fact, we have successfully amplified viral nucleic acid extracted from clinical material, both throat wash and nasal swab in the case of influenza and CSF for mumps, all of which have been sent through the post at ambient temperature.

A. INFLUENZA

The influenza genomic RNA segments have a common 3' terminal sequence[36] and a primer corresponding to these 12 nucleotides, the "universal" primer (Table 6–1), has been used extensively to synthesize cDNA from all segments in a single reaction. This is especially useful when subsequent analyses are to be performed on more than one segment. Alternatively, a segment-specific primer can be used for reverse transcription, for example, a "forward" amplification primer, as described in Table 6–1. In a typical experiment, the extracted RNA is reverse transcribed with the universal primer and an aliquot of the cDNA, using one tenth, used for amplification. In this way several amplifications, perhaps targeting different segments, can be performed from one cDNA sample. Using Amplitaq enzyme,®* our standard amplification conditions for influenza are to use 30 c of 94°C for 0.5 min, 55°C for 0.5 min, and 72°C for 1.0 min, with no additional incubation at the end of the 30 c. Reaction products are generally very clean since amplification is being performed on a nucleic acid sample of low sequence complexity with no high-level background of cellular DNA. While optimization of primer and Mg^{2+} concentrations has been performed for all primer pairs shown in Table 6–1, we have found that 0.2 μM primer and 2 mM Mg^{2+} typically provide optimal conditions. For direct sequencing, amplified DNA is subjected to a Sephadex®** G50 (Pharmacia) spun column and sequenced using ^{32}P-5' labeled primers. While exposure times are generally longer than if using α-^{32}P dATP in the sequencing reactions, the sequencing gels are of high quality and superior to other methods we have tried. Prior to cloning amplified DNA, the DNA is purified on a column of Sepharose®** 4B (Pharmacia) to remove excess primers and primer complexes. Primers for direct cloning of amplified DNA into M13 were designed that incorporate unique restriction sites suitable for insertion into the polylinker region of the M13mp18/19 vectors (Table 6–1). Sufficient nucleotide sequence data exist for a variety of influenza HAs such that a novel HA under study is likely to vary from an established sequence by only a limited number of nucleotides.[37] Thus restriction sites can be chosen that are unlikely (but with no guarantee) to exist within the region of interest in new isolates. For antigenic investigations this is the HA1 coding region, which is 1000 to 1100 nucleotides in size. In designing primers for direct cloning, the sequences flanking the appropriate HA1 coding region were examined for a close match to suitable restriction sites, and primers were synthesized that incorporated a site into the primer (Table 6–1). Some of these primers have only one to two mismatches with the HA sequence. Different sites were incorporated into forward and backward primers for forced cloning into

* Registered Trademark of Perkin-Elmer, Norwalk, CT.
** Registered Trademark of Pharmacia, Inc., Milwaukee, WI.

M13 so that the correct orientation of the insert was obtained for sequencing with our stock of HA-specific sequencing primers. More efficient cloning has been obtained with primers in which the cloning site is removed from the 5′ end of the primer. In a typical amplification for cloning, cDNA is generated using reverse transcriptase with either the universal primer or a standard (noncloning) forward PCR primer, and an aliquot of the cDNA is subjected to one round of PCR using cloning primers. If insufficient DNA is generated, an aliquot of the original cDNA is amplified using standard PCR primers which flank the cloning primers. After this amplification, an aliquot of the PCR product is reamplified using the cloning primers in nested PCR fashion. For example, to amplify the HA1 region of an H1 HA for cloning, cDNA is synthesized with either the universal primer or primer A/5/1, and PCR amplification is performed with Bam/A/52/1 and Eco/A/1109/2x (Table 6–1). If this amplification fails to provide sufficient quantities of DNA, the cDNA is amplified with A/5/1 and A/1117/2, and the amplified DNA is reamplified with Bam/A/52/1 and Eco/A1109/2x. Generally, there is sufficient virus in 100 μl infectious tissue culture fluid or allantoic fluid (10^7 to 10^9 pfu/ml) such that direct amplification with the cloning primers provides enough DNA for cloning or direct sequencing. For clinical material though, which is likely to contain only 10^0 to 10^5 pfu/ml, nested PCR as described above was usually required to generate sufficient DNA from 100 μl of sample for further analyses. Using this approach we have been able to successfully clone the HA1 region from virus in all clinical samples so far studied from which influenza virus was isolated, including both throat washes and nasal swabs, without the need to concentrate the virus in any way. All preamplification manipulations are performed in a separate laboratory from postamplification analyses.

B. MEASLES AND MUMPS

Unlike influenza, measles and mumps viruses do not have a segmented genome, and "universal" primers cannot be used. We chose to examine the F gene of measles and the F and SH genes of mumps and had specific primers made corresponding to published sequences.[38–40] Various combinations of primers, for both cDNA generation and PCR, were tried and those most commonly used by us are shown in Table 6–3.

Good laboratory practice is essential to avoid cross contamination in PCR. We routinely prepare PCR reaction mixes in bulk, usually sufficient for 10 or 20 reactions, and aliquot them into individual tubes. These "master mixes" are prepared in a laboratory separate from the one where the main virus work and post-PCR manipulations are carried out. When needed, master mixes are taken to the main laboratory where cDNA templates are added and PCR is carried out. Recently we adopted the so-called hot-start method whereby master mixes are prepared with a layer of histology wax separating dNTPs and primers from the *Taq* polymerase enzyme.[41] When the reaction tube is subjected to the first round of heating in the thermal cycler, the wax melts, the enzyme mixes with the other reagents, and the PCR begins. Hot-start PCR is thought to prevent premature enzyme activity which can produce primer dimers and other unwanted products. Our PCR protocols for measles and mumps are based on that of Cetus Corporation for the Amplitaq enzyme although we have titrated optimum magnesium, primer, and enzyme concentrations. Typically we find that 1.5 mM MgCl$_2$, 1 μM primer, and 2.5 U of enzyme are satisfactory. Amplification conditions must also be determined for different primers, but for those routinely used for measles and mumps we employ 30 c of 94°C for 0.5 min, 37°C for 0.5 min, and 72°C for 1.5 min. The

Table 6–3   Primers for measles and mumps amplification

Virus	Gene	Primer	Sequence (5′ → 3′)	Size of product
				(bp)
Measles	F	cDNA	ATTGAGGCAATCAGACAAGCAGGG	564
		PCR 1#	TTACGGGACCCCATATCTGCGGAG	
		PCR 2#	AGGGTCTTGATTAATGATCGTTCC	
Mumps	F	cDNA	ATCCTAGAGATCGGG	327
		PCR 1#	GTAGCACTGGATGGA	
		PCR 2#	ACTCACAGATTGGAG	
	SH and flanking regions	cDNA	GTCGATGATCTCATCAGGTAC	1129
		PCR 1#	GTCGATGATCTCATCAGGTAC	
		PCR 2#	AACGAGAATCCCCATGGAAACATA	

---

* Registered Trademark of Pharmacia, Inc., Milwaukee, WI.

reaction is completed by a final incubation at 72°C for 8.5 min. PCR products are then purified either by centrifugation through SizeSep™*-400 spun columns or by electrophoresis on a 0.8% agarose gel. Following the latter technique, the appropriate bands are excised from the gel and the DNA is eluted using a "freeze-squeeze" technique.[42] Purified PCR products are sequenced directly using α ^{32}P-dATP and T7 polymerase (Pharmacia). The protocol is based on that given by the enzyme supplier. This technique normally generates a sequence of between 100 and 200 nucleotides, which can be read off one gel. For our purposes of virus strain characterization, this gives sufficient information, and we do not need to clone PCR products prior to sequencing.

VI. DISCUSSION

A. PCR AND SEQUENCING

PCR has revolutionized sequence analysis of orthomyxoviruses and paramyxoviruses. Conventional sequencing is both intricate and labor intensive. Large quantities of virus must be grown in the laboratory and purified prior to extraction of the virion RNA. For a few model laboratory-adapted viruses, this is not too problematic; however, many viruses are difficult to grow and purify, especially those at low passage level. Additionally, direct sequencing of RNA templates is more demanding than corresponding analyses of DNA. The sensitivity of PCR allows analysis of virus derived from a single egg, a small sample of infectious culture fluid, or even from a single plaque of virus. Indeed, as described above, it has allowed sequence analysis of virus present in clinical material without the need to propagate virus in the laboratory. Analysis of virus currently causing disease is an important aspect of the control of influenza. While the major analytical approach remains the antigenic characterization of these viruses, sequence analysis of the HA genes of recent isolates, especially now that it can be performed easily and rapidly, has an important role to play in studying antigenic drift and assessing the impact of laboratory propagation on the selection of variants. Similarly, combining PCR and sequencing in the study of measles and mumps has identified distinct strains of wild-type viruses and confirmed the role of vaccine-virus in some adverse reactions.

B. PCR AND DIAGNOSIS OF INFLUENZA

Yamada et al.[16,17] demonstrated correspondence between the detection of influenza virus by PCR and conventional isolation methodology. However in a comparative study of rapid diagnostic methods, Chomel et al.[10] found that rapid tests, including the 24-WPC isolation method, were considerably more efficient than traditional isolation. The conclusion that might be drawn from these two studies is that the 24-WPC method would be considerably more efficient than PCR. However the relative viability of the samples tested in the two studies is unknown, and a direct comparison between 24-WPC and PCR on identical clinical samples would be required before such a conclusion can be reached.

The specificity of PCR is well established, and several workers have shown that when applied to influenza, specific types can be identified. The vast array of available sequence information on the influenza genome provides a wide choice of target sequences.[37] Individually, most of the genes of human influenza A viruses are sufficiently highly conserved that primers can be designed that are specific to that particular gene for all viruses within that type. A recent publication describes type-specific PCR primers for identification of influenzas A, B, or C, based on the NS genome segment. Their use in diagnosing influenza virus in clinical material was investigated using 21 specimens proven positive for influenza by culture/IF (immunofluorescence after 16- to 24-h culture); 19 of the samples were positive by PCR. For the two PCR negative samples, no virus could be isolated on tissue culture or in eggs.[43]

On the other hand, primers specific for the various hemagglutinin subtypes can be utilized, and their judicious use can enable determination of the subtype simultaneously with identification of the virus. Indeed, since serious respiratory disease can be caused by several different agents, it would be interesting to investigate the use of a cocktail of primers specific for a variety of possible respiratory tract viruses, e.g., influenzas A, B, and C, paramyxovirus, coronavirus, adenovirus, and respiratory syncytial virus. Confirmation of a positive PCR, when required, can be made by Southern blotting or by nested PCR. Indeed, in routine diagnosis such techniques are likely to be required to achieve sufficient sensitivity. Currently, diagnostic tests for influenza virus infection take sufficient time to perform that diagnosis is generally retrospective, but diagnosis is important nonetheless for epidemiological reasons. Thus current practices are generally of more benefit to the community than to the individual. Direct tests for the viral genome using PCR (1 to 2 days), or for virus antigen (1 day) or rapid 24-WPC virus isolation (2 to 3 days), would be more appropriate to diagnose current infection.

C. PCR AND DIAGNOSIS OF MEASLES AND MUMPS

PCR has so far been used only to a limited extent with measles and mumps viruses. Its value has been primarily in sequencing studies where PCR obviates the need to grow virus to high titer. This has facilitated the molecular characterization of virus strains and has been particularly useful in the study of vaccine-associated disease. However, the potential of PCR in diagnosis of measles or mumps infections is doubtful. Conventional assays work well, and the interpretation of PCR findings is not simple. Detecting fragments of measles and mumps genome in clinical material may not necessarily indicate current infection with these viruses.

VII. CONCLUSIONS

PCR is capable of being used as a diagnostic test for influenza, measles, and mumps. To establish its full potential, new and further comparative investigations need to be performed. Rapid alternative methods are currently available that have acceptable sensitivity and specificity. However, that should not obviate a requirement to investigate novel techniques, which may (or may not) ultimately prove to be superior in efficiency, in ease and speed of use, in cost, or in the specific information provided. In our opinion, the lability of RNA genomes, the requirement of a reverse transcription step, the care required to avoid contamination, and the existence of adequate alternative assays make the current usefulness of PCR as a diagnostic test for influenza, measles, and mumps debatable. Any advances in the application of PCR as a diagnostic tool are most likely to arise through improvements in automation of molecular assays.

REFERENCES

1. **Krug, R. M.,** *The Influenza Viruses,* Plenum Press, New York, 1989.
2. **Kingsbury, D. W.,** *The Paramyxoviruses,* Plenum Press, New York, 1991.
3. **Birrer, M. J., Udem, S., Nathenson, S., and Bloom, B. R.,** Antigenic variants of measles virus, *Nature,* 293, 67, 1981.
4. **Carter, M. J. and ter Meulen, V.,** Measles, in *Principles and Practice of Clinical Virology,* Zuckerman, A. J., Banatvala, J. E., and Pattison, J. R., Eds., John Wiley & Sons, Chichester, England, 1987, chap. 5.
5. **Leinikki, P.,** Mumps, in *Principles and Practice of Clinical Virology,* Zuckerman, A. J., Banatvala, J. E., and Pattison, J. R., Eds., John Wiley & Sons, Chichester, England, 1987, chap. 7.
6. **Schild, G. C. and Dowdle, W. R.,** Influenza virus characterisation and diagnostic serology, in *The Influenza Viruses and Influenza,* Kilbourne, E. D., Ed., Academic Press, New York, 1975, chap. 11.
7. **Minnich, L. and Ray, C. G.,** Comparison of direct immunofluorescent staining of clinical specimens for respiratory virus antigens with conventional isolation techniques, *J. Clin. Microbiol.,* 12, 391, 1980.
8. **Chomel, J. J., Thouvenot, D., Onno, M., Kaiser, C., Gourreau, J. M., and Aymard, M.,** Rapid diagnosis of influenza infection of NP antigen using an immunocapture ELISA test, *J. Virol. Meth.,* 25, 81, 1989.
9. **Woods, G. L. and Johnson, A. M.,** Rapid 24-well plate centrifugation assay for detection of influenza A virus in clinical specimens, *J. Virol. Meth.,* 24, 35, 1989.
10. **Chomel, J. J., Pardon, D., Thouvenot, D., Allard, J. P., and Aymard, M.,** Comparison between three rapid methods for direct diagnosis of influenza and the conventional isolation procedure, *Biologicals,* 19, 287, 1991.
11. **Perry, K., Brown, D. W. G., Parry, J. V., Panday, S., Pipkin, C., and Richards, A.,** The detection of measles, mumps and rubella antibodies in saliva using antibody capture radioimmunoassay, *J. Med. Virol.,* 40, 235, 1993.
12. **Norrby, E.,** Measles virus, in *Manual of Clinical Microbiology,* Lennette, E. H., Balows, A., Haustev, W. J., and Shadomy, H. J., Eds., American Society for Microbiology, Washington, DC, 1985, chap. 74.
13. **Norrby, E.,** Mumps virus, in *Manual of Clinical Microbiology,* Lennette, E. H., Balows, A., Haustev, W. J., and Shadomy, H. J., Eds., American Society for Microbiology, Washington, DC, 1985, chap. 75.
14. **Saiki, R. K., Gelfand, D. H., Stoffel, S., Scharf, S. J., Higuchi, R., Horn, G. T., Mullis, K. B., and Ehrlich, H. A.,** Primer directed enzymatic amplification of DNA with a thermostable DNA polymerase, *Science,* 239, 487, 1988.

15. **Zhang, W. and Evans, D. H.,** Detection and identification of human influenza viruses by the polymerase chain reaction, *J. Virol. Meth.,* 33, 165, 1991.

16. **Yamada, A., Imanishi, J., Nakajima, E., Nakajima, K., and Nakajima, S.,** Detection of influenza viruses in throat swab by using polymerase chain reaction, *Microbiol. Immunol.,* 35, 259, 1991.

17. **Yamada, A. and Imanishi, J.,** Detection of influenza B virus in throat swabs using the polymerase chain reaction, *Acta Virol.,* 36, 320, 1992.

18. **Robertson, J. S., Nicolson, C., Newman, R., Major, D., Dunleavy, U., and Wood, J. M.,** High growth reassortant influenza vaccine viruses: new approaches to their control, *Biologicals,* 20, 213, 1992.

19. **Robertson, J. S., Bootman, J. S., Newman, R., Oxford, J. S., Daniels, R. S., Webster, R. G., and Schild, G. C.,** Structural changes in the haemagglutinin which accompany egg adaptation of an influenza A(H1N1) virus, *Virology,* 160, 31, 1987.

20. **Wood, J. M., Oxford, J. S., Dunleavy, U., Newman, R. W., Major, D., and Robertson, J. S.,** Influenza A(H1N1) vaccine efficacy in animal models is influenced by two amino acid substitutions in the haemagglutinin molecule, *Virology,* 171, 214, 1989.

21. **Robertson, J. S., Nicolson, C., Bootman, J. S., Major, D., Robertson, E. W., and Wood, J. M.,** Sequence analysis of the haemagglutinin (HA) of influenza A(H1N1) viruses present in clinical material and comparison with the HA of laboratory-derived virus, *J. Gen. Virol.,* 72, 2671, 1991.

22. **Rajakumar, A., Swierkosz, E. M., and Schulze, I. T.,** Sequence of an influenza virus haemagglutinin determined directly from a clinical sample, *Proc. Natl. Acad. Sci. U.S.A.,* 87, 4154, 1990.

23. **Katz, J. M., Wang, M., and Webster, R. G.,** Direct sequencing of the HA gene of influenza (H3N2) virus in original clinical samples reveals sequence identity with mammalian cell-grown virus, *J. Virol.,* 64, 1808, 1990.

24. **Robertson, J. S., Bootman, J. S., Nicolson, C., Major, D., Robertson, E. W., and Wood, J. M.,** The haemagglutinin of influenza B virus present in clinical material is a single species identical to that of mammalian cell-grown virus, *Virology,* 179, 35, 1990.

25. **Takeuchi, K., Tanabayashi, K., Hishiyama, M., Yamada, Y. K., Yamada, A. K., and Sugiura, A.,** Detection and characterisation of mumps virus V protein, *Virology,* 178, 247, 1990.

26. **Elliott, G. D., Yeo, R. P., Afzal, M. A., Simpson, E. J. B., Curran, J. A., and Rima, B. K.,** Strain-variable editing during transcription of the P gene of mumps virus may lead to the generation of non-structural proteins NS1 (V) and NS2, *J. Gen. Virol.,* 71, 1555, 1990.

27. **Schulz, T. F., Hoad, J. G., Whitby, D., Tizand, E. J., Dillon, M. J., and Weiss, R. A.,** A measles virus isolate from a child with Kawasaki disease: sequence comparison with contemporaneous isolates from 'classical' cases, *J. Gen. Virol.,* 73, 1581, 1992.

28. **Gray, J. A. and Burns, S. M.,** Mumps meningitis following measles, mumps and rubella immunization, *Lancet,* 2, 98, 1989.

29. **Forsey, T., Mawn, J. A., Yates, P. J., Bentley, M. L., and Minor, P. D.,** Differentiation of vaccine and wild mumps viruses using the polymerase chain reaction and dideoxynucleotide sequencing, *J. Gen. Virol.,* 71, 987, 1990.

30. **Turner, P. C., Forsey, T., and Minor, P. D.,** Comparison of the nucleotide sequence of the SH gene and flanking regions of mumps vaccine virus (Urabe strain) grown on different substrates and isolated from vaccinees, *J. Gen. Virol.,* 72, 435, 1991.

31. **Yamada, A., Takeuchi, K., Tanabayashi, K., Hishiyama, M., and Sigiura, A.,** Sequence variation of the P gene among mumps virus strains, *Virology,* 172, 374, 1989.

32. **Brown, E. G., Furesz, J., Dimock, K., Yarosh, W., and Contreras, G.,** Nucleotide sequence analysis of Urabe mumps vaccine strain that caused meningitis in vaccine recipients, *Vaccine,* 9, 840, 1991.

33. **Taylor, M. J., Godfrey, E., Baczko, K., ter Meulen, V., Wild, T. F., and Rima, B. K.,** Identification of several different lineages of measles virus, *J. Gen. Virol.,* 72, 83, 1991.

34. **Yeo, R. P., Afzal, M. A., Forsey, T., and Rima, B. K.,** Identification of a new mumps virus lineage by nucleotide sequence analysis of the SH gene of ten different strains, *Arch. Virol.,* 128, 371, 1993.

35. **Sambrook, J., Fritsch, E. F., and Maniatis, T.,** *Molecular Cloning: A Laboratory Manual,* 2nd ed., Cold Spring Harbor Laboratory Press, Cold Spring Harbor, NY, 1989, Book 3, Appendix E.3.

36. **Robertson, J. S.,** 5′ and 3′ terminal nucleotide sequences of the RNA genome segments of influenza virus, *Nucl. Acids Res.,* 6, 3745, 1979.

37. **Webster, R. G., Bean, W. J., Gorman, O. T., Chambers, T. M., and Kawaoka, Y.,** Evolution and ecology of influenza A viruses, *Microbiol. Rev.,* 56, 152, 1992.

38. **Richardson, C., Hull, D., Greer, P., Hasel, K., Berkovich, A., England, G., Bellini, W., Rima, B., and Lazzarini, R.,** The nucleotide sequence of the mRNA encoding the fusion protein of measles virus (Edmonston strain): a comparison of fusion proteins from several different paramyxoviruses, *Virology,* 155, 508, 1986.

39. **Waxham, H. M., Server, A. C., Goodman, H. W., and Wolinsky, J. S.,** Cloning and sequencing of the mumps fusion protein gene, *Virology,* 159, 381, 1987.

40. **Elliott, G. D., Afzal, M. A., Martin, S., and Rima, B. K.,** Nucleotide sequence of the matrix, fusion and putative SH protein genes of mumps virus and their deduced amino acid sequences, *Virus Res.,* 12, 61, 1989.

41. **Chou, Q., Russell, M., Birch, D. E., Raymond, J., and Blouch, W.,** Prevention of pre-PCR mispriming and primer dimerization improves low-copy-number amplifications, *Nucl. Acids Res.,* 20, 1717, 1992.

42. **Tautz, D. and Renz, M.,** An optimised freeze-squeeze method for the recovery of DNA fragments from agarose gels, *Anal. Biochem.,* 132, 14, 1983.

43. **Claas, E. C. J., Sprenger, M. J. W., Kleter, G. E. M., van Beek, R., Quint, W. G. V., and Masurel, N.,** Type-specific identification of influenza viruses A, B and C by the polymerase chain reaction, *J. Virol. Meth.,* 39, 1, 1992.

44. **Robertson, J. S. and Forsey, T.,** unpublished results.

45. **Afzal, M. A., Pickford, A. R., Forsey, T., Heath, A. B., and Minor, P. D.,** The Jeryl Lynn vaccine strain of mumps virus is a mixture of two distinct isolates, *J. Gen. Virol.,* 74, 917, 1993.

Chapter 7

Detection and Analysis of Human Picornaviruses

Glyn Stanway and Timo Hyypiä

TABLE OF CONTENTS

I. INTRODUCTION

The picornavirus family includes several notable human and animal pathogens.[1] Poliomyelitis and type A hepatitis remain significant health problems, the common cold (frequently caused by rhinoviruses) is a major cause of morbidity and economic loss, and there are a number of other clinical manifestations of picornaviruses affecting a range of organ systems. Picornaviruses are divided, on the basis of physicochemical and genetic properties, into five main groups or genera: enteroviruses, rhinoviruses, hepatoviruses, aphthoviruses, and cardioviruses. The first three of these include the main human pathogens and will be the subjects of this review.

II. MEDICAL IMPORTANCE AND CONVENTIONAL DIAGNOSIS

Enteroviruses are a large group of clinically and biologically diverse serotypes. They are subgrouped into polioviruses, coxsackie A viruses (CAVs), coxsackie B viruses (CBVs), echoviruses, and enteroviruses 68 to 71 (Table 7–1). Virtually all the serotypes can cause infections of the central nervous system.[2] Polioviruses, which are now controlled by vaccination programs in most parts of the world, have been the subgroup most often associated with paralytic disease, while coxsackieviruses and echoviruses cause meningitis and encephalitis. CBVs are also found in association with myocarditis. Other clinical manifestations of enterovirus infections include rash illnesses, myalgia, respiratory infections, and epidemic conjunctivitis. In addition to the wide range of symptomatic cases, subclinical enterovirus infections are very common.

    Human rhinoviruses (HRVs), of the rhinovirus genus, are believed to be the most frequent cause of the common cold, and occasionally they are also responsible for infections of the lower respiratory tract, which can be severe or even fatal.[3] There are more than 100 serologically distinct HRV types, a major factor in the frequency of infection and a complication to the development of diagnostics and preventative or therapeutic measures. Hepatoviruses (hepatitis A virus; HAV), recently recognized as a distinct picornavirus genus, cause a common inflammatory liver disease which, although occasionally serious, usually has a good prognosis.

    Clinical symptoms and signs are not usually sufficiently typical in enterovirus infections to allow unambiguous diagnosis, and therefore specific identification of the causative agent is needed. Classical diagnostic methods for enteroviruses are based on isolation of the pathogen in cell cultures, followed by neutralization typing using antiserum pools, or by the observation of an increase in antibody titers

Table 7–1 **Human picornaviruses**

Group	No. of Serotypes	Examples of Clinical Diseases
Enteroviruses		
Polioviruses	3	Paralysis
Coxsackie A viruses	23	Hand, foot and mouth disease
Coxsackie B viruses	6	Epidemic myalgia, myocarditis
Echoviruses	31	Encephalitis, exanthema
Enteroviruses 68 to 71	4	Epidemic conjunctivitis, paralysis
Rhinoviruses	>100	Common cold
Hepatoviruses		
Hepatitis A virus	1	Hepatitis

between serum samples taken at the acute and convalescent phase of infection. The problem with these approaches is that they are both laborious and slow and therefore cannot provide a specific etiological diagnosis during the acute phase of infection. The same is true for HRVs, which are identified on the basis of their acid lability after cell culture isolation. The diagnosis of HAV infection is based on the demonstration of specific IgM-class antibodies in the blood during the acute phase of the illness. However, if epidemiological information on the virus strains is required, new methods would be needed since HAV cannot be isolated using standard techniques. Thus, for the reasons mentioned above, there is an obvious need to improve the speed, sensitivity, and accuracy of the diagnostic assays presently used for all the human picornaviruses.

During the past few years, an increasing amount of information has become available on the molecular characteristics of several picornaviruses.[4] This has significantly increased our understanding of their biological and pathogenetic behavior, but has also made available potentially useful tools for the diagnosis of picornavirus infections using molecular techniques. As in many other virus systems, PCR is perhaps the most promising of these new techniques, and a substantial amount of work has been expended in developing methodologies suitable for picornaviruses. The principles used in selection of the reagents for different diagnostic and epidemiological purposes as well as examples of their applications will be discussed in detail below.

III. MOLECULAR BIOLOGY OF PICORNAVIRUSES

Picornaviruses are small, naked, icosahedral particles which have an ssRNA molecule as their genetic material. Upon entry into the cell, this genome functions directly as a message for the production of a single polyprotein, which is cleaved by virus proteases to give 11 to 12 individual proteins.[4] In human picornaviruses, the genome is 7100 to 7500 nucleotides in length and is made up of a 5′ untranslated region (5′UTR, 600 to 750 nucleotides), an open reading frame (approximately 6600 nucleotides), a short 3′UTR (35 to 100 nucleotides), and a poly-A tract. Figure 7–1 shows a diagrammatic representation of the genome, indicating the regions encoding the final protein products. 1A to 1D are the capsid proteins (also called VP1–4), usually the most variable parts of the viruses when members of different genera and serotypes are compared. 2A to 2C and 3A to 3D, which include proteases, an RNA-polymerase, and proteins of unknown function, also show considerable variation but possess many common motifs. 2C and 3D, both of which are involved in RNA replication, are particularly well conserved. The 5′UTR contains several regions that are absolutely conserved between viruses of the same genus and have thus received the most attention for PCR as they open up the possibility of developing genus-specific reagents. The more variable parts of the genome may potentially allow serotype-specific detection although this possibility has been explored far less thoroughly.

IV. IMPLICATIONS OF PICORNAVIRUS GROUPS FOR PCR DETECTION

A large number of complete or substantially complete picornaviral nucleotide sequences have now been derived, giving a good coverage of the genera and subgroups.[4,5] From these, it is clear that the great majority of human picornavirus serotypes are closely related at the sequence level. This can be seen in

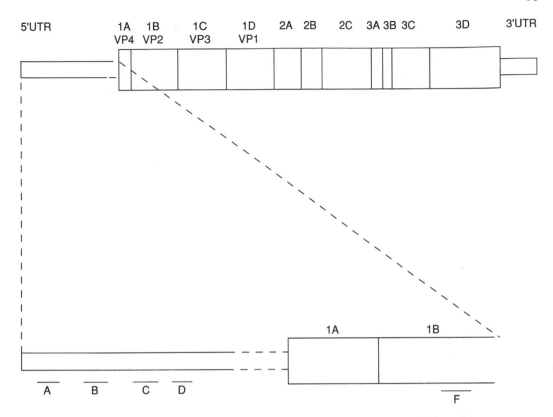

Figure 7–1 Diagrammatic representation of the picornavirus genome, which for all human members is 7100 to 7500 nucleotides in length. The wider box represents the open reading frame, and the regions encoding the proteins (1A to 3D) are indicated. Untranslated regions (UTRs) are represented as narrower boxes. The areas most frequently exploited for PCR of the major groups of human picornaviruses, rhinoviruses and enteroviruses, are shown on an expanded scale below the main diagram. Primers are commonly directed to the 5'UTR as it contains many conserved sequences (A to D, in Figure 7–3; Table 7–2). Enteroviruses and rhinoviruses can be differentiated by hybridizing the PCR product with genus-specific probes localized between the primers. Alternatively, the fact that rhinoviruses lack the part of the 5'UTR denoted by dotted lines means that differentiation can also be carried out on the basis of PCR product size, if a primer corresponding to, for example, D is used together with one corresponding to the conserved region F (Table 7–2).

Figure 7–2, where the amino acid sequence relationships between the 2C proteins of the sequenced serotypes are displayed. Similar relationships are seen throughout the genome. The data show that there is considerable overlap between enteroviruses and rhinoviruses; all of these being related by at least 45% amino acid sequence identity. Within the enteroviruses there are several serotypes, which are much more closely related. For instance, the three poliovirus serotypes are virtually identical to one another in this region, at least at the amino acid level, and the same is true of the CBVs so far analyzed. In contrast, the CAVs do not form a coherent group in sequence terms since some are closely related to either polioviruses or to CBVs.[6]

One enterovirus, echovirus 22, is highly anomalous.[7] Although it shares physical and some pathogenic properties consistent with its enterovirus classification and has a typical picornavirus genome organization, it differs substantially from all other sequenced picornaviruses (less than 30% sequence identity in 2C). Preliminary investigations suggest that echovirus 23 is similar to echovirus 22 in some regions of the genome. The evidence so far available implies that most, if not all, of the remaining echoviruses belong to the large entero-/rhinovirus cluster.[8] The same is true of at least the great majority of CAVs.[6] Similarly, PCR and sequence analysis of HRVs have so far revealed no atypical representatives.[4] The other human picornavirus belonging to the hepatovirus genus, HAV, is very different from the entero-/rhinovirus cluster (Figure 7–2). Thus, in sequence terms, three groups can be thought to comprise the main human serotypes: entero-/rhinoviruses, echovirus 22-like, and HAV.

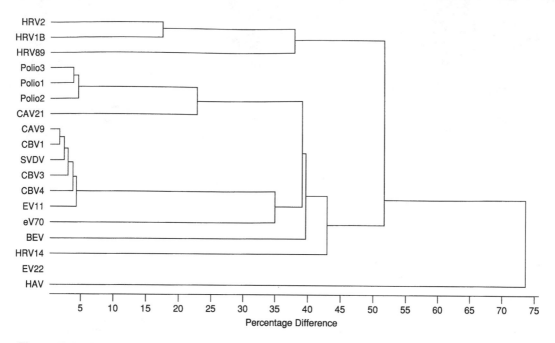

Figure 7–2   Dendrogram showing the relationships between groups of human picornaviruses. It is based on differences in the 2C polypeptide. HRV, human rhinovirus; Polio, poliovirus; CAV, coxsackie A virus; CBV, coxsackie B virus; SVDV, swine vesicular disease virus; EV, ECHO virus; eV, enterovirus; BEV, bovine enterovirus; HAV, hepatitis A virus.

Genetic relationships between serotypes have important implications for the design of PCR methodologies. Extensive nucleotide sequence identity has enabled the development of broadly reacting assays capable of detecting all the members of the large entero-/rhinovirus cluster described above.[9–15] Many of these take advantage of invariant or highly conserved sequences, which can be found in the 5′UTR of the genome. In addition, serotype-specific primers have been described.[14,16] The situation in the other groups is more straightforward because all HAV isolates belong to the same serotype and are comparatively homogeneous in genetic terms. The members of the echovirus 22 group also appear to be closely related, at least in certain regions of the genome.

V.   DEVELOPMENT OF PICORNAVIRUS PCR ASSAYS

In addition to the general complications associated with the application of PCR to diagnosis, for instance, the possibility of sample carryover giving rise to false-positive results, picornaviruses pose other problems. Serotype diversity, which can be overcome by using general primers that bind to conserved sequences, has already been described. Other problems arise as a result of the nature of the genetic material since ssRNA is much less stable than DNA, particularly in the nuclease-rich samples, for example, CSF, feces, and nasal lavages, that are routinely assayed. These samples also contain substances that may interfere with enzymic reactions. Finally, the RNA must first be reverse transcribed before PCR amplification. Several protocols have now been published that circumvent these difficulties, and picornavirus PCR is now well established.[9–16] When optimized assay conditions are used, the sensitivity of detection of enterovirus RNA with one set of reagents is in the range of 10 to 1000 genomes. When nested PCR is used, as few as 1 to 10 genomes still give a positive signal,[17] and a sensitivity of 2.5 genomic copies has been reported when PCR is followed by hybridization.[16] That this sensitivity is 20- to 100-fold better than that possible when conventional virus isolation is used illustrates the great potential of nested PCR in particular.

PCR was first successfully applied to picornaviruses from clinical material in 1988, by Gama et al., working with HRVs in nasal washes.[9,10] These authors found that inclusion of tRNA as a carrier using RNA isolation prior to PCR was beneficial and that the addition of the RNase inhibitor, vanadyl ribonucleoside complex, was essential if a positive result was to be obtained. Using primers directed to invariant sequences in the 5'UTR 380 nucleotides apart, it proved possible to detect all HRVs tested at a level of sensitivity of about 20 infectious virus particles.

Hyypiä et al.[12] described an assay system where sequences from the highly conserved 5'UTR were first used to amplify both entero- and rhinoviruses followed by differentiation between the groups by hybridization using oligonucleotide probes. It was shown that detection and differentiation of standard strains (except echovirus 22, which was detected with specific reagents) and clinical isolates was possible by combining the results obtained with the two probes because the rhinovirus probe was highly specific while the enterovirus probe reacted with some HRV strains, in addition to enteroviruses. Of the 29 clinical isolates tested, all except one were PCR positive, and a hybridization signal was obtained with all except three strains. No signal was observed in the three control virus preparations tested. A clinical echovirus 23 isolate among the reactive strains must represent either an atypical strain or a typing error since, according to the present knowledge on genetic relatedness, this virus should not be detected with the set of reagents used.

Olive et al.[13] selected one of their primers from the region coding for capsid proteins and another conserved primer from the 5'UTR. They were able to obtain amplification of three different HRV serotypes and the 26 enterovirus reference strains analyzed with this set of reagents. The different lengths of the amplification products originating from HRVs and enteroviruses could be used to differentiate between the groups. It was also possible to detect both HRVs and enteroviruses specifically in clinical material. Chapman et al.,[14] using conserved primers and a probe from the 5'UTR, detected all their 33 different enterovirus serotypes. Another pair of primers from the capsid encoding region was used for specific detection of coxsackievirus B3. Rotbart[15] used the same principles in the selection of reagents and developed an assay for enteroviruses that had a sensitivity of 10^3 TCID50. When 25 c were used, 9 of the 11 enteroviruses tested gave a positive signal by both gel electrophoresis and oligonucleotide hybridization. A further 25 PCR cycles gave a positive result with two strains negative in the primary assay, echoviruses 2 and 22, although this result must be treated with caution as other investigators consistently report negative results for echovirus 22 using enterovirus-specific primers.

In an extensive study of enteroviruses by Zoll et al.,[18] again using reagents recognizing conserved parts of the 5'UTR, PCR was applied to 66 reference strains. Sixty of these, excluding CAVs 11, 17, and 24 and echoviruses 16, 22, and 23, were positive by gel electrophoresis. When subsequent hybridization with a conserved probe was used, most of the PCR-positive serotypes reacted, the exceptions being CAVs 1, 12, 20, 21, and 22. The assay was further applied in the direct detection of enterovirus RNA from isolation positive clinical samples. All six throat wash samples, as well as the 12 stool specimens were detected positive. The sensitivity of the PCR was 0.1 fg of enterovirus cDNA, corresponding to approximately 10 molecules.

As discussed above, similar principles have been used in the selection of PCR primers and probes by most investigators. The broad reactivity of the reagents is obtained by utilizing the conserved parts of the genome in the 5'UTR or in the capsid protein encoding region. Alignment of four areas of the 5'UTR, applicable for primer selection, are shown in Figure 7–3. A set of reagents (A to D) that can be used for a nested PCR is presented in Table 7–2. Alternatively, primers A and B, or complementary sequences to primers C and D, can be used together with primer F from the capsid region to obtain amplification products of different lengths to differentiate between HRVs and enteroviruses (Figure 7–4).

VI. THE APPLICATION OF PCR TO SPECIFIC HEALTH PROBLEMS

A. POLIOMYELITIS

Paralytic poliomyelitis has been controlled in the developed world for over 30 years, due to the introduction of inactivated or attenuated vaccines. The vaccines contain three components, corresponding to the serotypes of the causative agents, polioviruses. Characterization of polioviruses is important for several reasons, not least because of the contribution this would make to the stated aim of the WHO to eradicate poliomyelitis from the world by the year 2000. A major aspect is the ability to differentiate between wild-type isolates and vaccine strains that circulate in the community as a result of campaigns using the Sabin, live attenuated vaccines. This gives an idea of the success of vaccination strategies in excluding wild-type strains and also enables an analysis of the etiology of paralytic disease when this occurs in areas well covered by a vaccine. Vaccine-associated paralysis can occur at very low frequency, and it is important to differentiate between this and wild-type cases.

Two methods have been described to answer these questions, both of which also enable the serotype of Sabin strains to be identified.[16,19] Yang et al.[16] used primers that recognize serotype-specific sequences

A

```
        50
Poliol  AGTACTCCGG TATTGCGGTA CCCTTGTACG CCTGTTT... TATACTCCC.
Polio2  AGTACACTGG TATTGCGGTA CCTTTGTACG CCTGTTT... TATACTCCC.
Polio3  AGTACACTGG TATCACGGTA CCTTTGTACG CCTGTTT... TATACTCCC.
CAV9    AGCACACTGG TATCACGGTA CCCTTGTGCG CCTGTTTTAT A.ACCCCAC.
CAV21   AGTAATCTGG TA.TCAGGTA CCTTTGTACG'CCTGTTTTAT ATCCCT....
CBV1    AGCACTCTGG TATCACGGTA CCTTTGTGCG CCTGTTTTAC ATCCCCTCC.
CBV3    AGCACTCTGG TATCACGGTA CCTTTGTGCG CCTGTTT... TATACCCCC.
CBV4    AGCACACTGG TATTCCGGTA CCTTTGTGCG CCTGTTTTAT AACCCCCCC.
EV70    AGTACTCCGG TACCCCGGTA CCCTTGTACG CCTGTTT... TATACTCCC.
HRV1B   TGTACTCTGT TATTCCGGTA ACTTTGTACG CCATTTTCCC TCCCTCCCC.
HRV2    GGTACTCTGT TATTACGGTA ACTTTGTACG CCAGTTT... ..TATCTCC.
HRV14   AGTACTCTGG TACT.ATGTA CCTTTGTACG CCTGTTTCTC CCCAACCACC
HRV89   TGTACTCTGT TATTACGGTA ACTTTGTACG CCAGTTTTTC CCACC.....
Con     -G-A--C-G- TA-----GTA -C-TTGT-CG CC--TTT--- ----------
```

B

```
        147
Pollol  CAGTACCACC ACGAACAAGC ACTTCTGTTT CCCCGG..TG ATGTCGTATA
Polio2  CAGTACCACC ACGAACAAGC ACTTCTGTTC CCCCGG..TG AGGCTGTATA
Polio3  CAGCGCCTCC GTGGGCAAGC ACTACTGTTT CCCCGG..TG AGGCCGCATA
CAV9    CAGTCACATC GTGACCAAGC ACTTCTGTCT CCCCGGACTG AGTATCAATA
CAV21   CAGTACCTCT ACGAACAAGC ACTTCTGTTT CCCCGG..TG AAATCATATA
CBV1    CAGCCATGTT TTGATCAAGC ACTTCTGTTA CCCCGGACTG AGTATCAATA
CBV3    CAGCCACGTT TTGATCAAGC ACTTCTGTTA CCCCGGACTG AGTATCAATA
CBV4    CAGCTGTGTT TTGGCCAAGT ACTTCTGTGT CCCCGGACTG AGTATCAATA
EV70    CAGTACCACC ACGAACACAC ACTTCTGTTT CCCCGG..TG AAGTTGCATA
HRV1B   CAGGTTGTCT AAGGTCAAGC ACTTCTGTTT CCCCGGTTGA CGTTGATAT.
HRV2    CAGATTACTG AAGGTCAAGC ACTTCTGTTT CCCCGG..TC AATGTTGATA
HRV14   AATGGTGTCT ATGTACAAGC ACTTCTGTTT CCCAGG..AG .CGAGGTATA
HRV89   CAGACTGTCA AAGGTCAAGC ACTTCTGTTT CCCCGG..TC AATGAGGATA
Con     -A-------- --G--CA--- ACT-CTGT-- CCC-GG---- -------AT-
```

Figure 7–3 Alignment of enterovirus and HRV sequences from selected parts of the 5′UTR. The A, B, C, and D represent highly conserved regions among all the viruses analyzed. The numbers refer to the poliovirus 1 sequence. The consensus sequence shows only nucleotides invariant in all viruses analyzed.

encoding the N-terminal region of VP1, which although variable between serotypes, is not under immune pressure and is well maintained during passage of the vaccine strains in humans. Primer design enabled a different-length product to be obtained for each of the three Sabin strains, thus allowing rapid strain identification. It was thought that the small size of the PCR product (97, 71, and 44 bp for Sabin 1, 2, and 3, respectively) enhanced the sensitivity; using strain-specific oligonucleotide probes, as few as 2.5 genomes could be detected.

The second method, introduced by Balanant et al.,[19] utilizes primers that amplify, from all poliovirus strains, a 480-nucleotide region encoding the N-terminal half of VP1. This product is then subjected to restriction enzyme digestion using 4-cutters, to generate RFLPs. It proved possible to differentiate between the original Sabin vaccine strains and wild-type isolates using this method. Similarly, vaccine-derived isolates could readily be identified, indicating that there is sufficient genetic stability to exploit this method widely. Additional information on the wild-type genotypes circulating in an area could also be obtained since closely related strains showed very similar profiles.

B. TYPE A HEPATITIS: ANTIGEN-CAPTURE PCR AND EPIDEMIOLOGY

Type A hepatitis is commonly observed in underdeveloped countries but is also a continuing problem elsewhere. The fact that there is a single serotype of HAV, allows reliable serological diagnosis. However,

C

```
          433
Poliol   TACATAAGAA  TCCTCCGGCC  CCTGAATGCG  GCTAATCCCA  ACCTCGGAGC
Polio2   TACCTGAGAG  TCCTCCGGCC  CCTGAATGCG  GCTAATCCTA  ACCACGGAGC
Polio3   TACATGAGAG  TCCTCCGGCC  CCTGAATGCG  GCTAATCCTA  ACCATGGAGC
CAV9     TAGCTGGTAG  TCCTCCGGCC  CCTGAATGCG  GCTAATCCCA  ACTGCGGAGC
CAV21    TACTCAAGAG  TCCTCCGGCC  CCTGAATGCG  GCTAATCCTA  ACCACGGAGC
CBV1     TAGTTGGTAA  TCCTCCGGCC  CCTGAATGCG  GCTAATCCTA  ACTGCGGAGC
CBV3     TAGTTGGTAG  TCCTCCGGCC  CCTGAATGCG  GCTAATCCTA  ACTGCGGAGC
CBV4     TAGTTGGTAG  TCCTCCGGCC  CCTGAATGCG  GCTAATCCTA  ACTGCGGAGC
EV70     TACCTGAGAG  TCCTCCGGCC  CCTGAATGCG  GCTAATCCCA  ACCACGGAGC
HRV1B    TCACTTTGAG  TCCTCCGGCC  CCTGAATGCG  GCTAACCTTA  AACCTGCAGC
HRV2     TCGCTTTGAG  TCCTCCGGCC  CCTGAATGTG  GCTAACCTTA  ACCCTGCAGC
HRV14    TGGTTGTGAG  TCCTCCGGCC  CCTGAATGCG  GCTAACCTTA  ACCCTAGAGC
HRV89    TCAGTGTGCT  TCCTCCGGCC  CCTGAATGTG  GCTAACCTTA  ACCCTGCAGC
Con      T---------  TCCTCCGGCC  CCTGAATG-G  GCTAA-C--A  A------AGC
```

D

```
          523
Poliol   AGTCCGTGGC  GGAACCGACT  ACTTTGGGTG  TCCGTGTTTC  CTTTTA.TTT
Polio2   AGTTCGTGGC  GGAACCGACT  ACTTTGGGTG  TCCGTGTTTC  CTTTTATTTT
Polio3   AGTCCGTGGC  GGAACCGACT  ACTTTGGGTG  TCCGTGTTTC  CTTTTA.TTC
CAV9     ACTCTGCAGC  GGAACCGACT  ACTTTGGGTG  TCCGTGTTTC  TTTTTATTCC
CAV21    AGTCTGTGGC  GGAACCGACT  ACTTTGGGTG  TCCGTGTTTC  CCTTTATATT
CBV1     ACTCTGCAGC  GGAACCGACT  ACTTTGGGTG  TCCGTGTTTC  ATTTTATTCC
CBV3     ACTCTGCAGC  GGAACCGACT  ACTTTGGGTG  TCCGTGTTTC  ATTTTATTCC
CBV4     ACTCTGCAGC  GGAACCGAGT  ACTTTGGGTG  TCCGTGTTTC  CTTTTATTCT
EV70     AGTCTGTGGC  GGAACCGACT  ACTTTGGGTG  TCCGTGTTTC  CTTTTATTTT
HRV1B    ATTGCGGGAT  GGGACCGACT  ACTTTGGGTG  TCCGTGTTTC  ACTTT.TTCC
HRV2     ATTGCGGGAT  GGGACCAACT  ACTTTGGGTG  TCCGTGTTTC  ACTTT..TTC
HRV14    ATTCCGGGAC  GGGACCGACT  ACTTTGGGTG  TCCGTGTTTC  TCATTTTTCT
HRV89    AGTGCGGGAT  GGGACCAACT  ACTTTGGGTG  TCCGTGTTTC  CTGTTTTTCT
Con      A-T--G----  GG-ACC-A-T  ACTTTGGGTG  TCCGTGTTTC  ---TT-----
```

Figure 7–3C and D

Table 7–2 **Primers selected from conserved parts of the 5′UTR (A to D) and capsid protein encoding region (F) of entero-/rhinoviruses**

Code	Primer
A(+)	5′-GTAC/ACT/CTTGTG/ACGCCA/TGTTT-3′
B(+)	5′-CAAGCACTTCTGTTTCCCCGG-3′
C(−)	5′-CATTCAGGGGCCGGAGGA-3′
D(−)	5′-GAAACACGGACACCCAAAGTA-3′
F(−)	5′-GGGACCTTCCACCACCANCC-3′

studies on virus epidemiology, which are important in developing a strategy for the control of the disease, have been hampered by the lack of simple and generally applicable virus-isolation procedures and by difficulty in distinguishing between virus strains. These limitations have led to the development of a PCR approach to enable amplification and analysis of HAV nucleic acid.[20,21]

Epidemiological analysis based on nucleotide sequencing of large numbers of virus isolates is particularly vulnerable to PCR contamination. This is acute in the case of HAV since the virus seems to be genetically stable, giving rise to rather fewer differences between strains than is seen in some picornaviruses. This means that contamination may not always be recognized. An important aspect of the

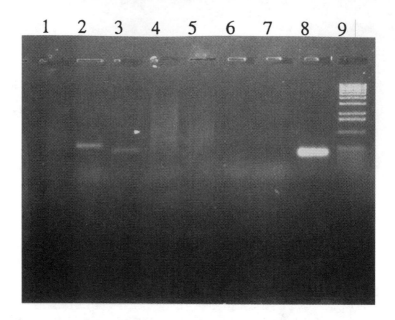

Figure 7–4   Example of picornavirus PCR. Lanes 1 to 4, analysis performed using the entero-/ rhinovirus general primers OL24 and OL68, corresponding to regions D and F (Figures 7–1 and 7–3; Table 7–2). Lanes 5 to 8, analysis performed using primers OL273 and OL274, based on the sequences of echoviruses 22 and 23; lanes 1 and 5, –ve control (no virus); lanes 2 and 6, CAV-9; lanes 3 and 7, HRV-85; lanes 4 and 8, echovirus 22; lane 9, molecular weight standard. Note the difference between the HRV-85 and CAV-9 product, which is consistently observed between all enteroviruses and rhinoviruses (13). OL273 and OL274 give a product only with echoviruses 22 and 23 of all human picornaviruses tested (data not shown).

work by Jansen et al.[20] was therefore the development of an antigen-capture PCR (AC/PCR). Virus present in clinical material (fecal suspension) was initially purified by immunoaffinity capturing using a specific monoclonal antibody bound to the walls of a reaction tube. The sample was washed thoroughly, then heat denatured to release the RNA so that it could be reverse transcribed and subjected to PCR. The regions amplified encode part of VP3 and the VP1/2A boundary, and sequence determination enabled the comparative analysis of several isolates. It was found that epidemiologically related strains were closely related in sequence, but unexpected relationships between strains causing geographically distant outbreaks were also observed.

Subsequently, an extensive comparative sequencing study, performed on 152 strains by Robertson et al.[21] (and see Chapter 5 by Robertson), indicated that there are four genotypes of HAV circulating in the human population. In the region of the genome studied (168 nucleotides around the VP1/2A boundary), these genotypes differed by 15 to 25%. The study also revealed that strains present in some geographical areas (e.g., the U.S.) belonged to the same genotype, while strains belonging to several genotypes could be isolated in other areas (e.g., Western Europe and Japan). These patterns of occurrence point to endemic transmission in the former case and to the import of HAV strains, presumably as a result of travel, in the latter.

It is probable that since it is based on two quite different steps, AC/PCR should reduce the possibility of PCR contamination. This advantage is augmented by the fact that the number of manipulations required, between clinical sample and final PCR product, is reduced significantly compared to the standard phenol extraction/ethanol precipitation procedure, which is widely used in picornavirus PCR. AC/PCR is presumably also applicable to other systems where antigenically well-defined viruses are being analyzed, although it will be less useful for other human picornaviruses because serotype diversity is so prevalent.

C. ETIOLOGICAL DIAGNOSIS OF VIRAL HEART DISEASE

In a number of reports, evidence of the involvement of, particularly, CBVs in myocarditis and cardiomyopathy has been presented. This has been mainly based on serological observations and isolation of viruses from sites other than heart muscle. Recently, availability of endomyocardial biopsy specimens and advances in molecular virology have made it possible to demonstrate viral nucleic acids directly from the cardiac samples and thus cast new light on the etiology of inflammatory heart disease. This has previously been carried out by spot hybridization and *in situ* hybridization, but recently increasing interest has been directed to applications of PCR for the purpose.

Jin et al.[22] used a set of primers and probes selected from the 5′UTR, which would detect CBVs and other related enteroviruses, to study inflammatory heart disease. When RNA purified from myocardial biopsy samples of 48 myocarditis and cardiomyopathy patients was analyzed, five of the specimens exhibited the presence of enteroviral sequences. A more specific PCR assay was applied for the same purpose by Weiss et al.,[23] who selected their primers and probe from the region encoding the nonstructural 2C protein of coxsackievirus B3. The sensitivity of the test was estimated to be 1 to 100 pfu of virus per gram of tissue, and the specificity was restricted to CBV-3 among the eight picornavirus strains analyzed. Of the five specimens from myocarditis patients, one gave a positive signal, while 11 myocardial samples from cases of idiopathic dilated cardiomyopathy were negative, as were all 21 samples from other heart diseases. Zoll et al.[18] found one of five samples from dilated cardiomyopathy patients to be positive by their broadly reacting PCR, whereas Grasso et al.[24] were unable to detect enterovirus RNA in samples from explanted hearts of 40 patients who had undergone cardiac transplantation.

In revealing viral sequences in the myocardium of inflammatory heart disease patients, as in many other diagnostic applications, PCR has the advantages of sensitivity and speed, and it makes possible simultaneous testing of large numbers of specimens. However, in addition to the contamination problems, it has the disadvantage of offering no additional information on the disease process when compared to *in situ* hybridization, which together with histopathology, reveals the localization of the infected cells. A combination of these two techniques, perhaps preceded by screening of the samples by PCR, would allow rapid and sensitive detection of viral sequences by nonradioactive probes in tissue sections.

D. RHINOVIRUSES AND THE COMMON COLD

Human rhinoviruses (HRVs) are among the most frequently occurring pathogens. The large number of serotypes (more than 100 are recognized) is a major factor in their high incidence as the immune reaction induced by one serotype is not cross protecting. This also greatly complicates the detection of HRVs because immunological techniques are cumbersome to apply. In addition, currently used detection methods rely on growth in cultured cells, but several serotypes are not readily thus propagated, and this may lead to an underestimation of the incidence of HRV infections. As specific anti-HRV drugs become available, it may be useful to have a rapid detection system, particularly in cases of lower respiratory tract infections, and such an assay would also be useful for epidemiological studies. PCR is potentially ideal for the purpose since it circumvents problems associated with serotype diversity and propagation difficulties. The development of an HRV PCR was described above.[9,10] Using this system, it has proved possible to analyze the incidence of HRVs associated with episodes of wheezing in children (unpublished). HRVs were detected in 3 out of 23 cases that were negative for respiratory syncytial virus. In all 3 cases, an HRV was not present in the nose when the child was healthy, implying that it had not been present as a commensal organism. In this study it proved possible to obtain a positive signal from nasal lavage but not nasal swabs. However, the advent of nested PCR may enable the less invasive sampling by nasal swabs to be employed.

HRV epidemiology, which may contribute to the development of anti-HRV therapies, would be facilitated by a more rapid method of typing the viruses. Two methods have been proposed that exploit the PCR product obtained from an assay using general primers. One of these, described by Torgersen et al.,[11] is based on restriction enzyme digestion to generate a pattern of fragment lengths characteristic of a serotype, while in the second, introduced by Bruce et al.,[25] the labeled PCR product is used to probe a standard filter containing reference serotypes. Preliminary results were encouraging although both methods may prove to be difficult to apply to wild-type isolates in view of the large number of HRV serotypes.

As in other systems, PCR is now being used as a rapid means of generating sequence data for comparative and epidemiological purposes. Using HRV reference strains, it has proved possible to develop a procedure to analyze highly variable parts of the virus genome, particularly those encoding antigenic sites, exploiting primers that bind to highly conserved flanking sequences (Gama et al., unpublished). A similar approach is being taken to study wild-type strains (Khan et al., unpublished). This sort of analysis should enable a detailed comparison of HRVs, which is necessary for an understanding of diversity, a key feature in the epidemiology. General primers should also be applicable to the study of molecular relationships in other serotypically diverse picornavirus groups, for instance, CAVs.[6]

E. OTHER DIAGNOSTIC APPLICATIONS

As discussed above, in many of the described picornavirus PCR assays the method has already been successfully applied to the analysis of clinical material, including nasal washings,[9] stool,[20] and cerebrospinal fluid (CSF).[13] In addition to these reports there are other studies, concerning particularly diagnostic applications for enteroviruses, that deserve to be summarized. Since enteroviruses replicate in the gut and are secreted in stools for a few weeks after the acute infection, analysis of these samples offers perhaps one of the most sensitive diagnostic approaches. However, as in the case of virus isolation, it is important to keep in mind that subclinical infections are common, and positive findings from sites other than the affected tissue are only suggestive of the disease association. Abebe et al.[26] have shown that the sensitivity of PCR, when compared to enterovirus isolation from stool samples, is 69% after 30 c and 100% when a nested PCR is used. The results support the idea that conventional virus isolation can be supplemented by molecular methods.

During the acute phase of enterovirus infection, there is usually a short period of time when virus can be isolated from the blood. This would be suitable sample material for standardized assays when the specimen has been taken early after the onset of symptoms. In accord with this idea, Thoren et al.[27] showed that 7 out of 12 acute-phase sera from patients with enterovirus meningitis were positive by nested PCR. Meningitis is one of the common complications of enterovirus infections, and CSF is usually available for diagnostic purposes in these cases. Rotbart[28] showed that all of their enterovirus isolation-positive CSF specimens were also found to be positive by PCR, as were also some of the isolation-negative samples. In addition, PCR was positive in a case of persistent enterovirus infection.[29]

One of the attractive applications of PCR is the search for viral sequences in diseases with unknown etiology. Picornavirus PCR has been used, in addition to the analysis of inflammatory heart disease, for studying muscle biopsies from patients with idiopathic inflammatory myopathies. In a careful investigation using PCR reagents for a number of viruses, including picornaviruses, Leff et al.[30] were unable to observe any evidence of viral involvement. In another study, patients with postviral fatigue syndrome were tested by Gow et al.,[31] and the authors were able to detect enteroviral RNA sequences in muscle tissue of 53% of the patients but also in 15% of the controls. Although in the absence of confirmatory reports this finding should be treated with care, it hopefully will encourage further detailed investigations of other patient groups.

VI. CONCLUSIONS

For all the human picornaviruses, it has proved possible to overcome problems inherent in serotype diversity and the nature of the genomic material, to devise efficient protocols for the application of PCR to detection and analysis. In this review we have tried to show how the sensitivity and rapidity of the methods are already making them attractive alternatives to conventional diagnostic procedures. In some cases, for example, HAV epidemiology, PCR is the only viable approach to data generation. All the signs are that picornavirus PCR will become increasingly well used for detection and epidemiological studies. Its ability to circumvent many problems encountered by conventional techniques should improve the speed and accuracy of diagnosis, thus giving a more detailed picture of the role of picornaviruses in disease, potentially benefitting patient care and enhancing the prospects for development of novel therapeutic agents.

REFERENCES

1. **Rueckert, R. R.,** Picornaviridae and their replication, in *Virology,* Fields, B. N. and Knipe, D. M., Eds., Raven Press, New York, 1990, 507.
2. **Grist, N. R., Bell, E. J., and Assaad, F.,** Enteroviruses in human disease, *Progr. Med. Virol.,* 24, 114, 1978.
3. **Stanway, G.,** Rhinoviruses, in *Encyclopaedia of Virology,* Webster, R. G. and Granoff, A., Eds., Academic Press, San Diego, CA, in press.
4. **Stanway, G.,** Structure, function and evolution of picornaviruses, *J. Gen. Virol.,* 71, 2483, 1990.
5. **Palmenberg, A. C.,** Sequence alignments of picornaviral capsid proteins, in *Molecular Aspects of Picornavirus Infection and Detection,* Semler, B. L. and Ehrenfeld, E., Eds., American Society for Microbiology, Washington, D.C., 1989, 211.
6. **Hyypiä, T. and Stanway, G.,** Biology of coxsackie A viruses, *Adv. Virus Res.,* 42, 343, 1993.

7. **Hyypiä, T., Horsnell, C., Maaronen, M., Khan, M., Kalkkinen, N., Auvinen, P., Kinnunen, L., and Stanway, G.,** A novel picornavirus group identified by sequence analysis, *Proc. Natl. Acad. Sci. U.S.A.,* 89, 8847, 1992.

8. **Auvinen, P. and Hyypiä, T.,** Echoviruses include genetically distinct serotypes, *J. Gen. Virol.,* 71, 2133, 1990.

9. **Gama, R. E., Hughes, P. J., Bruce, C. B., and Stanway, G.,** Polymerase chain reaction amplification of rhinovirus nucleic acids from clinical material, *Nucl. Acids Res.,* 16, 9346, 1988.

10. **Gama, R. E., Horsnell, P. R., Hughes, P. J., North, C., Bruce, C. B., Al-Nakib, W., and Stanway, G.,** Amplification of rhinovirus specific nucleic acids from clinical samples using the polymerase chain reaction, *J. Med. Virol.,* 28, 73, 1989.

11. **Torgersen, H., Skern, T., and Blaas, D.,** Typing of human rhinoviruses based on sequence variations in the 5′ non-coding region, *J. Gen. Virol.,* 70, 3111, 1989.

12. **Hyypiä, T., Auvinen, P., and Maaronen, M.,** Polymerase chain reaction for human picornaviruses, *J. Gen. Virol.,* 70, 3261, 1989.

13. **Olive, D. M., Al-Mufti, S., Al-Mulla, W., Khan, M. A., Pasca, A., Stanway, G., and Al-Nakib, W.,** Detection and differentiation of picornaviruses in clinical samples following genomic amplification, *J. Gen. Virol.,* 71, 2141, 1990.

14. **Chapman, N. M., Tracy, S., Gauntt, C. J., and Fortmueller, U.,** Molecular detection and identification of enteroviruses using enzymatic amplification and nucleic acid hybridization, *J. Clin. Microbiol.,* 28, 843, 1990.

15. **Rotbart, H. A.,** Enzymatic RNA amplification of the enteroviruses, *J. Clin. Microbiol.,* 28, 438, 1990.

16. **Yang, C.-F., De, L., Holloway, B. P., Pallansch, M. A., and Kew, O. M.,** Detection and identification of vaccine-related polioviruses by the polymerase chain reaction, *Virus Res.,* 20, 159, 1991.

17. **Hyypiä, T.,** Etiological diagnosis of viral heart disease, *Scand. J. Infect. Dis.,* in press.

18. **Zoll, G. J., Melchers, W. J. G., Kopecka, H., Jambroes, G., van der Poel, H. J. A., and Galama, J. M. D.,** General primer-mediated polymerase chain reaction for detection of enteroviruses: application for diagnostic routine and persistent infections, *J. Clin. Microbiol.,* 30, 160, 1992.

19. **Balanant, J., Guillot, S., Candrea, A., Delpeyroux, F., and Crainic, R.,** The natural genomic variability of poliovirus analysed by a restriction fragment polymorphism assay, *Virology,* 184, 645, 1991.

20. **Jansen, R. W., Siegl, G., and Lemon, S. M.,** Molecular epidemiology of human hepatitis A virus defined by an antigen-capture polymerase chain reaction method, *Proc. Natl. Acad. Sci. U.S.A.,* 87, 2867, 1990.

21. **Robertson, B. H., Jansen, R. W., Khanna, B., Totsuka, A., Nainan, O. V., Siegl, G., Widell, A., Margolis, H. S., Isomura, S., Ito, K., Ishizu, T., Moritsugu, Y., and Lemon, S. M.,** Genetic relatedness of hepatitis A virus strains recovered from different geographical regions, *J. Gen. Virol.,* 73, 1365, 1992.

22. **Jin, O., Sole, M. J., Butany, J. W., Chia, W.-K., McLaughlin, P. R., Liu, P., and Liew, C.-C.,** Detection of enterovirus RNA in myocardial biopsies from patients with myocarditis and cardiomyopathy using gene amplification by polymerase chain reaction, *Circulation,* 82, 8, 1990.

23. **Weiss, L. M., Movahed, L. A., Billingham, M. E., and Cleary, M. L.,** Detection of coxsackievirus B3 RNA in myocardial tissues by the polymerase chain reaction, *Am. J. Pathol.,* 138, 497, 1991.

24. **Grasso, M., Arbustini, E., Silini, E., Diegoli, M., Percivalle, E., Ratti, G., Bramerio, M., Gavazzi, A., Vigano, M., and Milanesi, G.,** Search for coxsackievirus B3 RNA in idiopathic dilated cardiomyopathy using gene amplification by polymerase chain reaction, *Am. J. Cardiol.,* 69, 658, 1992.

25. **Bruce, C. B., Gama, R. E., Hughes, P. J., and Stanway, G.,** A novel method of typing rhinoviruses using the product of a polymerase chain reaction, *Arch. Virol.,* 113, 83, 1990.

26. **Abebe, A., Johansson, B., Abens, J., and Strannegård, Ö.,** Detection of enteroviruses in faeces by polymerase chain reaction, *Scand. J. Infect. Dis.,* 24, 265, 1992.

27. **Thoren, A., Robinson, A. J., Maguire, T., and Jenkins, R.,** Two-step PCR in the retrospective diagnosis of enteroviral viraemia, *Scand. J. Infect. Dis.,* 24, 137, 1992.

28. **Rotbart, H. A.,** Diagnosis of enteroviral meningitis with the polymerase chain reaction, *J. Pediatr.,* 117, 85, 1990.

29. **Rotbart, H. A., Kinsella, J. P., and Wasserman, R. L.,** Persistent enterovirus infection in culture-negative meningoencephalitis: demonstration by enzymatic RNA amplification, *J. Infect. Dis.,* 161, 787, 1990.

30. **Leff, R. L., Love, L. A., Millar, F. W., Greenberg, S. J., Klein, E. A., Dalakas, M. C., and Plotz, P. H.,** Viruses in idiopathic inflammatory myopathies: absence of candidate viral genomes in muscle, *Lancet,* 339, 1192, 1992.

31. **Gow, J. W., Behan, W. M. H., Clements, G. B., Woodall, C., Riding, M., and Behan, P. O.,** Enteroviral RNA sequences detected by polymerase chain reaction in muscle of patients with postviral fatigue syndrome, *Br. Med. J.,* 302, 692, 1991.

Chapter 8

Relevance of PCR to Flavivirus Research

Deepak A. Gadkari

TABLE OF CONTENTS

I. INTRODUCTION

The polymerase chain reaction (PCR)[1] has become a well-established technique in the field of virology. PCR can be used for two purposes: (1) rapid diagnosis of viral infections and (2) amplification of specific regions of the viral genome for sequencing/expression of the amplified product to obtain specific proteins. The former approach has immediate practical utility while the latter will give us data that will be useful in answering more basic questions.

II. FLAVIVIRUSES

The Flaviviridae family contains a major genus *(Flavivirus)* that comprises more than 65 different viruses, the majority of which are transmitted by arthropod vectors.[2] These viruses have been classified into six antigenic complexes based on neutralization with polyclonal antibodies.[3] Flaviviruses are approximately 45 nm in diameter, and virions contain three structural proteins: envelope (E), membrane (M), and core (C). Flaviviruses are enveloped viruses and are inactivated by treatment with lipid solvents like ether, chloroform, and deoxycholate. The virions contain a linear, single-stranded RNA of positive polarity. RNA and protein synthesis preferentially occur in the perinuclear region, and viruses mature within the cisternae of the endoplasmic reticulum.

    More than 20 viruses of the *Flavivirus* genus are pathogenic to man. A few cause disease in domestic and wild animals. The mechanisms for the maintenance of vector-borne flaviviruses in nature are transovarial and transtadial transmission (in mosquitoes and ticks) and natural reservoirs, infecting hosts such as monkeys, pigs, birds, and rodents. For the majority of vector-borne flaviviruses, the human is an incidental host.

Yellow fever, Japanese encephalitis, and dengue viruses have great public health importance due to their high rate of infection and severity of disease. Case fatality rates over 40% have been reported in Japanese encephalitis cases.[4] Table 8–1 summarizes important flavivirus diseases affecting humans with regard to their geographic distribution, age group affected, and vectors involved.

III. NEED FOR FLAVIVIRUS DIAGNOSIS

Several flaviviruses cause large outbreaks or epidemics, with thousands of cases being reported. Isolation of some flaviviruses from blood collected during the acute phase is fairly easy. Yellow fever (YF), Kyasanur forest disease (KFD), and dengue viruses fall into this category. However, virus isolation from patient's blood or cerebrospinal fluid (CSF) in Japanese encephalitis (JE), Murray valley encephalitis, and St. Louis encephalitis (SLE) cases is difficult, rare, and time consuming. Most of the flavivirus infections occur in tropical countries. Transportation of clinical specimens requires maintenance of a "cold chain". High ambient temperatures may result in inactivation of virus. Virus isolation attempts generally fail if the cold chain is not properly maintained.

Cerebral malaria, Reye's syndrome, bacterial meningitis, and tubercular meningitis or encephalitis due to other viruses cause symptoms similar to those caused by JE virus. Encephalitis due to SLE virus infection is generally seen in the older age group and is likely to be misdiagnosed as stroke. Some cases of tick-borne encephalitis show symptoms similar to poliomyelitis. To rule out such confusions in diagnosis, especially in the early stages of disease, rapid and reliable tests are necessary.

Serology plays an important part in flavivirus diagnosis. Such tests have great utility in testing large numbers of samples collected during serological surveys or epidemics. For diagnosis, however, a paired sample of acute and convalescent is required. IgG class antibodies show significant cross reactions with antigenically related viruses, and often conclusive diagnosis cannot be made.

No specific treatment exists for flavivirus infection. Treatment is mostly symptomatic. However, if a correct and rapid diagnosis is made, supportive treatment can be given to patients, and excessive administration of antibiotics can be avoided. In severe encephalitic cases, control of fever, convulsions, and cerebral edema should be given priority in order to save the patient's life.

Although flaviviruses are distributed worldwide, the members of this genus are restricted to certain geographical areas (Table 8–1). However, new foci of flavivirus activity are continually being found. For example, JE is endemic in different parts of India. In recent years, two new foci (Haryana and Maharashtra states) have appeared, with JE cases reported for the first time (Reference 5 and NIV, Pune, unpublished data). Local medical practitioners were not aware of such cases and were handling JE cases for the first time. Rapid diagnosis is essential in such instances to make the medical faculty aware of flavivirus infections and the consequences for proper management of patients.

Mosquitoes are vectors for many flavivirus infections of humans. Mosquito breeding habitats range from domestic water containers to various types of water bodies, including paddy fields. If flavivirus activity is detected, public awareness can be raised to abolish such mosquito breeding places wherever feasible. It is possible to predict an epidemic by studying vector density; by isolating or detecting virus from the vector species; and by monitoring anti-flavivirus antibodies in the population and amplifying hosts. Short- and/or long-term measures to control the vector population can then be undertaken. Thus, although specific treatments are not available for the patients, rapid and accurate diagnosis plays an important role in controlling flavivirus infections.

IV. DIAGNOSTIC TESTS FOR FLAVIVIRUS INFECTIONS

A. TESTS FOR ANTIBODY DETECTION

As mentioned earlier, virus isolation from clinical samples is not routinely achieved for most flavivirus infections. Diagnosis therefore mainly relies on serological tests or antigen detection, as described below.

1. Hemagglutination Inhibition (HI) Test[6]

This test is used to detect and quantitate antibodies in serum specimens. Most mosquito- and tick-borne flaviviruses cause agglutination of goose or chick red blood cells (RBCs). When a known antigen is allowed to react with the test sample, the presence of antibodies reactive to this antigen inhibits the agglutination of RBCs. This is one of the most widely used tests in flavivirus diagnosis. A large number of serum samples can be tested simultaneously against several antigens. The drawbacks of this test are

Table 8–1 **Major flavivirus diseases of man**

Disease	Major Vector	Places	Age Group Affected	Virus Isolation from Clinical Cases	Amplifying Host (Reservoir)
Yellow fever	*Aedes aegypti*	Africa, South and Central America, U.S.,[a] Europe[a]	All	Easy	Monkeys
Japanese encephalitis	*Culex tritaeniorhynchus*	South East and Far East Asia, Japan[a]	(High in ages 3 to 15 years)	Difficult	Pigs, birds
St. Louis encephalitis	*Culex pipiens* *Culex tarsalis*	U.S., Jamaica	Adult	Difficult	Birds
West Nile	*Culex* (several spp.)	Africa, Middle East, South East Asia (Europe, former Soviet Union)	Children	Relatively easy	Birds
Murray Valley encephalitis	*Culex annulirostris*	Australia, New Guinea	Children	Difficult	Water birds
Dengue	*Aedes aegypti*	Tropical and subtropical countries	All	Easy	None
TBE	*Ixodes ricinus* *Ixodes persulcatus*	Eastern Europe	Adult	Easy	Rodents, large and small mammals
KFD	*Ixodid ticks*	India	Adult	Easy	Monkeys, bats, rodents

[a] Disease controlled.

cross reactions with antigenically related viruses; the need to pretreat samples (acetone extraction) to remove nonspecific agglutinins; and the requirement for paired samples, acute and convalescent. A fourfold rise in antibody level in convalescent sample compared to that in acute sample is indicative of infection.

2. Complement Fixation Test[7,8]

This test is mainly used to detect antibodies in serum specimens. A quick complement fixation test has also been reported for detection of an unidentified antigen.[9] When an antigen-antibody reaction occurs, it fixes the complement, which is then not available for lysis of the indicator system (sheep RBCs and anti-sheep RBCs). Although this test has more specificity than the HI test and is able to detect early antibodies, it is more difficult to perform and requires more reagents.

3. Single Radial Hemolysis[10]

This is a variation of the complement fixation test. Agar gel containing antigen-coated erythrocytes and complement is allowed to solidify on a slide. Test sera are added in wells punched in the gel. Presence of virus-specific antibodies is indicated by hemolysis of erythrocytes in the gel. Antibodies diffuse radially, and the zone of lysis is directly proportional to the antibody concentration.

4. IgM Antibody Capture Assay[11,12]

This is probably the most sensitive and specific test for indirect diagnosis of a recent flavivirus infection. Virus-specific IgM antibodies can be detected in serum or CSF samples collected during the acute phase of the disease. The sample does not require pretreatment, and due to the high sensitivity of the test, IgM antibodies from CSF specimens can also be detected and diagnosis can be made with a single sample. A slight modification of this test where, instead of enzyme-conjugated antibody, RBCs are added as the indicator system has also been reported (the hemadsorption immunosorbent test; HIT[13]).

B. TESTS FOR ANTIGEN/VIRUS DETECTION
1. ELISA

The test is used to detect antigen from arthropod vectors. A simple sandwich assay is followed using mono- or polyclonal antibodies as capture and probe.[14,15] This is a fairly sensitive method for detection of certain flaviviruses, and large numbers of samples can be handled.

2. Reverse Passive Hemagglutination

This test has been recently reported for detection of JE virus in CSF samples. RBCs are coated with antibodies, preferably monoclonal, and mixed with the suspected sample. Agglutination is observed if the sample contains sufficient quantities of homologous antigen.

3. Immunofluorescence[17,18]

This test can be performed on field-collected arthropod vectors or on mosquitoes inoculated with suspected sample. For detection of viral antigen from mosquitoes, either head squashes or dissected salivary glands are taken for direct or indirect immunofluorescence.

4. Virus Isolation

A variety of primary cell cultures and continuous cell lines are in use for flavivirus isolation. Infant mice, 1 to 3 days old, are also excellent hosts for flavivirus isolation. Inoculation of test sample in mosquitoes has also been reported for isolation of JE virus.[19] Once the virus is isolated, either from clinical material or from the vector species, identification is confirmed by checking virus reactivity with a panel of known antisera in serological tests and in neutralization tests *in vitro* (plaque-reduction test) or *in vivo,* employing adult or infant mice.

V. FLAVIVIRUS RNA

Studies on the molecular aspects of flaviviruses have progressed rapidly since elucidation of the complete nucleotide sequence of the prototype, YF virus.[20] Several unresolved issues — number of nonstructural proteins; single initiation vs. internal initiations for virus protein synthesis — were clarified after the complete genome sequence was known. Subsequently, several flavivirus RNA genomes were completely

Table 8–2. **Availability of nucleotide sequencing data on some flaviviruses**

Virus	Strain	Sequencing Data	Ref.
Yellow fever	17 D	Complete	20
	Asibi	Complete	21
Japanese encephalitis			
	JaGArS 982	Complete	22
	Nakayama	C → N term of NS3	23
	SA14, Wild type	Complete	24,25
	SA14–14–2, Attenuated	Complete	24,25
West Nile		Complete	26
Kunjin	MRM 61 C	Complete	27
Dengue 1	Nauru Island	C → NS1	28
	Singapore S275/90	Complete	29
Dengue 2	1409/Jamaica	Complete	30
	PR159, S1	Complete	31
	16681, Wild type	Complete	32
	16681, Attenuated	Complete	32
Dengue 3	H 87	Complete	33
Dengue 4	Caribbean, 814669	Complete	34
St. Louis encephalitis	SLI-7	5′ → N term of NS3	35
Murray Valley encephalitis	Australia	Complete	36
Tick-borne encephalitis			
	Far Eastern	Complete	37
	Western, Neudoefl	Complete	38
Louping Ill	369/T2	Structural proteins	39
Langat	TP21	Structural proteins	40

or partially sequenced (Table 8–2).[21–40] All the flaviviruses sequenced so far have been shown to possess a similar genome organization. The genomes share the following common features: single plus stranded RNA of about 11 kb; 5′-capped RNA; noncoding sequences at both 5′- and 3′ ends; structural protein coding genes at the 5′ end, covering about one fourth of the genome; the remainder of the protein coding region codes for the nonstructural (NS) proteins. The genomic RNA is devoid of a poly-A tail at the 3′ end, with the exception of certain tick-borne encephalitis (TBE) virus isolates.[41] There is a long open reading frame coding for more than 3400 amino acids. Internal initiation of protein synthesis has not been confirmed. The polypeptides are generated as a result of post- or cotranslational processing. Uncleaved polyproteins have not yet been unambiguously identified from virus-infected cells. The order of the genes is 5′-C, preM/M, E, NS1, NS2A, NS2B, NS3 NS4A, NS4B and NS5–3′ (Figure 8–1).

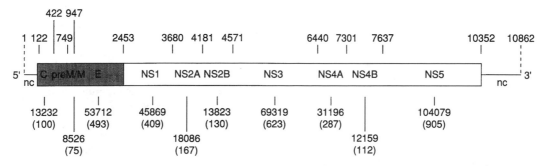

Figure 8–1 Genome organization in flaviviruses. The numbers above the genome are the starting nucleotides in the prototype yellow fever (YF) virus RNA for each protein. The numbers below the line are the deduced molecular weights of proteins (number of amino acids). Glycosylation is not taken into account. Data on YF virus are from Reference 20. The C, M, and E are structural proteins, and NS1 through NS5 are nonstructural proteins. nc, noncoding sequences.

VI. PCR FOR FLAVIVIRUSES

There is a limited amount of published work on application of PCR for flavivirus detection, and most of the data concern the dengue viruses. Detection of flaviviruses by PCR involves (1) extraction of RNA from the sample; (2) synthesis of cDNA using random or specific primer(s), and amplification of cDNA using a suitable pair of primers; (3) identification of the amplified product.

A. RNA EXTRACTION

Viral RNA can be extracted from serum from the viremic host, CSF, host tissues, vectors, and infected cultured cells or culture fluid. Care is required to avoid RNase contamination, and endogenous RNase activity must be inactivated. Physical degradation of RNA can be minimized by avoiding overvortexing during the extraction process. Tardieuex and Poupel[42] have extracted viral RNA from midgut cells of mosquitoes fed on dengue-virus-infected blood by lysing the cells with NP40 in the presence of proteinase K, followed by phenol treatment. Deubel et al.[43] and Laille et al.[44] have also used NP40 to lyse cells and, after removal of cell nuclei, extracted the RNA by phenol treatment in the presence of 1% SDS. Extraction of dengue virus RNA from serum samples collected from dengue patients during the viremic phase has been achieved by guanidine isothiocyanate (GITC) treatment of serum or by treatment with proteinase K in the presence of SDS, followed by phenol extractions.[45] Eldadah et al.[46] used an elaborate method for obtaining RNA from infected mouse brain tissue. Cells were lysed with GITC, and the RNA was pelleted through cesium chloride. Morita et al.,[47] in their attempt to simplify the PCR protocol, detergent treated (1% NP40) dengue-virus-infected cell culture fluids for 1 min at room temperature. Without further treatment, this preparation was used for reverse transcription and PCR. This simplified procedure was as efficient in obtaining the final signal as that obtained from purified RNA preparations.

In our laboratory GITC or sodium iodide has been used to lyse JEV-infected porcine kidney cells and ground mosquito suspensions. The RNA was bound to silica particles.[48] After several quick washes, the bound RNA was eluted in distilled water.

B. REVERSE TRANSCRIPTION AND PCR (RT-PCR)

These reactions can be performed in a single tube with RT and PCR reagents combined. The thermocycler is programmed for the RT reaction for 10 to 40 min and subsequently the cycles required for PCR amplification. Many workers, however, prefer to carry out the RT reaction separately. The entire RT reaction is then supplemented with the PCR reagents. Henchel et al.[45] used degenerate oligonucleotide primers for cDNA synthesis. For detection of any of the four dengue virus serotype primers complementary to the NS1 gene were selected. This region codes for the same amino acids in all four dengue subtypes but the code is degenerate. A mixture of primers representing all four nucleotides at position 12 and C or T at position 15 were used for the RT reaction. The authors have termed these "universal primers". The advantage of this strategy is that synthesis of cDNA can be achieved from samples collected from dengue patients, irrespective of subtype etiology. After the RT reaction, a second set of degenerate primers representing the region towards the 5′ side (upstream) of the first set of primers was added. These primers were degenerate at positions 3, 9, and 12. Duebel et al.[43] used a mixture of primers representing all four subtypes from the envelope protein coding gene. The primer pairs selected were from fairly conserved regions (within the subtypes). However, the region amplified has previously been shown to be variable, and it codes for identifiable antigenic epitopes, including a potential glycosylation site.

Morita et al.[47] used four different protocols for the detection of dengue virus from infected culture supernatants. A standard PCR was performed with the RNA obtained after detergent treatment without further processing. Also RNA obtained after phenol extraction was used. A rapid PCR was performed, with lysis, RT, and PCR carried out in a single tube. RT was performed at 55°C for 10 min, and the PCR conditions were 92°C for 1 min, 55°C for 1 min, and 72°C for 1 min. The cycle was repeated 25 to 35 times. All protocols gave satisfactory results. The advantages with rapid PCR were that reaction tubes required minimal handling and that the results could be obtained in 2 to 3 h.

C. IDENTIFICATION OF PCR-AMPLIFIED PRODUCT

With access to nucleotide sequence data and the primer sequences, the PCR product size can be predicted. Analysis of the product on agarose gels and comparison with known size markers is a fair indication of correct amplification. However, agarose gel electrophoresis often reveals more than one DNA band. In such instances authentication of the amplicon becomes obligatory. The simplest way is to use a nested, internal primer pair to reamplify a portion of the primary PCR product. Four different JE virus strains isolated from the Indian subcontinent have been amplified using primers flanking the E gene. The PCR product of one of the strains was reamplified using nested primers (Figure 8–2, lanes 1 and 2), giving a product of 705 bp.

Many workers have authenticated the PCR amplification of products by hybridization. This method has become particularly useful in identifying dengue virus subtypes. The universal primers used by Henchel et al.[45] were able to amplify any of the four dengue subtypes. However, identity of a specific subtype was possible only after hybridization with type-specific radiolabeled oligonucleotide probes. Duebel et al.[43] used a mixture of primers for amplification of dengue virus subtypes. The subsequent identification of the subtype was revealed again by hybridization with the type-specific probes. This strategy was exploited by Laille et al.[44] who demonstrated the rare occurrence of dual infection with dengue 1 and dengue 3 viruses in six confirmed cases.

Direct nucleotide sequencing of the PCR product, or cloning followed by sequencing, has been reported by several workers. Strain variation is a commonly observed phenomenon for the flaviviruses. Monoclonal antibodies (MAbs) to several flaviviruses are available. Detection of strain variations based on differential reactivities of MAbs (or lack of their reactivity) is possible. Differences between YF virus vaccine strains and the wild-type strain were observed by MAb studies.[49] It was possible to postulate that sequence changes in the E protein of YF virus were responsible for attenuation. Several strains of JEV were classified into two groups on the basis of their reactivities with MAbs.[50]

Though such studies do have some advantages, they are restricted to analysis of virus structural or envelope proteins. The PCR technique has an advantage in that it can be used to amplify any portion of the viral genome. Further analysis of the product can lead to detection of changes at the nucleotide level (and therefore at the corresponding amino acid position if the amplified region represents protein coding region). Such variations were detected in the E protein coding gene of YF virus vaccine strains 17DD and 17D-213, when compared to their parent wild-type Asibi strain.[51] The amino acid changes observed can be further studied to check their relevance to the process of virus attenuation.

Two dengue virus strains isolated from patients with hemorrhagic fever were thought to be new subtypes based on the data obtained in complement fixation, agar gel diffusion, and neutralization tests. However, PCR amplification followed by nucleotide sequencing revealed that the isolate TH-36 was indeed dengue type 2 and that TH-Sman was dengue 1 subtype as there was >97% nucleotide sequence homology with the respective dengue virus subtypes.[52]

Lewis et al.[53] have described a protocol for direct sequencing of large (>2000 bp) PCR products obtained from all four dengue subtypes. Though 0.1 plaque forming unit (pfu) was required to amplify a region of 510 bp, a much larger quantity of virus (10,000 pfu) was required to obtain a product of 2370 bp sufficient for direct sequencing.

The double-stranded DNA product, obtained after PCR amplification of two strains of dengue 3 virus, was converted to a single-stranded form using lambda exonuclease.[54] A phosphorylated primer served as a substrate for the exonuclease, resulting in complete digestion of one strand. By selecting an appropriate

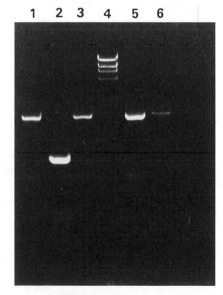

Figure 8–2 RT-PCR amplification of envelope (E) protein coding gene of Japanese encephalitis (JE) virus strains. The primer pair used flanked the E gene: primer 1 (–), nucleotides 2582 to 2602 and primer 2 (+), nucleotides 942 to 965. The expected size of the PCR product was 1660 bp. Four JE virus strains shown were isolated from Vellore, Southern India, P20778 (lane 1), Bankura, Eastern India 733913 (lane 3), Goa, Western India, 826309 (lane 5), and Sri Lanka, 691004 (lane 6). Lane 2 shows reamplification of the product from lane 1 using a nested set of primers, primer 3 (–), nucleotides 2478 to 2507 and primer 4 (+), nucleotides 1802 to 1831. The expected PCR product using primers 3 and 4 was 705 bp. Lane 4 shows the profile of lambda DNA digested with Hind III on 1% (w/v) agarose gel in TAE buffer containing 0.04% ethidium bromide. The nucleotide numbers correspond to the JEV strain sequenced by Sumiyoshi et al.[22]

primer for phosphorylation, a single-stranded template for sequencing can be prepared from either of the two strands. The authors claimed that better sequencing results were obtained using a single-stranded rather than a double-stranded template.

In the sequencing studies cited above, the authors have concluded that direct nucleotide sequencing of the PCR product is error free and gives extremely reliable data. They argue that using appropriate reaction conditions, the error rate of the reverse transcriptase and *Taq* polymerase is so low that in the large pool of amplified product, errors are not detected. This was shown by sequencing multiple clones or amplicons.

VII. APPLICATIONS AND UTILITY

PCR, in flavivirus research, certainly has advantages. Though several simple tests exist for direct or indirect diagnosis of flavivirus infections, they each have limitations. Detection of virus-specific IgM antibodies is a rapid diagnostic test. However, virus-specific IgM cannot be detected in a significant proportion of confirmed cases of JE.[55] Although detection of virus in CSF by reverse passive hemagglutination (RPHA) has been demonstrated,[16] this test has not yet become routine for flavivirus diagnosis. PCR, therefore, has great potential as a major diagnostic tool in flavivirus infections.

The sensitivity of PCR was checked by Morita et al.[47] For the detection of dengue viruses, a positive PCR signal was obtained when the virus load was 3 to 8 pfu. This indicated a superior sensitivity of the plaque assay over PCR. A similar level of sensitivity was observed by Henchel et al.[45] when the PCR results were compared to plaque assay. Interestingly, the presence of high levels of virus-neutralizing antibodies in the test sample did not decrease the sensitivity of the PCR. However, isolated virus is required for a plaque assay, and subtype identification of dengue viruses is not possible unless neutralization or other serological tests are performed using specific antibodies. As has been demonstrated, PCR can overcome these obstacles by careful selection of two specific primers and/or hybridization probes for identification of the amplified product. The sensitivity of the PCR can be enhanced using biotinylated nucleotides during amplification; nested PCR; booster PCR;[56] and the subsequent hybridization of the amplicon with radioactive or nonradioactive probes.

PCR has another advantage in allowing rapid detection of nucleotide changes in the genome of isolates. Direct sequencing of amplicons has bypassed the need to clone them. Strain variation and subtype identification done in this way is much easier than by RNA fingerprinting. Sequencing of the extreme 5′ and 3′ ends of flavivirus RNA is difficult due to secondary structures. Mandl et al.[57] described a novel approach to overcoming this. The RNA has a type I cap at the 5′ end and can be circularized using RNA ligase. The 5′ and 3′ junction region is then RT-PCR amplified using primers from the known 5′ and 3′ portions of the viral RNA. The PCR amplicon covers the junction region. Sequencing of the product thus obtained will give the nucleotide sequence of the extreme 5′ and 3′ ends of the RNA.

Cross contamination of samples is a major cause of false-positive results. False-negative results need to be checked carefully. Henchel et al.[45] have reported that low virus titers in the test sample did not cause false-negative results but repeated freezing and thawing of the sample might have done so. We have reproducibly amplified a 1.66-kb fragment of JE virus RNA covering the entire E protein gene. However, when tissue-culture-grown virus (10^8 pfu/ml) was exposed to 45°C for 45 min prior to RNA extraction, a PCR signal was not obtained, though the plaque titer remained precisely the same as the unheated sample (unpublished observations). The cause of this has not been discovered. Co-purification of other DNAs, RNAs, and impurities in the extracted samples may also play a role in inefficient amplification. Urea, heparin, hemoglobin, phenol, and sodium dodecyl sulfate have been shown to interfere with the assay.[58–60] Primer sequences and their lengths may also affect the PCR results. The optimal experimental parameters may therefore change from virus to virus, even within members of the same family.

As a routine diagnostic tool, PCR will turn out to be far more expensive than the available serological, immunological, and virological tests. Isolation of flaviviruses from clinical specimens or vectors will continue to be the standard technique since virus isolates will be required for further research and reference purposes. However, PCR will be a helpful tool in selecting samples for virus-isolation attempts.

ACKNOWLEDGMENTS

I wish to thank Dr. Cecilia for reading the manuscript and suggesting improvement. I am also grateful to Sadhana Kode and S. S. Bedekar for computer assistance.

REFERENCES

1. **Saiki, R. K., Scharf, S., Faloona, F., Mullis, K. B., Horn, G. T., Erlich, H. A., and Arnheim, N.,** Enzymatic amplification of β-globin genomic sequences and restriction site analysis for diagnosis of sickle cell anaemia, *Science,* 230, 1350, 1985.

2. **Wengler, G.,** Flaviviridae, in *Classification and Nomenclature of Viruses, Fifth Report of the International Committee on the Taxonomy of Viruses,* Francki, R. B. I., Fauquet, C. M., Knudson, D. L., and Brown, F., Eds., *Arch. Virol.,* Suppl. 2, 223, 1991.

3. **Calisher, C. H., Karabatsos, N., Dalrymple, J. M., Shope, R. E., Porterfield, J. S., Westaway, E. G., and Brandt, W. E.,** Antigenic relationships between flaviviruses as determined by cross-neutralisation tests with polyclonal antisera, *J. Gen. Virol.,* 70, 37, 1989.

4. **Umenai, T., Krzysko, R., Bektimorov, T., and Assaad, F.,** Japanese encephalitis: Current worldwide status, *Bull. WHO,* 63, 625, 1985.

5. **Sharma, S. N. and Panwar, B. S.,** An epidemic of Japanese encephalitis in Haryana in the year 1990, *J. Commun. Dis.,* 23, 204, 1991.

6. **Clarke, D. H. and Casals, J.,** Technique of haemagglutination and haemagglutination-inhibition with arthropod borne viruses, *Am. J. Trop. Med. Hyg.,* 7, 561, 1958.

7. **Casals, J. and Palacious, R.,** The complement fixation test in the diagnosis of virus infections of the central nervous system, *J. Exp. Med.,* 74, 409, 1941.

8. **Sever, J. L.,** Application of a microtechnique to viral serological investigations, *J. Immunol.,* 88, 320, 1962.

9. **Pavri, K. M. and Shaikh, B. H.,** A rapid method of specific identification of Japanese encephalitis-West Nile subgroup of arboviruses, *Curr. Sci.,* 35, 455, 1966.

10. **Gaidamovich, S. Y. and Melnikova, E. E.,** Passive hemolysis-in-gel with togaviridae arboviruses, *Intervirology,* 13, 16, 1980.

11. **Burke, D. S. and Nisalak, A.,** Detection of Japanese encephalitis virus immunoglobulin M antibodies in serum by antibody capture radioimmunoassay, *J. Clin. Microbiol.,* 15, 353, 1982.

12. **Gadkari, D. A. and Shaikh, B. H.,** IgM antibody capture ELISA in the diagnosis of Japanese encephalitis, West Nile and dengue virus infections, *Indian J. Med. Res.,* 80, 613, 1984.

13. **Gunasegaran, K., Lim, T. W., Ahmed, A., Aaskov, J. G., Lam, S. K., and Pang, T.,** Haemadsorption immunosorbent technique for the detection of dengue immunoglobulin M antibody, *J. Clin. Microbiol.,* 23, 170, 1986.

14. **Hall, R. A., Kay, B. H., and Burgess, G. W.,** An enzyme immunoassay to detect Australian flaviviruses and identify the encephalitis subgroup using monoclonal antibodies, *Immunol. Cell. Biol.,* 65, 103, 1987.

15. **Tsai, T. F., Bolin, R. A., Montoya, M., Bailey, R. E., Francy, D. B., Jozan, M., and Roehrig, J. T.,** Detection of St. Louis encephalitis virus antigen in mosquitoes by capture enzyme immunoassay, *J. Clin. Microbiol.,* 25, 370, 1987.

16. **Ravi, V., Premkumar, S., Chandramukhi, A., and Kimura-Kuroda, J.,** A reverse passive haemagglutination test for detection of Japanese encephalitis virus antigen in cerebrospinal fluid, *J. Virol. Meth.,* 23, 291, 1989.

17. **Pang, T., Lam, S. K., Chow, C. B., Poon, G. K., and Ramalingam, S.,** Detection of dengue virus by immunofluorescence following inoculation of mosquito larvae, *Lancet,* i, 1271, 1983.

18. **Ilkal, M. A., Dhanda, V., Rodrigues, J. J., Mohan Rao, C. V. R., and Mourya, D. T.,** Xenodiagnosis of laboratory acquired dengue infection by mosquito inoculation and immunofluorescence, *Indian J. Med. Res.,* 79, 587, 1984.

19. **Mourya, D. T., Ilkal, M. A., Mishra, A. C., George Jacob, P., Pant, U., Ramanujam, S., Mavale, M. S., Bhat, H. R., and Dhanda, V.,** Isolation of Japanese encephalitis virus from mosquitoes collected in Karnataka state, India from 1985 to 1987, *Trans. R. Soc. Trop. Med. Hyg.,* 83, 550, 1989.

20. **Rice, C. M., Lenches, E. M., Eddy, S. R., Shin, S. J., Sheets, R. L., and Strauss, J. H.,** Nucleotide sequence of yellow fever virus: implications for flavivirus gene expression and evolution, *Science,* 229, 726, 1985.

21. **Hahn, C. S., Dalrymple, J. M., Strauss, J. H., and Rice, C. M.,** Comparison of the virulent Asibi strain of yellow fever virus with the 17 D vaccine strain derived from it, *Proc. Natl. Acad. Sci. U.S.A.,* 84, 2019, 1987.

22. **Sumiyoshi, H., Mori, C., Fuke, I., Morita, K., Kuhara, S., Kondou, J., Kikuchi, Y., Nagamatu, H., and Igarashi, A.,** Complete nucleotide sequence of the Japanese encephalitis virus genome RNA, *Virology,* 161, 497, 1987.

23. **McAda, P. C., Mason, P. W., Schmaljohn, C. S., Dalrymple, J. M., Mason, T. L., and Fournier, M. J.,** Partial nucleotide sequence of the Japanese encephalitis virus genome, *Virology,* 158, 348, 1987.

24. **Aihara, S., Chunming, R., Yong-Xin, Y., Lee, T., Watanabe, K., Komiya, T., Sumiyoshi, H., Hashimoto, H., and Nomoto, A.,** Identification of mutations that occurred on the genome of Japanese encephalitis virus during the attenuation process, *Virus Genes,* 5, 95, 1991.

25. **Nitayaphan, S., Grant, J. A., Chang, G. J., and Trent, D. W.,** Nucleotide sequence of the virulent SA-14 strain of Japanese encephalitis virus and its attenuated vaccine derivative, SA-14–14–2, *Virology,* 177, 541, 1990.

26. **Castle, E., Leidner, U., Nowak, T., and Wengler, G.,** Primary structure of the West Nile flavivirus genome region coding for all non-structural proteins, *Virology,* 149, 10, 1986.

27. **Coia, G., Parker, M. D., Speight, G., Byrne, M. E., and Westaway, E. G.,** Nucleotide and complete amino acid sequences of Kunjin virus: definitive gene order and characteristics of the virus-specified proteins, *J. Gen. Virol.,* 69, 1, 1988.

28. **Mason, P. W., McAda, P. C., Mason, T. L., and Fournier, M. J.,** Sequence of dengue-1 virus genome in the region encoding the three structural proteins and the major nonstructural protein NS1, *Virology,* 161, 262, 1987.

29. **Fu, J., Tan, B.-H., Yap, E.-H., Chan, Y.-C., and Tan, Y. H.,** Full length cDNA sequence of dengue type 1 virus (Singapore strain S 275/90), *Virology,* 188, 953, 1992.

30. **Deubel, V., Kinney, R. M., and Trent, D. W.,** Nucleotide sequence and deduced amino acid sequence of the nonstructural proteins of dengue type 2 virus, Jamaica genotype: comparative analysis of the full length genome, *Virology,* 165, 234, 1988.

31. **Hahn, Y. S., Galler, R., Hunkapiller, T., Dalrymple, J. M., Strauss, J. H., and Strauss, E. G.,** Nucleotide sequence of dengue 2 RNA and comparison of the encoded proteins with those of other flaviviruses, *Virology,* 162, 167, 1988.

32. **Blok, J., McWilliam, M., Butler, H. C., Gibbs, A. J., Weiller, G., Herring, B. L., Hemsley, A. C., Aaskov, J. G., Yoksan, S., and Bhamarapravati, N.,** Comparison of a dengue 2 virus and its candidate vaccine derivative: sequence relationships with the flaviviruses and other viruses, *Virology,* 187, 573, 1992.

33. **Osatomi, K. and Sumiyoshi, H.,** Complete nucleotide sequence of dengue type 3 genome RNA, *Virology,* 176, 643, 1990.

34. **Mackow, E., Makino, Y., Zhao, B., Zhang, Y. M., Markoff, L., Buckler-White, A., Guiler, M., Chanock, R., and Lai, C. J.,** The nucleotide sequence of dengue type 4 virus: analysis of genes coding for nonstructural proteins, *Virology,* 159, 217, 1987.

35. **Trent, D. W., Kinney, R. M., Johnson, B. B., Vorndam, A. V., Grant, J. A., Duebel, V., Rice, C. M., and Hahn, C.,** Partial nucleotide sequence of St. Louis encephalitis virus RNA: structural proteins, NS1, NS2a and NS2B, *Virology,* 156, 293, 1987.

36. **Lee, E., Fernon, C., Simpson, R., Weir, R. C., Rice, C. M., and Dalgarno, L.,** Sequence of the 3′ half of the Murray Valley encephalitis virus genome and mapping of the nonstructural proteins, NS1, NS3 and NS5, *Virus Genes,* 4, 197, 1990.

37. **Pletnev, A. G., Yamshchikov, V. F., and Blinov, V. M.,** Nucleotide sequence of the genome and complete amino acid sequence of the polyprotein of tick-borne encephalitis virus, *Virology,* 174, 250, 1990.

38. **Mandl, C. W., Heinz, F. X., Stockl, E., and Kunz, C.,** Genome sequence of tick-borne encephalitis virus (Western subtype) and comparative analysis of nonstructural proteins with other flaviviruses, *Virology,* 173, 291, 1989.

39. **Shiu, S. Y. W., Ayres, M. D., and Gould, E. A.,** Genomic sequence of the structural proteins of Louping Ill virus. Comparative analysis with tick-borne encephalitis virus, *Virology,* 180, 411, 1991.

40. **Mandl, C. W., Iacono-Eonners, L., Wallner, G., Holzmann, H., Kunz, C., and Heinz, F. X.,** Sequence of the genes encoding the structural proteins of the low-virulence tick-borne flaviviruses Langat TP21 and Yelantsev, *Virology,* 185, 891, 1991.

41. **Mandl, C. W., Kunz, C., and Heinz, F. X.,** Presence of poly (A) in a flavivirus: significant differences between the 3′ noncoding regions of the genomic RNAs of tick-borne encephalitis virus strains, *J. Virol.,* 65, 4070, 1991.

42. **Tardieux, I. and Poupel, O.,** Use of DNA amplification for rapid detection of dengue viruses in midgut cells of individual mosquitoes, *Res. Virol.,* 141, 455, 1990.

43. **Deubel, V., Laille, M., Hugnot, J. P., Chungue, E., Guesdon, J. L., Drouet, M. T., Bassot, S., and Chevrier, D.,** Identification of dengue sequences by genomic amplification: rapid diagnosis of dengue virus serotypes in peripheral blood, *J. Virol. Meth.,* 30, 41, 1990.

44. **Laille, M., Deubel, V., and Sainte-Marie, F. F.,** Demonstration of concurrent dengue 1 and dengue 3 infection in six patients by the polymerase chain reaction, *J. Med. Virol.,* 34, 51, 1991.

45. **Henchel, E. A., Polo, S. L., Vorndam, V., Yaemsiri, C., Innis, B. L., and Hoke, C. H.,** Sensitivity and specificity of a universal primer set for the rapid diagnosis of dengue virus infections by polymerase chain reaction and nucleic acid hybridisation, *Am. J. Trop. Med. Hyg.,* 45, 418, 1991.

46. **Eldadah, Z. A., Asher, D. M., Godec, M. S., Pomeroy, K. L., Goldfarb, L. V., Feinstone, S. M., Levitan, H., Gibbs, C. J., and Gajdusek, D. C.,** Detection of flaviviruses by reverse transcriptase polymerase chain reaction, *J. Med. Virol.,* 33, 260, 1991.

47. **Morita, K., Tanaka, M., and Igarashi, A.,** Rapid identification of dengue virus serotypes by using polymerase chain reaction, *J. Clin. Microbiol.,* 29, 2107, 1991.

48. **Boom, R., Sol, C. J. A., Salimans, M. M. M., Jansen, C. L., Werthem-van Dillen, P. M. E., and Noordaa, J.,** Rapid and simple method for purification of nucleic acids, *J. Clin. Microbiol.,* 28, 495, 1990.

49. **Sil, B. K., Dunster, L. M., Ledger, T. N., Wills, M. R., Minor, P. D., and Barrett, A. D. T.,** Identification of envelope protein epitopes that are important in the attenuation process of wild type yellow fever virus, *J. Virol.,* 66, 4265, 1992.

50. **Ghosh, S. N., Sathe, P. S., Sarthi, S. A., Cecilia, D., Dandawate, C. N., Athavale, S. S., and Pant, U.,** Epitope analysis of strains of Japanese encephalitis virus by monoclonal antibodies, *Indian J. Med. Res.,* 89, 368, 1989.

51. **Post, P. R., Santos, C. N. D., Carvalho, R., Cruz, A. C. R., Rice, C. M., and Galler, R.,** Heterogeneity in envelope protein sequence and N-linked glycosylation among yellow fever virus vaccine strains, *Virology,* 188, 160, 1992.

52. **Shiu, S. Y. W., Jiang, W. R., Porterfield, J. S., and Gould, E. A.,** Envelope protein sequences of dengue virus isolates TH-36 and TH-Sman, and identification of a type specific genetic marker for dengue and tick-borne flaviviruses, *J. Gen. Virol.,* 73, 207, 1992.

53. **Lewis, J. G., Chang, G. J., Lanciotti, R. S., and Trent, D. W.,** Direct sequencing of large flavivirus PCR products for analysis of genome variation and molecular epidemiological investigations, *J. Virol. Meth.,* 38, 11, 1992.

54. **Lee, E., Nestorowicz, A., Marshall, I. D., Weir, R. C., and Dalgarno, L.,** Direct sequence analysis of amplified dengue virus genomic RNA from cultured cells, mosquitoes and mouse brain, *J. Virol. Meth.,* 37, 275, 1992.

55. **Thakare, J. P., Gore, M. M., Risbud, A. R., Banerjee, K., and Ghosh, S. N.,** Detection of virus specific IgG subclasses in Japanese encephalitis patients, *Indian J. Med. Res.,* 93, 271, 1991.

56. **Ruano, G., Fenton, W., and Kidd, K. K.,** Biphasic amplification of very dilute samples via booster PCR, *Nucl. Acids Res.,* 17, 5407, 1989.

57. **Mandl, C. W., Ileinz, F. X., Puchhamemer-Stöckl, E., and Kunz, C.,** Sequencing the termini of capped viral RNA by 5′ and 3′ ligation and PCR, *BioTechniques,* 10, 984, 1991.

58. **Bentler, E., Gelbart, T., and Kuhl, W.,** Interference of heparin with polymerase chain reaction, *BioTechniques,* 9, 166, 1990.

59. **Khan, G., Kangro, H. O., Coates, P. J., and Heath, R. B.,** Inhibitor effects of urine on the polymerase chain reaction for cytomegalovirus DNA, *J. Clin. Pathol.,* 44, 360, 1991.

60. **Brisson-Noel, A., Aznar, C., Chureau, C., Nguyen, S., Pierre, C., Bartoli, M., Bonete, R., Pialoux, G., Gicquel, B., and Garrigue, G.,** Diagnosis of tuberculosis by DNA amplification in clinical practice evaluation, *Lancet,* 338, 364, 1991.

Chapter 9

Polymerase Chain Reaction for Rubella Virus Diagnosis

William F. Carman and Elizabeth A. B. McCruden

TABLE OF CONTENTS

I. INTRODUCTION

Rubella is caused by a positive-stranded RNA virus that probably spreads by droplets and direct contact. The exact route of the virus through the body until the rash appears is unclear but it seems likely that monocytes or other white blood cells play a role. The rash is probably immune mediated. The infection is important because of the life-threatening or crippling disease it causes in infants whose mothers are infected during the first 15 weeks of gestation. In this chapter, we will outline the epidemiology, diagnosis, and genome structure of rubella, suggest clinical situations where PCR may be useful, and provide methods that have been assessed previously.

A. EPIDEMIOLOGY

Since the introduction of the MMR (measles, mumps, rubella) vaccine, rubella incidence has declined substantially in the U.S. and Europe. Temporary increases in incidence will still occur, dependent upon vaccine coverage, but already doctors entering practice in the U.S. no longer expect to see cases. In Britain, there is still a small annual epidemic; the difference between the two countries is due to past differences in vaccine policy. In the U.S., young children have been vaccinated for many years, with more recent recommendations[1] that high-school-age children and susceptible postpartum women also be immunized. In the U.K., it was thought that only teenage girls and susceptible postpartum women need be vaccinated. It has since been recognized that this policy is unlikely to lead to eradication of congenital rubella, leading to the introduction of vaccination for children of both sexes at 15 months of age in the U.K.

Introduction of vaccination has led to changes in incidence. In the U.S., there was only one case of fetal disease (congenital rubella syndrome; CRS) in 1989,[1] compared with 13 in 1986. In the U.K., 7 CRS cases were seen in 1990,[2] a decline from 37, 19, and 8 cases in 1987, 1988, and 1989, respectively. Only

3 terminations for maternal rubella were performed in the U.K. in 1989, compared with 63 and 36 in 1987 and 1988, respectively.[2] Cases of U.K.-laboratory-confirmed infection were the lowest recorded in 1990, and this can be expected to decline as the vaccinated cohort become the predominant population. Only 1.2% of antenatal clinic attenders in the U.K. are susceptible. It is hoped that limiting the spread of virus in the community by vaccination will make it rare for pregnant women to come into contact with an infected case and that if they do, they will be immune. In the underdeveloped world, where rubella is not regarded as a major pathogen, the incidence is unknown, but it is probably high with regular epidemics.

B. CLINICAL IMPORTANCE

Rubella is important because of the effects on the fetus, which are widespread and lead to either abortion or crippling abnormalities. After birth, infection is asymptomatic, or mild and self-limiting disease may occur.

The first 15 weeks of pregnancy, when the fetal organs are forming, are the most critical.[3] If a pregnant woman has a primary infection (an earlier sample has been shown to be IgG negative) with rash between 12 days after her last menstrual period[4] and before 11 weeks gestation, she has an almost 100% chance of having a diseased fetus. The risk of disease in infected fetuses drops to about one third in those aged from 12 to 16 weeks and is very small after this gestational age. Similarly, the fetal-infection rate itself declines from nearly all to two thirds to one third at the same ages.[3] As a result, before 11 or 12 weeks, the woman would be offered a termination of pregnancy. From 12 weeks on, the outcome of maternal rubella is very difficult to predict.

One early study[3] showed that asymptomatic infections carry no risk of fetal infection; however, other studies[5,6] have shown that fetal infection can occur and that it can lead to CRS. Nevertheless, some fetuses do not become infected and disease will not always occur after infection. Even in the first trimester, if the pregnancy is particularly important to the family, a case can be made for trying to assess whether the fetal tissues are infected. Appearance of a rubella rash within 12 days of the last menstrual period does not lead to fetal infection, but one cannot fully rule out the possibility that preconception rubella in the mother might lead to fetal infection.

C. IMMUNITY AND VACCINATION

Rubella immunity can be established by many techniques that measure either total antibody or specific IgG. Antibody due to vaccination or past infection is considered to be protective, except for rare examples, to be discussed later. Problems can be encountered with some tests if the level is close to or below the standard "protective" level (15 IU/l in the U.K.). If there is any doubt and the patient is not a pregnant woman, a repeat dose of vaccine can be administered. Vaccination during pregnancy is contraindicated because the vaccine is live; however, it is clear that inadvertent vaccination during pregnancy is not an indication for termination as there are no reports of fetal disease being caused by it although the fetus can be infected.[7]

D. DIAGNOSIS

The subject has recently been comprehensively reviewed by Cradock-Watson.[8] Acute infection is usually diagnosed by demonstration of specific IgM in the serum. Care is needed, however, as IgM may not be detectable in the first 48 h after rash onset; in those circumstances, a second sample should be requested. Acute infection may also be confirmed by demonstrating a fourfold or greater rise in antibody titer, using hemagglutination inhibition or specific IgG tests.

A pregnant woman might have a history of a rash some weeks before and be unsure whether it occurred before or after her last menstrual period. The finding of rubella IgM in her serum would not help to resolve this as it persists for about 6 months after infection. IgM assays can give equivocal results, and false-positive results attributable to the presence in serum of rheumatoid factors can be seen in some assay formats.

Virus isolation in tissue culture is mainly reserved for diagnosis of congenital rubella. This may not be successful if samples have been in transit for longer than 72 h. Originally, virus was grown in monkey kidney; after passage to increase viral titer, a challenge dose of an enterovirus was added. If interference with enteroviral growth occurred, detected by a lack of enteroviral cytopathic effect, this was considered evidence of the presence of rubella virus. This cumbersome technique has been replaced with growth in cells such as RK-13 or vero, both semicontinuous lines, in which either cytopathic effect can be observed or antigen can be detected by specific fluorescence using labeled antirubella antibody. The cytopathic effect takes weeks to appear and may require multiple transfers of the virus into fresh tissue culture. Not many laboratories would offer this service because of the ease and wide availability of rubella IgM detection assays.

Prenatal diagnosis of fetal infection is difficult. IgM has been detected in fetal blood samples[6,9] occasionally as early as 15 weeks of pregnancy, but is worth looking for only after 18 weeks (one group[6] showed reliability after 21 weeks). Furthermore, in many fetuses, blood can, unfortunately, be withdrawn under ultrasound guidance from the umbilical vein only after 20 weeks. This would be of little benefit because of the age limits at which termination of pregnancy is desirable. Even if the fetal serum contains IgM, this does not necessarily indicate fetal damage, nor does the absence of IgM preclude a damaged infant. Nevertheless, one study[6] showed that 12 of 18 fetuses were positive for IgM when sampled between 20 and 26 weeks. A clinical decision was made depending on the gestational age. One fetus gave a false negative because IgM had not yet been generated; unfortunately, a diseased infant was born. If the mother is infected very early in the pregnancy, IgM may not appear in the fetus.[4]

Amniotic fluid[10,11] or chorionic villus biopsies[12] can be cultured, but this is not particularly sensitive (from 70 to 90%) as it is dependent on the amount of infectious virus present. Amniotic fluid was cultured in three patients with confirmed maternal infection.[10] The pregnancy of the patient from whom rubella was cultured was terminated, and virus was found in many organs of the fetus. The other two babies, where amniocentesis was negative, were normal and uninfected at birth. Culture of chorionic villus biopsies gives variable results compared with hybridization assays[12,13] (see below).

RNA detection has been tried both in cultured cells and directly on fetal tissues. In a study designed to compare the efficacy of monoclonal antibody staining of rubella antigen to detection of nucleic acid by hybridization in infected tissue cultures, first-trimester abortuses from infected mothers were examined.[13] Of the 40 positives, half had RNA detectable without antigen detected by antibody staining. This was attributed to defective interfering particles. Hybridization assays directly on tissues have been shown to be as sensitive as culture,[12] but this is unlikely to be sensitive enough, suggesting that PCR may be useful.

PCR would be predicted to be more sensitive than culture as the virus does not need to be infectious. PCR may not be entirely satisfactory, however, as it may give rise to false negatives if the infection is not distributed throughout the tissues. For this reason, a negative PCR result cannot exclude rubella virus infection. False positives may occur because of the possibility that maternal blood will contaminate the fetal tissues. We do not know for how long viral RNA can be detected in maternal blood by PCR following acute rubella; it is possible that low levels may remain for some months.

E. REINFECTIONS

Infection does not always lead to disease. In particular, it seems that reinfections seldom result in damaged fetuses. This gives rise to difficult decisions in cases where reinfections have occurred. The diagnosis of rubella usually rests on the finding of a rise in antibody titer or the appearance of IgM. Rises in IgG may occur in immune individuals during epidemics but probably do not, on their own, reflect an infectivity risk to the fetus.[5]

Maternal reinfections may occur after both natural infection and vaccination. Three prospective studies[5,14,15] have been performed to address the risk to the fetus after reinfection. In the first two, no fetal infections were seen. In the third,[15] 3 of 10 reinfections in the first trimester led to fatal infection. An infant who was infected at 12 weeks postconception had no disease; other fetuses have been aborted without diseased organs being detectable. Of 34 confirmed reinfections that continued to term, only 3 (of 31 tested) were infected. Interestingly, this study found that fetal infection occurred only in asymptomatic, unvaccinated mothers with reinfection; nevertheless, symptomatic reinfection in the mother is associated occasionally with fetal infection. Fetal disease, in contrast to infection, in confirmed maternal reinfections have been described, but only as isolated case reports.[8]

Reinfections are detected because of a rise in IgG titer or the presence of IgM. If IgM is not present, or the patient gives a history of vaccination or past infection, it would be appropriate to test the avidity of the IgG.[16] The principle of this is that the immune response late after infection is associated with high-avidity antibody; such antibody binds strongly to antigen on the solid phase of an assay. In contrast, IgG that has been recently synthesized (during a primary infection) will be easily stripped off the antigen using mild protein denaturants. The assay is performed with and without such denaturants in the solutions, and a ratio is calculated. The ratio is compared to known controls and gives some indication of the time lapse since infection. This gives useful information, but may not be completely reliable. IgG_1 avidity can be useful,[17] as can the presence of specific IgG_3; the latter is always found in primary infection, but only occasionally in reinfections. This subject has been reviewed recently.[18]

F. PERSISTENT INFECTION

The association with arthritis, transient or persistent, has been the subject of a number of studies.[19,20] PCR may be appropriate in searching for viral persistence in such cases, as there is no diagnostic serological test for this condition.

II. SUGGESTED INDICATIONS FOR PCR IN RUBELLA INFECTION

1. Reinfection of mother is suspected.
2. Rubella appears in the second trimester. In this situation, the fetus has a lower chance of being infected than in first-trimester cases, but there is still some risk of disease.
3. Rubella appears in the first trimester, and termination of pregnancy cannot be considered on ethical grounds. Most of these fetuses will be infected and have disease.
4. The mother had rubella before conception occurred. (It is known, however, that maternal disease after 12 days from the last menstrual period is predictive of fetal infection.)
5. The date of the last menstrual period before conception is unclear in relation to the timing of the rash in a patient found to have rubella IgM.
6. Serology in the mother is unclear.
7. Persistent infection is suspected.

A. MOLECULAR BIOLOGY OF RUBELLA VIRUS AND PRIMER CHOICE

The genome contains between 10,000 and 11,000 bp of positive-polarity single-stranded RNA. There are two mRNA molecules, the longer of which is genomic and codes for the nonstructural proteins. The 24S RNA, at the 3′ end of the genome, encodes the structural proteins in the form of a post-translationally cleaved polyprotein. Comparative sequence data are available from the structural proteins only and, in particular, the envelope glycoproteins, E1 and E2. E1 is reasonably conserved at the nucleotide level. There are too few data available on core sequences to design reliable primers.

E1 primers have been used reliably, and it seems that this is the region of choice with our present knowledge. If further information is generated on the nonstructural regions, better primers may become apparent. In fact, one group found that the palindromic sequence at the 3′ end of E1 made for very difficult amplification, and complicated procedures to allow priming were necessary.[19] Primers that have been used, with their positions on the genome, are given in Figure 9–1. Other areas of the genome would make good primer sites, but have not yet been employed. Multiple alignment of the available sequences should allow design of suitable primers, either from completely conserved regions or from conserved sequence islands flanking variable regions, perhaps those that are antigenically important. Sequencing of PCR amplicons from such variable regions would allow molecular epidemiological comparison of rubella virus strains.

A[a]	773	CAA GCG AGT AAG CCA GCG AG	754
B[a]	650	ATG GCA CAC ACA CCA CTG CT	669
Ru1[b] 1	888	GCT TTC ACC TAC CTC TGC ACT GCA	1911
Ru2[b]	1990	TGC TTT GCC CCA TGG GAC CTC GAG	2013
Ru3[b]	2310	GGC GAA CAC GCT CAT CAC GGT	2290
Ru4[b]	2478	GTC GGG CGG GAC CTG GAC CTC GAG	2455
Sense[c]	840	ATG GGC GAG GAG GCT TTC 857	
Antisense[c]		TAG TAC AAG CAT TTG GCA C	
R1[d]	1073	AAC TTC AGC CCC AAG GGG CC	1054
R2[d]	651	CAA CAC GCC GCA CGG ACA AC	670

Figure 9–1   Primers used for amplification of rubella RNA. The numbers are the locations on the E1 protein sequence (start and finish) of each primer. a: Carman et al.;[24] b: Eggerding et al.,[23] c: Cusi et al.;[21] and d: Ho-Terry et al.[22]

B. TISSUE PREPARATION

Three groups, given below as methods 1,[13,20] 2,[21] and 3,[22] have published protocols for the preparation of either tissue culture RNA or tissue RNA.

1. Method 1

Tissue is solubilized in 4 M guanidinium isothiocyanate at 60°C and extracted with phenol-chloroform and chloroform at 60°C. After precipitation with ethanol, the pellet is digested with proteinase K (200 µg/ml) in 0.1 M Tris-HCl, pH 7.4, 50 mM NaCl, 10 mM EDTA, and 0.2% sodium dodecyl sulfate (SDS) at 37°C for 30 to 60 min. Following extraction with phenol-chloroform and chloroform at 60°C, the nucleic acids are precipitated with ethanol, washed twice with 75% ethanol to ensure complete removal of SDS, and resuspended in 300 µl of 10 mM Tris-HCl, 1 mM EDTA, pH 8.0 (TE), containing 100 U/ml of RNasin.

The RNA is purified by selective elution from an anionic exchange resin immobilized in a micropipette tip (Diagen). A Diagen tip 20 is equilibrated by washing with 400 mM NaCl, 50 mM 3 MOPS, 1% ethanol, pH 7.0, RNA in 300 µl of TE with RNasin is adjusted to 250 mM NaCl and 35 mM MOPS, pH 7.0, and is applied to the resin. After extensive washing of the resin with equilibration buffer to remove proteins, nucleotides, and other impurities, RNA is eluted in 1.05 M NaCl, 50 mM MOPS, 2 M urea, 15% ethanol, pH 7.0. DNA remains bound to the resin and RNA is precipitated with isopropanol at −20°C. After washing twice with 75% ethanol, the RNA is taken up in sterile water at a rate of 5 µl per 10 mg tissue. Recovery of RNA from Diagen tips is approximately 70 to 80%. The final concentration of RNA obtained is 2 to 5 µg/µl.

2. Method 2

This method was applied to tissue culture samples. Cells are harvested by centrifugation, and cytoplasmic extracts are prepared by resuspending the washed cells in lysis buffer (150 mM NaCl, 10 mM Tris-HCl, pH 7.4, 1.5 mM MgCl$_2$, 0.2% Nonidet P-40) containing 1% diethylpyrocarbonate. Nuclei are removed by sedimentation at 800 g for 5 min, and cytoplasmic RNA is extracted from the cell lysates after centrifugation. RNA is extracted twice with phenol-chloroform, precipitated as usual and collected by centrifugation, washed with 70% ethanol, and dissolved in water at 1 mg/ml. RNA can also be analyzed directly from lysed infected cell extracts. Infected cells are suspended in 500 µl of water containing 100 µg/ml proteinase K, incubated for 1 h at 30°C, and then heated to 100°C for 5 min.

3. Method 3

Infected cells are lysed by repeated freeze thawing (3×) and clarified by centrifugation at 10,000 g for 10 min. The virus is concentrated from lysed cells and the tissue culture medium by centrifugation at 50,000 g for 3.5 h. RNA is purified by phenol-chloroform extraction.

To test the need for RNA extraction, 8 µl of concentrated virus preparation is added to an equal volume of RSB lysis buffer (10 mM Tris-HCl, pH 7.5, 10 mM NaCl, 5 mM MgCl$_2$ containing 2% [v/v] Triton X-100). Samples are left on ice for 5 to 10 min prior to reverse transcription and amplification by protocol 2, described below.

To assess the need for high-speed centrifugation, 8 µl of a cell suspension was added to 8 µl RSB lysis buffer. After 5 to 10 min on ice, the cellular debris was pelleted by centrifugation for 1 min in a microfuge. The supernatant was collected, reverse transcribed, and amplified as described in protocol 2.

C. REVERSE TRANSCRIPTION (RT) AND PCR
1. Method 1

Extracted RNA (2 µl) is denatured in 10 µl of 10 mM Hepes, pH 6.9, 0.1 mM EDTA, and 5 µM hexanucleotide random primer at 90°C for 2 min. The sample is quenched in ice, and Moloney murine leukemia virus RT (20 U in 10 µl of 100 mM Tris-HCl, pH 7.5, 150 mM KCl, 6 mM MgCl$_2$, 20 mM DTT, 1 mM of each dNTP, and 20 U of RNasin) is added at room temperature; further incubation is done at 37°C for 90 min. The reaction mixture is extracted with phenol-chloroform and chloroform and is used directly as cDNA in the PCR without precipitation.

Every RT run includes (1) negative controls with no RNA or RNA extracted from normal placenta and (2) positive controls with RNA extracted from purified rubella virus or rubella-infected cells or RNA transcribed from a vector carrying a cloned rubella E1 cDNA insert.

The final conditions for PCR are 2.5 µl cDNA, 10 mM Tris-HCl, pH 8.3 (at 25°C), 50 mM KCl, 1.5 mM MgCl$_2$, 0.001% gelatin,* 25 pmol of each primer, and 1.25 U of Taq polymerase in a final volume of 50 µl. Initial denaturation is at 99°C for 3.5 min. Enzyme is added at 70°C. Amplification is usually 40 c of denaturation at 95°C for 15 s followed by annealing/extension at 70°C for 100 s. The final extension at 70°C is for 8 min.

2. Method 2

Total cytoplasmic RNA (1 µg) is reverse transcribed in the presence of the antisense oligonucleotide primer. RNA is heated to 95°C for 2 min and cooled on ice before reverse transcription is initiated. First-strand synthesis of cDNA is carried out in a total volume of 20 µl containing 50 mM KCl, 10 mM Tris-HCl, pH 8.3, 1.5 mM MgCl$_2$, 0.001% (w/v) gelatin,* 500 µM of each dNTP, 20 U of RNasin, 200 U of RT, and 20 pmol of antisense oligonucleotide primer. The reaction mixture is incubated at 42°C for 1 h and then heated for 5 min at 95°C. PCR amplification can be carried out on the entire RT reaction mixture. PCR is performed in 100 µl containing 50 mM KCl, 10 mM Tris-HCl, pH 8.3, 1.5 mM MgCl$_2$, 0.001% (w/v) gelatin, 200 µM dNTPs, primers at 1 µM, and 2.5 U of Taq polymerase. Cycling is performed at 95°C for 90 s, 55°C for 90 s, and 72°C for 120 s for a total of 35 c; 1 µl of the product of this reaction is nested using standard protocols.

3. Method 3

Three different methods for optimum reverse transcription of purified RNA and subsequent amplification of cDNA were directly compared. In the first protocol the RNA was reverse transcribed in a standard RT solution (40 µl) containing 50 mM Tris-HCl, pH 8.3, 10 mM MgCl$_2$, 60 pmol sense primer, 3.75% (v/v) dimethylsulfoxide (DMSO), 2 U RNasin, 500 µM of each dNTP, and 10 U of avian myeloblastosis virus (AMV) RT. After 1 to 3 h incubation at 42°C, 40 pmol sense primer and 100 pmol antisense primer are added, and the solution is adjusted to a concentration of 16.6 mM ammonium sulfate.

In the second protocol, RNA was reverse transcribed under PCR conditions (67 mM Tris, pH 8.8, 6.7 mM MgCl$_2$, 16.6 mM ammonium sulfate, 10 mM 2-ME, 60 pmol sense primer, dNTPs), 3.75% DMSO and RT; KCl was added (range of 44 to 95 mM was assessed). After 2 to 3 h incubation at 42°C, 40 pmol sense and 100 pmol antisense primers were added as well as BSA, dNTP, and Taq in a final volume of 100 µl.

The third protocol was the same as the second except DMSO was excluded. Forty cycles were performed; samples were incubated at 95°C for 1 min, 60°C for 1 min, and 70°C for 1 min. For the first cycle only, the samples were incubated at 95°C for 5 min.

D. RESULTS USING DESCRIBED PROTOCOLS

Table 9–1 summarizes the analysis of a number of samples taken from first-trimester rubella-infected mothers tested by method 1. Some of the samples are prenatal (chorionic villus, amniotic fluid) and others are from aborted fetuses and placentas. It can be seen that there was about 75% concordance between cell culture (as detected by PCR after culture) and direct PCR. Thus, direct PCR would have picked up an extra 25% of cases. There were no false negatives by PCR. Note that although numbers were small, no fetal infection was found in the asymptomatic infection, reinfection, and preconception samples.

Using method 3, no amplified DNA was detected by protocol 1. However, when samples were reverse transcribed in a standard amplification solution supplemented with KCl, positive results were obtained (protocol 2). Final KCl concentrations from 44 to 96 mM give similar results. The exclusion of DMSO from the reverse transcription reaction (protocol 3) decreased the yield of amplified product.

A DNA fragment was also produced after reverse transcription and DNA amplification of crude cell lysates. However, the ability to detect rubella RNA in crude cell lysates appeared to decrease with an increase in cell concentration; although negative results were obtained when amplifications were performed on nucleic acid from 10^4 cells, a 1/10 dilution of this sample resulted in a positive result. The results were most reliable when free virus was used rather than infected cells. This indicates that DNA amplification is not impaired by the residual RSB, Triton X-100, or KCl from the lysis and reverse transcription steps.

---

* Gelatin may be omitted.

Table 9–1. **Detection of rubella-specific nucleic acid sequences by polymerase chain reaction (PCR)**

Patient Category	Nos.	Types of Infection	No. of Specimens Examined per Patient	Direct PCR Positive	Negative	PCR after Co-cultivation Positive	Negative
a A	1	Rubella	3	0	3	0	3
	2		5	0	5	0	5
	3–4		1	0	1	0	1
B	5–12	Rubella	1	1	0	1	0
C	3	Rubella	3	1	2	3	0
	14		2	2	0	1	1
	16		2	1	1	2	0
D	17	Rubella	11	1	10	0	11
	18–20		1	1	0	0	1
b E	1–4	Infection not related to rubella	1	0	1	0	1
F	5–6	Second-trimester rubella infection	1	0	1	0	1
G	7–9	Reinfection	1	0	1	0	1
	10		8	0	8	0	8
H	11	Asymptomatic infection	3	0	3	0	3
	12	Contact	4	0	4	0	4
	13	Preconception	1	0	1	0	1
	14	Vaccination	1	0	1	0	1

Note: The specimens examined include chorionic villus samples, cells concentrated from amniotic fluid, aborted fetal organs, and aborted placentas. a: Patients with serologically confirmed first-trimester rubella, where testing for rubella RNA sequences by direct PCR and by PCR after co-cultivation showed (A) concordant negative results: (B) concordant positive results: (C) concordant positive results with respect to patients but not in all specimens: (D) discordant results, positive by direct PCR but negative by PCR after co-cultivation. b: Patients (E) with viral infections other than rubella or (F, G, and H) with rubella infection but not during the first trimester of pregnancy.

From Ho, L. and Terry, G. M., in *Diagnosis of Human Viruses by Polymerase Chain Reaction Technology,* Becker, Y. and Darai, G., Eds., Springer-Verlag, Berlin, 1992, 231. With permission.

E. SENSITIVITY

An assessment of sensitivity was not made with method 1. Method 2 was thought to be able to detect a single RNA molecule (as measured by dilution experiments). These authors found that not all their products amplified with the same efficiency. For example, the 591-bp target was variably identified after a single round, whereas the 321-bp fragment was regularly identifiable. However, nesting the 591-bp product to generate the 321 bp resolved this problem. Southern blot hybridization with an internal probe or nesting and staining with ethidium bromide each detected rubella in 5 fg of infected cell RNA. Method 3 detected 2 ng of rubella RNA by ethidium bromide staining of a single round of PCR.

III. SUMMARY

PCR for the diagnosis of rubella seems to be a useful procedure. However, problems arise in interpretation. It is well known that a first-trimester maternal primary infection has a high chance of harming the fetus, and therapeutic abortion is usually the preferred approach. Even if the fetus is shown to be PCR negative, one cannot be sure that the fetus is uninfected, and few clinicians and, probably, mothers would be happy to let the pregnancy go to term. On the other hand, if fetal tissues from a reinfection or a preconception maternal infection can be shown to be PCR negative, there is a much higher chance that the baby will be normal, and it would be consoling to the mother to have confirmation of this early in the pregnancy.

REFERENCES

1. Rubella prevention: recommendations of the Immunization Practices Advisory Committee (ACIP), *Morbid. Mortal. Week. Rep.,* 39/RR-15, 1, 1990.
2. **Miller, E., Waight, P. A., Vurdien, J. E., White, J. M., Jones, G., Miller, B. H. R., Tookey, P. A., and Peckham, C. S.,** Rubella surveillance to December 1990: a joint report from the PHLS and National Congenital Rubella Surveillance Programme, *Commun. Dis. Rep.,* 1, R33, 1991.
3. **Miller, E., Cradock-Watson, J. E., and Pollock, T. M.,** Consequences of confirmed maternal rubella at successive stages of pregnancy, *Lancet,* 8032, 781, 1982.
4. **Enders, G., Nickerl-Pacher, U., Miller, et al.,** Outcome of confirmed pericoconceptional maternal rubella, *Lancet,* 1, 1445–1447, 1988.
5. **Cradock-Watson, J. E., Ridehalgh, K. S., Anderson, M. J., and Pattison, J. R.,** *J. Hyg. Cambridge,* 87, 147, 1981.
6. **Daffos, F., Forestier, F., Grangeot-Keros, L., Capella Pavlovsky, M., Lebon, P., Chartier, M., and Pillot, J.,** Prenatal diagnosis of congenital rubella, *Lancet,* 2, 8393, 1, 1984.
7. **Tookey, P. A., Jones, G., Miller, B. H. R., and Peckham, C. S.,** Rubella vaccination in pregnancy, *Commun. Dis. Rep.,* 1, R86, 1991.
8. **Cradock-Watson, J. E.,** Laboratory diagnosis of rubella: past, present and future, *Epidemiol. Infect.,* 107, 1, 1991.
9. **Enders, G. and Jonatha, W.,** Prenatal serological diagnosis of intrauterine cytomegalovirus infection, *Infection,* 15, 162, 1987.
10. **Skvorc-Ranko, R., Lavoie, H., St.-Denis, P., Villeneuve, R., Gagnon, M., Chicoine, R., Boucher, M., Guimond, J., and Dontigny, Y.,** Intrauterine diagnosis of cytomegalovirus and rubella infections by amniocentesis, *Can. Med. Assoc. J.,* 145, 649, 1991.
11. **Levin, M. J., Oxman, M. N., Moore, M. G., et al.,** Diagnosis of congenital rubella *in utero, N. Engl. J. Med.,* 290, 1187, 1974.
12. **Ho-Terry, L., Terry, G. M., Londesborough, P., Rees, K. R., Wielaard, F., and Denissen, A.,** Diagnosis of fetal rubella infection by nucleic acid hybridization, *J. Med. Virol.,* 24, 175, 1988.
13. **Ho, L. and Terry, G. M.,** Diagnosis of prenatal rubella by polymerase chain reaction, in *Diagnosis of Human Viruses by Polymerase Chain Reaction Technology,* Becker, Y. and Darai, G., Eds., Springer-Verlag, Berlin, 1992, 231.
14. **Morgan-Capner, P., Hodgson, J., Hambling, M. H., Dulake, C., Coleman, T. J., Boswell, P. A., Watkins, R. P., Booth, J., Stern, H., Best, J. M., and Banatvala, J. E.,** Detection of rubella-specific IgM in subclinical rubella reinfection in pregnancy, *Lancet,* 1, 244, 1985.
15. **Morgan-Capner, P., Miller, E., Vurdien, J. E., and Ramsay, M. E. B.,** Outcome of pregnancy after maternal reinfection with rubella, *Commun. Dis. Rep.,* 1, R57, 1991.
16. **Hedman, L. and Seppala, I.,** Recent rubella virus infection indicated by low avidity of specific IgG, *J. Clin. Immunol.,* 8, 214, 1988.
17. **Thomas, H. I. J. and Morgan-Capner, P.,** Rubella-specific IgG subclass avidity ELISA and its role in the differentiation between primary rubella and rubella reinfection, *Epidemiol. Infect.,* 101, 591, 1989.
18. **Thomas, H. I. J. and Morgan-Capner, P.,** The use of antibody avidity measurement for the diagnosis of rubella, *Rev. Med. Virol.,* 1, 41, 1991.
19. **Chantler, J. K., Ford, D. K., and Tingle, A. J.,** Persistent rubella infection and rubella-associated arthritis, *Lancet,* 1, 1323, 1982.
20. **Spruance, S. L., Metcalf, R., Smith, C. B., Griffiths, M. M., Wood, J. R.,** Chronic arthropathy associated with rubella vaccination, *Arthritis Rheum.,* 20, 741, 1977.
21. **Cusi, M. G., Cioé, L., and Rovera, G.,** PCR amplification of GC-rich templates containing palindromic sequences using initial alkali denaturation, *BioTechniques,* 12, 502, 1992.
22. **Ho-Terry, L., Terry, G. M., and Londesborough, P.,** Diagnosis of foetal rubella virus infection by polymerase chain reaction, *J. Gen. Virol.,* 71, 1607, 1990.
23. **Eggerding, F. A., Peters, J., Lee, R. K., and Inderlied, C. B.,** Detection of rubella virus gene sequences by enzymatic amplification and direct sequencing of amplified DNA, *J. Clin. Microbiol.,* 29, 945, 1991.
24. **Carman, W. F., Williamson, C., Cunliffe, B. A., and Kidd, A. H.,** Reverse transcription and subsequent DNA amplification of rubella virus RNA, *J. Virol. Meth.,* 25, 21, 1989.

Chapter 10

PCR Technology for Lyssavirus Diagnosis

Noël Tordo, Hervé Bourhy, and Débora Sacramento

TABLE OF CONTENTS

I. INTRODUCTION TO RABIES DISEASE

A. HISTORY

Rabies disease was first mentioned in the Eshnunna code, 23rd Century BC. Up to late in the 19th century, the treatment remained basically irrational, reflecting the terror provoked by the disease itself: hot or cold baths, salt or red-hot iron application to wounds, and magic potions.[1] The scientific approach only began with Pierre-Victor Galtier (1879) and subsequently Louis Pasteur (1880). Pasteur serially passaged in rabbit brains a cow isolate from the suburbs of Paris, and obtained a fixed strain with high virulence and a constant short incubation period. Dessiccation of the rabbit spinal cords progressively attenuated their virulence. Serial inoculations of dogs with spinal cord extracts of increasing virulence rendered them refractory to rabies. In 1885, this method of rabies prophylaxis was successfully used on Joseph Meister, who had been severely bitten several days earlier by a rabid dog.[2–4] One century later, the control of rabies

0-8493-4833-1/95/$0.00+$.50
© 1995 by CRC Press Inc.

remains one of the priorities of the WHO, notably in the developing world, and its eradication a challenge for the future.

B. TAXONOMY

Viruses responsible for rabies encephalitis are classified in the *Lyssavirus* genus of the *Rhabdoviridae* family.[5] The genus was first divided into four serotypes on the basis of serological relationships and antigenic reactivity with monoclonal antibodies.[6-9] This grouping was recently confirmed and extended by sequence comparison of the nucleoprotein genes.[10] Six genotypes were characterized (Figure 10–1): the first four matched the corresponding serotypes; while the last two delineated sub-types of the European Bat Lyssaviruses (EBL), which were previously unclassified. The prototype viruses are: (1) rabies; (2) Lagos bat; (3) Mokola; (4) Duvenhage; (5) EBL1; and (6) EBL2. Members of genotype 1 are rabies viruses. Members of genotypes 2 to 6 are rabies-related viruses. Almost all the genotypes are known to cause human disease. The term sero-genotype 1 to 6 will be used in this review.

C. MOLECULAR STRUCTURE AND REPLICATION MECHANISM

The lyssavirus is bullet shaped (diameter: 75 nm; length: 100 to 300 nm). It consists of an internal helically coiled ribonucleocapsid surrounded by a lipid envelope of cellular origin.[11] The ribonucleocapsid comprises the RNA genome, which is strongly associated with the N nucleoprotein, and two other viral proteins involved in replication: the M1 phosphoprotein and the L polymerase. Two viral proteins are associated with the envelope: the inner M2 membrane protein, and the transmembrane G glycoprotein. A recent study on vesicular stomatitis virus (VSV) has re-evaluated the position of the M2 protein — it is more likely to be internal of the ribonucleocapsid coil.[12] The only external viral peptide is the ectodomain of the G protein, which is presumed to bind the viral receptor on the host cell. When the virus penetrates the cell by pinocytosis its membrane fuses with that of the lysosomal vacuole. This liberates into the cytoplasm, the ribonucleocapsid, which is capable of transcription — the encapsidated RNA genome is the template for the viral polymerase.

PROTOTYPE/SEROTYPE/		GENOTYPE/	GEOGRAPHIC DISTRIBUTION/	ANIMAL SPECIES
Rabies	1	1	world except... Australia, New Zealand Antartica, Japan, Hawai Great-Britain, Scandinavia	Man <u>carnivore</u>, cattle <u>bat (insectivorous, frugivorous, vampire)</u>
Lagos bat	2	2	Nigeria, Guinea, Senegal Ethiopia, Central Afr. Rep. South Afr. Rep., Zimbabwe	<u>frugivorous bat</u> dog, cat
Mokola	3	3	Nigeria, Cameroon Ethiopia, Centrafr. Rep. Zimbabwe	Man shrew, rodent dog, cat
Duvenhage	4	4	South Afr. Rep., Zimbabwe	Man <u>insectivorous bat</u>
EBL1 European bat lyssavirus 1	?	5	Europe	Man <u>insectivorous bat (Eptesicus, Pipistrellus)</u>
EBL2 European bat lyssavirus 2	?	6	Europe	Man <u>insectivorous bat (Myotis)</u>

Figure 10–1 Taxonomy of the *Lyssavirus* genus using antigenic and genetic criteria. The animal species serving as principal reservoirs or vectors are underlined.

The lyssavirus genome is an unsegmented negative-stranded RNA (Figure 10–2). It has been cloned and sequenced and shown to code for a 3′ short leader RNA, then the N, M1, M2, G, L proteins.[13–17] The messenger RNAs are capped and polyadenylated, and are not encapsidated. The genes are separated by variable non-transcribed intergenic regions which are generally short, except for the G-L intergene — the Ψ pseudogene. During transcription, the messenger RNAs are transcribed in a cascade with a decreasing rate from the 3′ to the 5′ termini.[11] After protein synthesis, the (–) genome is replicated into a (+) antigenome, which serves in turn to amplify the (–) genomes for the progeny virions. The full length (+) and (–) strands are always found in ribonucleocapsid structures. Thus, negative strand replication is combined with the encaspidation of the RNA.

D. EPIDEMIOLOGY

Rabies virus is maintained and transmitted by several animal species which serve as reservoirs as well as vectors.[18] These animals should be distinguished from other species which, although susceptible to infection, constitute epidemiological cul-de-sacs (man for example). The combination of the vector and virus results from adaptation at the molecular, physiological, and behavioral levels. This equilibrium can be modified by natural or artificial selection pressure, resulting in the (dis)appearance of vector species. For example, the systematic elimination of stray dogs during the first half of this century led to the elimination of canine rabies in Europe. However, rabies spread among the fox population. More recently, the virus adapted to the Eastern European racoon dog, a feral species originally imported for its fur from Asia into Eastern Europe.

Rabies is present world-wide except in Oceania, Antarctica, and several islands such as U.K., Japan, and some Antilles islands (Figure 10–1). It exists as two major epidemiological forms: (1) urban rabies with the stray dog as reservoir and vector; and (2) sylvatic rabies involving wildlife. Urban rabies has presently disappeared from developed countries but remains a serious health problem in developing ones. It is responsible for about 90% of the estimated 35,000 human deaths, due to rabies, per year in the world.[19,20] Sylvatic rabies occurs world-wide.

Two main groups of wild vector species can be distinguished: terrestrial mammals and chiropters. Terrestrial vectors in Europe are the red fox, the raccoon dog (Eastern), and the wolf (Yugoslavia). In North America the vectors are the red fox (North-East), the arctic fox (far North), the raccoon (East), the skunk (Central and Western), and the mongoose (Carribean Islands). In Asia they are the arctic fox (Siberia), the red fox (Middle-East), and the wolf (Afghanistan, Iran, and Iraq); while in Africa they include the mongoose (South), the bat-eared fox, and the jackal. The chiropteran vectors are either hematophagous (vampire-bats from the North of Mexico to the North of Argentina), or frugivorous (Sub-Saharan Africa), or insectivorous (America, Europe, and South-Africa).

It is interesting to note that sero-genotypes can have a very sharp (sero-genotypes 2, 5, 6) or diverse (sero-genotype 1) vector range (see Figure 10–1). In addition, it is curious that the bat viruses from America (vampires or insectivorous) belong to sero-genotype 1. In contrast, bat viruses from Europe (insectivorous) and Africa (frugivorous or insectivorous) are exclusively rabies-related viruses of sero-genotypes 2, 4, 5, 6.

E. PATHOLOGY

Rabies is a lethal disease affecting the central nervous system (CNS) of mammals, and a zoonotic one that can be transmitted to man.[21,22] The virus is usually transmitted by a bite or by wound licking. However, direct penetration of the virus through mucosal surfaces is also possible: aerosols and saliva constitute an important mode of transmission for bats in caves. Depending on the infection site, the virus can directly enter the sensory and motor nerve endings, or enter after persistence or growth in epithelial or muscular cells.[23–26] The viral particles are transported by retrograde axonal flow and replicate in the neuronal perikaryon. After a peripheral infection, the CNS is sequentially infected: the brain stem first, then the thalamus and then the cortex. Late in infection, the virus returns to peripheral tissue by anterograde axoplasmic flow and reaches non-neural tissues: salivary and lachrymal glands; basal hair cells; buccal, nasal and intestinal mucosa; myocardium; kidney; pancreas; lungs; and interscapular fat. These external sites of viral particle production differ slightly from one species to another. The mode of transmission of the virus is influenced by the particular site of replication.

The incubation period of the virus varies depending on the individual infected; on the animal species; on the site of infection; and on the virus itself (1 to 3 months in 85% of human cases). The symptomatic period is rapid (<1 week) and violent, and generally starts with ache and irritation at the site of the bite. By this stage the virus is already being excreted in the saliva. Various symptoms are then observed:

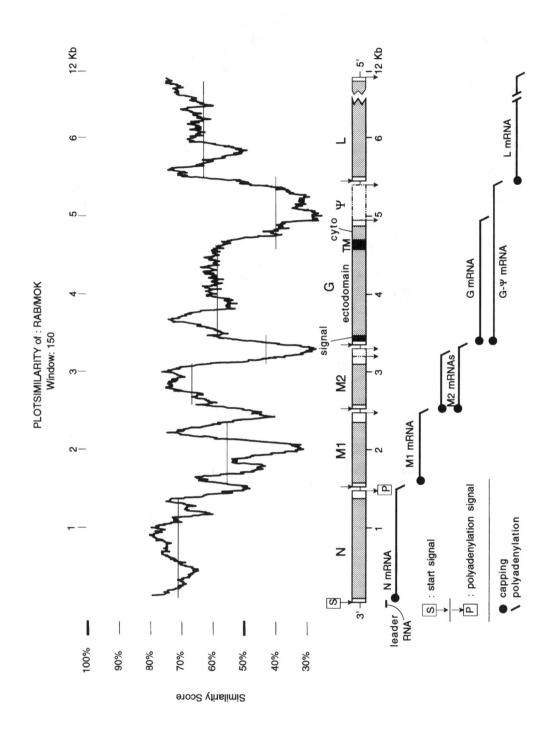

Figure 10–2 Similarity profile along the 6500 first 3′ nucleotides of the rabies (PV strain) and the Mokola (MOK5 isolate) genomes. The profile was calculated by comparing successive windows of 150 nucleotides. The different domains of the G glycoprotein (signal peptide, ectodomain, transmembrane domain, cytoplasmic domain) are shown. At the bottom of the Figure, the transcription step of the genome is indicated. The mRNAs are produced in cascade from the 3′ to the 5′ genomic direction. A phenomenon of alternative termination (or polyadenylation) is observed for the M2 and L cistrons, each of them being capable of producing both a short and a large mRNA.

hyperactivity, anxiety, and mental confusion. In animals, a feverish period generally precedes a paralytic one. In addition, the timber of the voice is frequently modified and behavioral disorders appear (loss of fear of human threat, aggressiveness) which help transmission of the disease. In humans, spasms provoked by visual, auditory, or tactile stimuli are the classical clinical form, although a paralytic form has also been reported.[27] The most typical human symptom is hydrophobia, a spasm of the larynx following sight, sound, or contact with water. Except for three reported cases, the disease is always fatal. Death follows a paralysis of the cardio-respiratory system. Nothing is known concerning the deep-seated mechanism of death. Rare histological perturbations are seen, as if the virus killed the organism, without killing the cell, by inducing neurone or brain functional alterations.[28] Negri bodies, dense cytoplasmic inclusions, are observed in neurones of the CNS in 60 to 70% of the cases. These are composed of numerous viral ribonucleocapsids. Interestingly, these histopathological modifications are slightly different for different lyssaviruses.

F. PROPHYLAXIS

The relatively few human cases in the developed world stands in sharp contrast with the increasing number in developing countries. The lower incidence in the industrialized world results from the disappearance of the urban reservoir following strict sanitary and medical prophylaxis: systematic elimination of stray dogs or potentially contaminated wild animals; obligatory vaccination of owned dogs. Oral vaccination of wild or stray animals with baits is the most suitable means of human protection.

For humans, in the absence of any suitable therapy (interferon, ribavirin, and immunoglobulin have been tried), post-exposure vaccination remains the only defense; extensive washing of wounds with soap is the first efficient prophylactic measure. Pre-exposure vaccination should be proposed for individuals professionally exposed to rabies (laboratory workers, veterinarians, wildlife agents, fire-fighters) and recommended for people in enzootic areas (hunters). In post-exposure treatment, the decision of whether or not to vaccinate is taken by the physician, who must consider several scenarios.[6] Briefly, vaccination is recommended when the biting animal is unknown. In this case, local and parenteral immunoglobulin treatment may complement vaccination. If the biting animal can be captured and killed, the diagnosis of rabies may be confirmed or excluded in the laboratory. In the case of a dog or cat, if it can be followed up and remains well after 10 days then post-exposure treatment can be discontinued.

G. IMMUNOLOGY AND VACCINES

The establishment of immunity against rabies virus is a complex phenomena involving numerous effectors:[29] helper and cytotoxic T lymphocytes, B lymphocytes, and circulating antibodies (Abs). Several antigenic sites (epitopes) have been located on the viral proteins. Because of its external position, the G protein ectodomain is the only antigen able to elicit the synthesis of neutralizing Abs which bind to the virion spikes. Consequently, the G protein is sufficient to protect against a virulent challenge performed intracerebrally. The internal antigens (ribonucleocapsid) only protect against a peripheral challenge. However, the central role of the N protein in the development of the helper T cell response, associated with a lesser genetic variability than the G protein, promotes it as the better candidate to broaden the protective spectrum of the vaccines beyond their original serotypes. The recent discovery of superantigen properties of the N protein may increase its vaccine potential.[30]

During the last century, the Pasteur Institute's vaccine as well as the vaccination schedules, have undergone continuous technical improvement for safety and efficiency.[31,32] Today live vaccines (virus attenuated by multiple passages; apathogenic antigenic mutants; recombinant pox- or adenoviruses expressing rabies antigens) are exclusively reserved for veterinary use. They are particularly suitable for oral vaccination of wild or stray animals using baits. Human vaccines consist of rabies virus inactivated by physical (UV; heat) or chemical (β-propiolactone) treatments. Virus production methods have progressively evolved in parallel with the criteria for their acceptability, notably concerning residual cellular DNA (<100 pg/dose).[33] The virus can be grown in adult nervous tissue (mostly sheep); suckling mouse brain; duck embryos; and in Vero and human diploid cell cultures. For financial reasons, numerous developing countries still use the more crude methods, despite frequent neurologic accidents (1/1000). The current development of less expensive cell culture technology (BHK-21), as well as the proposed modification of the criteria for vaccine acceptability should help to reduce these problems.[33,34]

In contrast to these technical developments, human and veterinary rabies vaccines are always produced from viral strains derived from isolates made more than 50 years ago, frequently from Pasteur's cow isolate itself.[32] All of them belong to the sero-genotype 1. It has been experimentally established that they provide absolutely no protection against a challenge with viruses from sero-genotypes two[35] and three.[36] With sero-genotype five,[37] it was shown that protection was dependent on the vaccine strain used. In summary, the current rabies vaccine is unable to protect against all lyssaviruses.

H. DIAGNOSIS

In the absence of constant typical symptoms, differential clinical diagnosis of rabies from other viral encephalitis is difficult. Only laboratory diagnosis can be certain. Two types of diagnosis are distinguished: ante- and post-mortem.

In animals, ante-mortem diagnosis mostly consists of a veterinary survey to see if the biting animal could have originated the human infection. Methods to detect virus in saliva (generally 1 to 5 days before the symptoms) are only now under development. In humans, ante-mortem diagnosis is performed on saliva, serum, cerebrospinal fluid (CSF), corneal impressions, and skin biopsies from the neck by virological or immunological techniques. However, ante-mortem diagnosis remains poorly sensitive with numerous false negative results.

Post-mortem laboratory diagnosis is systematically performed on the cortex, the hippocampus, and the bulb. There are a variety of laboratory methods. The choice must take into account: (1) the urgency of the anti-rabies treatment for exposed contacts; (2) the working conditions (field or laboratory); and (3) the objective (routine diagnosis, typing, expertise). The selection criteria include rapidity, specificity, sensitivity, detection spectrum, and cost.

During the last 3 decades, advances in knowledge of the virus, as well as technical progress have largely improved the diagnostic tools.[38-41] Histological detection of the cytoplasmic Negri bodies (Seller's technique), which is poorly specific, has been gradually replaced by virus isolation and/or immunological detection of viral ribonucleocapsid antigens. The former method requires the maintenance of virus infectivity, while the latter one is convenient for a damaged sample. Virus isolation by the suckling mouse inoculation test (MIT) is very sensitive and still frequently used in developing countries. However, the delay before obtaining the results (at least 7 days) is clearly too slow for routine diagnosis. The tissue culture inoculation test (RTCIT; 20 h) is preferred. Nowadays, three very efficient techniques are advised and routinely employed for rabies diagnosis:[38,41] (1) direct detection of viral nucleocapsid inclusions on brain smears with a fluorescent polyclonal antibody (FAT); (2) immuno-capture of viral ribonucleocapsids in brain extracts by an ELISA technique using a polyclonal antibody linked to peroxidase (rapid rabies enzyme immuno-diagnosis — RREID); and (3) viral isolation in neuroblastoma cell cultures (RTCIT) detected as in (1), by an immunofluorescent test.

The sensitivity of the three routine techniques is excellent with isolates from sero-genotype 1, but may vary from poor to inefficient with the other sero-genotypes, depending both on the sero-genotype itself and on the laboratory equipment. To solve this problem, the RREID technique has been recently improved:[42] the resulting RREID-lyssa test, uses pluri-specific polyclonal antibodies directed against the sero-genotypes 1, 3, and 5 and a streptavidin-biotin amplification method.

II. RABIES AND PCR

A. LITERATURE SURVEY

Since the first use of PCR to detect rabies RNA in 1990, at least ten other descriptions have been published. Two were concerned with diagnosis;[43,44] four with typing of the virus using restriction fragment length polymorphisms (RFLP);[43,45-47] four with molecular epidemiology, by correlating the genome variability with its geographical location, or with its host;[10,47-49] and one with the intrinsic variability of the isolate.[50] Two of the PCR tests were used taxonomically,[10,46] and PCR was also used to detect virus in experimental rabies infections.[26,51] Several reviews compare these different approaches.[32,52–54] It is of note that the above work mostly focused on nucleoprotein gene amplification, except for one study which targeted the glycoprotein gene,[50] and two which used the non-coding Ψ pseudogene as a target.[43,48]

B. EVOLUTION OF THE LYSSAVIRUS GENOME

The current knowledge of the *Lyssavirus* genus indicates that sero-genotypes 1 (rabies) and 3 (Mokola), are the more phylogenetically distant.[10,55] A pairwise comparison of their respective genomes is illustrative of the maximum genetic variability among lyssaviruses of medical interest[43,52] (Figure 10–2). The similarity profile shows the great variability in the non-protein coding regions, particularly the M2-G intergene and the G-L intergene (or Ψ pseudogene) regions, two areas with a strong evolutionary plasticity among unsegmented negative strand genomes (Order *Mononegavirales*).[56] The protein coding regions themselves vary to different degrees (Figure 10–2): the central core of the N protein, probably including the RNA genome binding site, is conserved among different lyssaviruses.[10] In contrast, the central part of the M1 protein is very divergent. The signal, transmembrane and cytoplasmic domains of the G protein are less conserved than the ectodomain. The L protein, although not totally sequenced, seems to be a concatenation of conserved catalytic domains joined by divergent areas, as previously noted among the polymerases of Mononegavirales.[57] It is predictable that peculiar L domains, notably those presumed to be responsible for RNA synthesis, would be more conserved than the N protein.[57–59] From the mean values, M1 is more variable than the G ectodomain, which is more variable than M2 then N when rabies and Mokola, i.e. distant sero-genotypes (1 and 3) are compared. In contrast, M2 is more variable than M1 between closely related strains (of same sero-genotype 1).[60] This suggests that the mutation rate of each protein is not constant but changes with evolutionary distance, a feature to keep in mind when defining the regions of interest for amplification.

C. PURPOSE OF EACH GENOME REGION

For routine diagnosis, the challenge is the detection of minute traces of infection by the broadest spectrum of lyssaviruses. This requires a sensitive and broadly reactive test. A genome target for PCR that meets this criterion should be highly expressed and conserved, at least in the primer binding sites. The latter criterion suggests the N and L genes as targets.[43,52] The need for sensitivity suggests that the 3′ encoded genes are suitable targets, since they are more intensively transcribed into mRNA, owing to the decreasing transcription rate in the 3′ to 5′ direction.[11] Therefore, it is logical that the N gene (specifically the conserved central region), which fulfills both criteria, should have been the target in all described diagnostic trials.[43,44] However, experimental data suggest that the messenger RNAs are more sensitive to nucleases than (–) and (+) full length RNA associated with nucleocapsids.[43] This reduces the disadvantage of targeting the L gene for diagnosis. Indeed, the extreme stability of several motifs within the L gene should help improve and widen the use of PCR for epidemiological purposes from the lyssavirus genus to other rhabdoviruses.[57–59]

Beside routine diagnosis, the characterization of the infecting virus is important for a reference laboratory. Here, the specificity of the PCR, and the ability to use it as a typing test are important. For example, the oral vaccination campaigns for wildlife rabies elimination, currently underway in North Western Europe, require the development of efficient tools to rapidly differentiate vaccine strains from autochthonous isolates, and to classify the latter. Two principal PCR options exist: either using primer pairs for a very specific amplification; or using ubiquitous or generic primers, and subsequently typing the amplicon. For the latter, three methods are conceivable: (1) molecular hybridization with specific probes; (2) RFLP analysis of the amplicon; and (3) definitive genome characterization by sequencing. The most suitable genomic region for typing depends both on the method used for identification, and on the evolutionary distance between the viruses to be identified. The large non-protein coding regions such as the Ψ pseudogene are susceptible to mutation free of external selective pressure, and are the most sensitive targets for differentiating closely related isolates (same sero-genotype, same or neighboring biotype) by RFLP or sequencing,[43,48] and for distinguishing distant viruses of different sero-genotypes (rabies and Mokola) by hybridization.[43] The conserved protein regions are also convenient for typing viral genomes by RFLP or sequencing.[10,45–47,49] In summary, the lyssavirus genome appears as a microscope for evolution with multiple magnifications; one should always choose that best adapted to the object observed.

For taxonomic purposes, the N gene is the best candidate for a comparative evaluation of molecular tests, because it has been characterized as an antigen in taxonomic studies.[7–9,61] The G ectodomain is also interesting since numerous isolates have been classified through their reactivity pattern with anti-G MAbs. Obviously, comparative study of the sequence of antigenic epitopes of both genes allows an understanding of the molecular basis of the antigenicity and cross-protection.[62,63]

D. RULES FOR CHOOSING PRIMERS

As for all RNA virus, PCR amplification of rabies genes requires two steps: (1) reverse transcription of RNA into cDNA; and (2) PCR amplification of the cDNA. The cDNA step may use either specific or random priming. The advantage of specific priming is that it allows selection of either the (+) or (−) RNA strand as a target, as well as the precise region transcribed into cDNA. The inconvenience is that subsequent PCR primers have to be in genomic regions to located 3′ side from the cDNA primer. Random priming avoids this inconvenience, but is not specific as it allows transcription of any viral RNA strand as well as non-viral RNAs that may be a majority population in the sample. For sequence purposes it is advisable to focus on a single RNA strand (i.e., to use specific priming) because there may be sequence differences during replication between (+) and (−) strands, such as composition (mismatches) and length (editing, polyadenylation). From this perspective, the unique (−) strand genome is a better target than the (+) strand molecules, which have two possible origins, either from transcription (leader and messenger RNAs), or from replication (antigenome).[11] Specific priming on the (−) strand genome target was used in the original rabies PCR method, which was developed to allow the amplified segment to serve successively for diagnosis, typing (RFLP), and molecular epidemiology (sequencing).[43] It is presently being compared with random priming for diagnosis use. Preliminary results suggest that the latter is more sensitive.

For primer design, the tendency has been to select those capable of amplifying the greatest number of lyssaviruses. These have been located in very stable sequence regions. The knowledge of the stable regions of the genome has evolved in parallel with growing understanding of the lyssavirus genome. This has come from comparisons between vaccine strains from the sero-genotype 1,[17,60] between sero-genotypes 1 and 3,[43,55] and between all sero-genotypes.[10] Today, the continuing increase of new sequences of wild isolates of sero-genotype 1 allows refinement of primer sets by taking into account mismatches present in certain groups of isolates.[10,47–50] This is a first step toward the design of type-specific primer sets. Up to now, however, published typing methods have described RFLP and sequence analysis of a common amplified fragment, obtained from generic primers.

When working with unknown wild isolates it is always possible, no matter how great the global conservation of the primer target, that local mismatches occur. To reduce the risk, particular attention should be paid to the 3′ end of the primer. If located in a coding region, the 3′ end should be an invariable residue of a strongly conserved codon (avoiding the wobble position). Besides the tryptophan codon, which is highly conserved, it has been noted that glycine and basic amino acids are strongly conserved in proteins with catalytic activities, such the L polymerases.[57] Cysteines are very stable in structural proteins like the G glycoprotein.[55] In the non-coding regions, the start and stop transcription signals flanking each gene are functionally conserved,[11] but almost identical and therefore ill-suited for differentiating between different genes.

E. THE DIFFERENT PRIMER SETS

The above rules have been partially observed in the design of characterized primer sets (their exact structure is available in the articles mentioned; see Figure 10–3). To amplify the N gene, the upstream primer mostly encompassed the N transcription start signal and the initiator codon.[10,44–47,49] The downstream one either flanked the stop codon to amplify all the coding region,[44,45,47,49] or mapped in a conserved region of the M1 gene to allow study of the non-coding region of the N gene.[10,46] For diagnosis, three proposed primer sets matched internal areas of the N protein and produced small fragments (330 to 520 bp) promising a high amplification rate, even with degraded samples.[43,44,51] The specific advantage of the N_1-N_2 set is that it flanks the more conserved region of the N gene (position 600 to 1200 on Figure 10–1) and was designed from the Mokola sequence (even though this virus is rare).[43] This choice of template-primer heterogeneity allows amplification of variable isolates. However, the N_1-N_2 primer set detects sero-genotype 3, but not sero-genotypes 5 and 6.[43] Within sero-genotype 1, this set is more efficient for European isolates. The N_7-N_8(M1) primer set (1500 residues) is efficient for sero-genotypes 1, 4, 5, and 6, but inefficient for sero-genotype 3.[46]

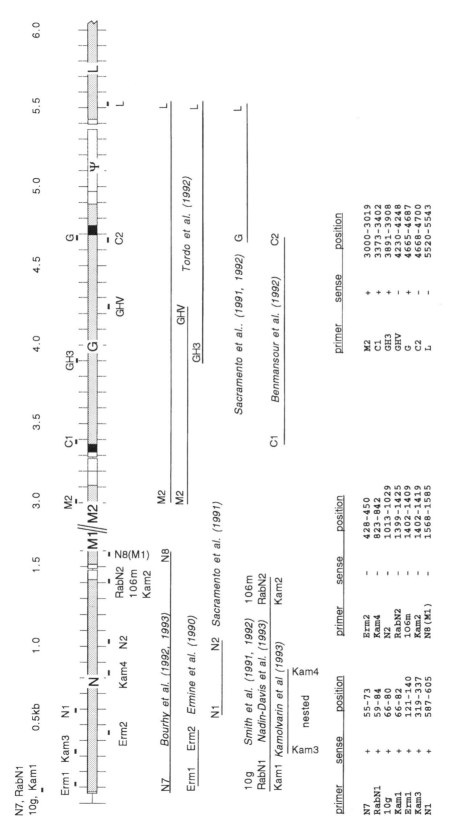

Figure 10–3 Summary of the different primer sets that have been tried for amplifying lyssavirus genes by PCR. The primer location is shown at the top of the Figure and is precise at the bottom, by reference to the genome sequence of the PV vaccine strain of rabies virus.[13-15] The middle of the Figure shows a schema of the amplicons of different authors and the respective references.

The amplification of the G glycoprotein and, above all, the Ψ pseudogene regions is more complex due to their greater variability. A primer set was designed with the aim of focusing on the coding region of the glycoprotein of the sero-genotype 1.[50] Interestingly, two M2 and L primers were found in very conserved domains of the flanking M2 and L genes.[43,52] Although both primers are efficient on most lyssavirus tested, their use in the same set may be difficult because they are far apart (2500 residues). An alternative involves two smaller overlapping amplifications in M2-GHV and GH3-L with intermediate primers (unpublished results). A suitable G primer was found within the conserved ectodomain of the glycoprotein, to specifically focus on the highly divergent genomic area encompassing the transmembrane and cytoplasmic domains of the G protein, and the consecutive Ψ pseudogene.[43,48,52] Although only moderately similar between rabies and Mokola genomes (5 mismatches in 23 residues), the G primer terminates in five conserved nucleotides (a tryptophan codon and the two first residues of a glycine codon). Consequently, the G-L set is able to amplify isolates of sero-genotypes 1 and 3, despite the fact that no significant conservation is observed between the Ψ pseudogenes of any sero-genotype.[43] Beside the primer sets used for amplification, numerous others have been designed for sequence analysis of amplicons. The constraints of conservation at this level are generally less important.

III. PROTOCOLS

A. GENERAL REMARKS

The reader interested in experimental details of the procedures is referred to the original articles where the protocols have been extensively described. The following will only highlight the main options and particular experimental details.

In the progressive PCR method originally described by Sacramento,[43] particular attention was paid to simplifying and standardizing methods, and combining commercially available enzymes, kits, and apparatus, following the manufacturers recommendations. The goal was to develop standard methods for use in a classical diagnostic and/or reference laboratory. Because of the extreme sensitivity of PCR, it is crucial that special care be taken to avoid cross-contamination of samples and carry-over of amplified products, both which can result in false-positive results. Guidelines for avoiding false positive with PCR have been described elsewhere. In general, this can be minimized by careful laboratory technique and must be monitored by the use of negative controls (uninfected brain), as well as with reactions where the nucleic acid sample is replaced by water which serves to detect contamination of both specimens and solutions. All reagents and laboratory supplies used for extractions need to be devoid of RNase.

B. ANIMAL SAMPLE COLLECTION

In the majority of cases, the laboratory diagnosis of rabies is performed on brain specimens. Every effort should be made to limit the handling of samples, thus avoiding contamination between them. Simple techniques have been devised to minimize these problems, based on the collection of an internal brain sample by introduction of a disposable plastic pipette into the skull via the occipital foramen[64] or via the retro-orbital route.[65] These techniques are 100% similar in sensitivity with skull opening and are particularly useful for working in field conditions. Brain samples are kept deep frozen at −80°C in plastic tubes.

In many laboratories, the virus is passaged on suckling mouse brain or cell cultures either for diagnosis (MIT or RTCIT) or for amplification. It has been shown that less than 15 to 20 passages in either system do not modify the consensus sequences of the fragment amplified (unpublished results). However, no study has been performed to evaluate the intrinsic variability of the viral population (quasi-species). However, ten passages by peripheral route (intra-muscular) are sufficient to modify the consensus sequence.

C. RNA EXTRACTION

The RNA can be prepared from cell cultures washed with PBS, or from brain samples (approximately 2 g) crushed to homogeneity with a plastic pestle into extraction buffer (for example 1% SDS, 1% Nonidet P40, 1 mM EDTA pH 8.0, 50 mg/ml dextran sulfate). Then the RNA is extracted by phenol-chloroform[43,45] or by guanidinium thiocyanate,[47] or by a combination of both methods,[44] or by centrifugation on a continuous cesium trifluoroacetate gradient.[50] After precipitation and washing the RNA is resuspended in pyrolyzed water.

D. cDNA SYNTHESIS AND PCR AMPLIFICATION

For cDNA synthesis, total (or cytoplasmic) RNA (1 to 2 µg) from infected material is annealed with the primer (20 to 100 pmol) for 3 min at 65°C, chilled on ice, adjusted to 10 to 50 µl with cDNA buffer and incubated for 45 to 90 min at 37 to 42°C. The reverse transcription medium slightly varies according to different authors (KCl or $CaCl_2$; MMLV or AMV reverse transcriptase; with or without RNasin). However, 50 mM Tris-HCl pH 8.3, 75 mM KCl, 3 mM $MgCl_2$, 10 mM DDT, 1 mM each dNTP, 0.5 U/µl RNasin 20 U/µl MMLV reverse transcriptase can be considered as a starting standard reaction mixture.

The resulting RNA/cDNA hybrid is diluted 5-,[44,45,47] 20-[50] to 100-fold[43] in PCR buffer, which is also slightly variable from author to author: 10 mM Tris-HCl pH 8.3, 50 mM KCl, 2 mM $MgCl_2$, 0.2 mM each dNTP, 1 µM each primer, 0.02 U/µl Taq DNA polymerase are basically required in a final 100 µl volume; DMSO (10%), gelatin (0.01 to 0.1%), or Triton X-100 (0.1%) are added occasionally. The mixture is covered with mineral oil and amplification is conducted in a thermocycler.

The number and setting of the consecutive denaturation (D), annealing (A), and elongation (E) steps should be empirically adjusted taking into account: (1) the length and composition of the primer set; (2) the primer complementary with its matching region on the genome of the isolate studied; (3) the size of the amplified segment; and (4) the abundance of the target RNA. The temperatures, D (94°C) and E (72°C) are usually constant because they are optimal for DNA strand denaturation and Taq polymerase activity, respectively. In contrast, A may vary from 37°C (15mer primers) to 55°C (30mer primers) and is strongly influenced by the primer sequence. Considering the time length, D (0.5 to 1 min) and A (1 to 2 min) are less stringent, except that an excess D (94°C) may prematurely impair the Taq polymerase activity. In contrast, E should be progressively extended from about 1 min for 0.4 kb amplifications (primer set N_1-N_2) to 2.5 min for a 2.5 kb (primer set G-L). In parallel, the number of cycles should be extended from 30 to 40. Finally, pauses and control of the temperature modulation (ramping) may be inserted between the steps. The ultimate E step is generally performed for 10 min to complete the synthesis of all amplicons.

Often, the cDNA synthesis and the PCR amplification are performed as two different reactions. However, considering the similarity of the cDNA and PCR buffers, it is possible to find a consensus buffer where both steps take place successively in a single tube (unpublished results). This method could be helpful in routine diagnosis for two reasons: (1) because handling the reaction mixture is reduced; and (2) because it promises to be more sensitive, combining the RT/PCR amplification of (+) and (−) strands. In contrast, this method may not be optimal for research purposes: (1) a high concentration of $MgCl_2$ may impair the fidelity of the Taq polymerase; (2) sequence differences between (+) and (−) strands may exist.

E. METHODS FOR THE DETECTION OF AMPLICONS

Amplified products can be directly visualized after 1 to 3% agarose gel electrophoresis in TAE buffer containing ethidium bromide (10 mg/ml) under UV light (312 nm). They can be indirectly detected either after Southern blotting of the gel, or by direct dot blotting, using a DNA probe internal to the amplified segment. These probes are synthetic oligodeoxynucleotides, or are prepared from restriction digest of various lyssavirus DNA clones.[13–16] Radioactive labeling of the probe can be of random priming or be at the 5′ end (with T4 DNA kinase). Non-radioactive labeling with digoxigenin-dUTP can also be used. Conditions for electrophoresis, hybridization, washing, exposure, immunological detection of digoxigenin, enzymatic revelation with alkaline phosphatase etc., are available in the commercial kits or in general laboratory manuals.[66]

F. METHODS FOR ANALYSIS OF AMPLICONS: RESTRICTION FRAGMENT LENGTH POLYMORPHISM AND NUCLEOTIDE SEQUENCING

For typing lyssavirus isolates, aliquots of the amplified fragments (2 to 10 µl) are digested by a selected restriction enzyme panel and analyzed by restriction fragment length polymorphism (RFLP) by agarose gel electrophoresis in TAE buffer containing ethidium bromide (10 mg/ml).

Nucleotide sequencing of the PCR product was performed either after cloning of individual molecules (quasi-species)[47,50] or by direct sequencing of the amplified segment (consensus sequence).[10,43,47,49] This was by the dideoxy chain termination technique[67] using Taq[47,49] or T7[10,43,47,50] polymerases. A very simple method[43] has been developed: (1) run the PCR products on a 0.7 to 2% NuSieve GTG (FMC) agarose gel electrophoresis with ethidium bromide, to separate the amplified fragment from the (possible) non-specific products and the excess primers; (2) excise the amplified fragment from the gel; (3) melt 10 µl of gel (containing the purified fragment) and directly sequence it with a T7 kit under normal conditions, except that the sequence reactions are maintained at 37°C instead of room temperature, to avoid agarose polymerization; (4) concentrate sequence reactions to 1 to 2 µl by lyophilization; (5) load onto a denaturing sequencing gel (for example: 60 cm × 0.1 mm; gradient from 0.5 × TBE to 5% acrylamide-urea to 5 × TBE to 7% acrylamide-urea; electrophoresis for 5 h at 50 W); and (6) fix (10% ethanol, 10% acetic acid), heat-dry, and autoradiograph. In this way, the results are available for computer analysis three days after sample receipt. In addition, the piece of gel containing the amplified segment can be kept at −20°C for months.

IV. DISCUSSION

A. DIAGNOSTIC METHODS THROUGH DETECTION OF VIRAL (AMPLIFIED) TRANSCRIPTS

Rabies virus genome transcription and replication produce exclusively RNA molecules.[1] Experiments to show their presence among the total RNAs of the sample, either on dot or Northern blots are not, in their current form,[68] sensitive enough. Whether radio- (^{32}P) or non-radio-labeled (digoxigenin) DNA probes are used, only the more heavily infected samples are detected. However, following reverse transcription and cDNA amplification by PCR, rabies virus RNA can easily be detected.

At the Pasteur Institute of Paris, 100 suspect samples received in the National Reference Centre for rabies were checked in systematic blind trials by PCR (N_1-N_2 or N_7-N_8 (M1) primer sets)[43] and by the 3 routine techniques (FAT; RTCIT; RREID).[38,41] A 100% correlation of the results (22 positive samples) was obtained.[43] However, it was found that blotting of the PCR products and subsequent hybridization (dot or Southern) with either radio- (^{32}P) or non-radiolabeled (digoxigenin) probes was an obligatory step for reliable results. Indeed, direct observation of the amplified segment on agarose gel with ethidium bromide resulted in several false positives among highly degraded samples, due to non-specific bands co-migrating with the expected product amplicon. From the technical point of view, dot blots are preferable to Southern blots because of their simplicity and rapidity, and because they are as sensitive with non-radiolabeled (digoxigenin) probes as with radio-labeled (^{32}P) ones.[43,54] This allows detection of as little as 1 pg i.e., 10^5 molecules of amplified product. Indeed, in our experience, probing of Southern blots with a non-radiolabeled probe may not be sensitive enough. On the other hand, the digoxigenin-label is favored because it is more stable than the ^{32}P-labeled one (several months rather than 2 weeks).

During a collaborative work involving different teams in Bangkok (Thailand), 206 dog and 3 rabid human brain isolates were studied.[44] Over 96 dogs were positive with classical techniques (95 by FAT and MIT, 1 by MIT only), but only 36 (37.5%) were found positive by a single-step PCR (involving the whole N gene). However, all the 96 (100%) were found positive after a double-step PCR with nested primers (see Figure 10–3). The three rabid human brain samples were positive with either single- or double-step PCR. In contrast with the previous study,[43] very few false-positives were reported during a single-step PCR, and absolutely none after the nested PCR. In addition, it was shown that sample decomposition has clearly less effects on nested PCR than on FAT. The sensitivity of the technique was estimated to be about 8 pg of rabies-specific RNA.

Both of the above articles clearly promote PCR as an efficient alternative protocol for rabies diagnosis, although a more systematic use is required to improve its sensitivity and reproducibility. The unequal sensitivity of the single-step PCR in the two articles (100%[43] to 37.5%[44] of positive results) could be a result of differing kinetics of the two enzymatic amplification reactions. It could explain the greater number of false-positive samples in the former one.

B. PCR AND ROUTINE TECHNIQUES FOR RABIES DIAGNOSIS

In Thailand, where sample degradation is a real problem because of the climatic conditions, nested PCR appears both more sensitive and stable than FAT. Assuming that routine laboratory conditions for PCR become available there, this method should then become basic for diagnosis. In Paris, a greater sensitivity of PCR was not observed: no sample positive by PCR was found negative with classical techniques. Therefore, PCR must exhibit other advantages to be promoted from a sophisticated confirmatory tool to a reference method for routine use. The following discussion attempts to illustrate this:

- **Rapidity:** As a possible basis for medical decisions (see Introduction), a good rabies diagnostic method must be rapid. The delay of availability of results for the different methods are compared in

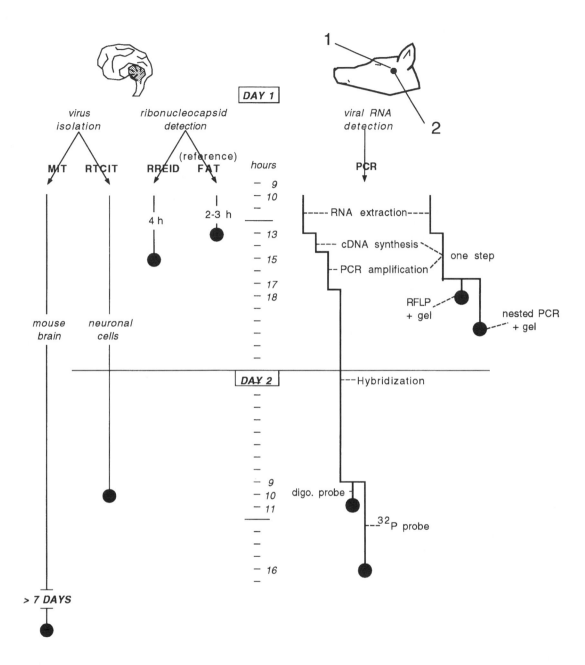

Figure 10–4 Comparative evaluation of the delay for the routine diagnosis techniques (FAT, Fluorescent Antibody Test; RREID, Rapid Rabies Immune-Enzyme Immuno-diagnosis; RTCIT, Rapid Tissue Culture Inoculation Test; MIT, Mouse Inoculation test) and the PCR method for diagnosis. Improvements of PCR method (single step for cDNA and PCR, RFLP, nested PCR) are suggested.

Figure 10–4.[38,41,52] The suspected samples are received and autopsied every morning before 10 a.m. The FAT is completed in 2 to 3 h, the RREID in 4 h. Their results are available early to middle afternoon. The RTCIT requires an overnight amplification on neuronal cell cultures and any positivity is revealed the next morning, by FAT. In comparison, the current PCR technique is slower: the internal brain collection is rapid, but total RNA is hardly available before early afternoon (after the extractions and precipitation); the cDNA synthesis (1 h 30 min) and the PCR (3 h) are only completed by late afternoon. Whatever the methods used next, the results are only available next morning, in synchrony with RTCIT (1) if a nested PCR (3 h) is started the agarose gel can only be run (30 min) the next morning; (2) if dot blotting (30 min) is performed; hybridization is left overnight, and the immuno-enzymatic revelation of digoxigenin-labeled probes is also only completed the next morning. Several ways of reducing this time handicap of PCR are currently under investigation in different laboratories: (1) trials of various nucleic acid extraction procedures; (2) combination of cDNA synthesis and PCR amplification in a single step resulting in an increased sensitivity since the N gene and the N messenger RNAs are simultaneously amplified; (3) direct characterization of the amplified products by restriction fragment length polymorphism (RFLP), in order to avoid the hybridization step. However, it is unlikely that PCR will soon become faster than the FAT technique, which will remain the reference method until PCR is able to surpass its sensitivity and speed.

- **Cost:** In 1988, a comparative estimation of the laboratory cost of the different diagnosis techniques was performed.[41] The routine cost range was FAT (1 unit monies (um)) cheaper than RTCIT (1.5 um) cheaper than RREID (5 um) cheaper than PCR (7 um) cheaper than MIT (20 um). The investment cost range was FAT (1 um) equal to RRIED (1 um) equal to PCR (1 um) cheaper than RTCIT (4 um) equal to MIT (4 um). This has already placed the PCR method in a competitive position, which can probably be bettered, as the price of *Taq* polymerase falls.

- **Polyvalence and specificity:** An important requirement of a diagnostic technique is its ability to detect all infecting Lyssaviruses. The parallel use of N_1-N_2 and N_7-N_8(M1) primer sets was shown to be efficient for rabies, Mokola and European Bat Lyssavirus isolates.[10,43,46] With a growing knowledge of Lyssavirus genetics, other primer sets polyvalent for the whole genus or specific to peculiar genotypes will be designed. This extension of the diagnostic spectrum is a specific advantage of PCR. Among the classical techniques, immuno-capture of viral nucleocapsids by ELISA has been recently extended to viruses divergent from serotype 1 (RRIED Lyssa).[42]

- **Ante-mortem diagnosis:** "Ante-mortem" diagnosis during clinical surveys or during the quarantine of biting animals is another important aspect of PCR. Here, other techniques, (FAT, RTCIT) are only poorly efficient and their results inconsistent. Work is in progress for the early detection of viruses in the secretion (saliva) of rabid animals. Improvements will be necessary before this method is established as a routine procedure.

C. COMPARISON OF TYPING BY RESTRICTION FRAGMENT LENGTH POLYMORPHISM (RFLP) AND MONOCLONAL ANTIBODIES

During the last decade, various panels of anti-N and anti-G monoclonal antibodies (MAbs) have been defined for typing rabies isolates.[7–9,61] Typing with anti-N MAbs is rapid when directly applied to a brain print by fluorescent antibody tests. However, it is less powerful for differentiating isolates than using anti-G MAbs, in correlation with the respective antigenic variability of each protein. On the other hand, reactivity to anti-G MAbs can be determined by neutralizing tests, which require a previous cell culture adaptation of the isolate. In summary, typing with MAbs requires more than one method, and the infectivity of the isolate has often to be preserved.

In comparison, molecular typing by PCR-RFLP offers numerous advantages. From a technical point of view it is easy (only one PCR amplification) and quick (always completed the day of sample reception), and it does not require a live, virulent sample. Also, the genome is directly analyzed without the alterations accompanying the cell culture adaptation; and the genomic region sampled can be changed as necessary, unlike antibody tests that are exclusively focused on murine immunologic epitopes. In addition, modification of restriction enzyme sites may accompany (be linked to) the emergence of viral variants, and thus identify samples of interest for further sequence studies.

Different panels of restriction enzymes have already been defined for discriminating between viral variants based on their geography, host, or taxonomy.[43,45–47] The G-L primer set is of particular interest for its capacity to amplify the very variable Ψ pseudogene of most rabies and rabies-related viruses. This creates a powerful typing tool efficient for very divergent as well as very close isolates: (1) the former are differentiated by hybridization on Southern blot using typical (for example genotype specific) Ψ probes; and (2) the latter by RFLP with selected restriction enzymes.[43] Thus, a panel of five basic enzymes (BamHI, HindIII, PstI, RsaI, TaqI) is sufficient to differentiate the principal rabies vaccine strains PV, ERA, SADB19, HEP, CVS, and PM, both between each of them and from the wild variants currently circulating in French foxes.[43,48] This can only be achieved with two successive panels of 8 anti-N and 6 anti-G MAbs (Sureau, P., personal communication). Such tools are extremely useful for surveying residual cases during oral vaccination campaigns of wildlife. Figure 10–5 also illustrates that a panel of three enzymes (DdeI, PstI, and PvuII) is convenient for identifying by RFLP of the N gene [N_7-(N_8)M1

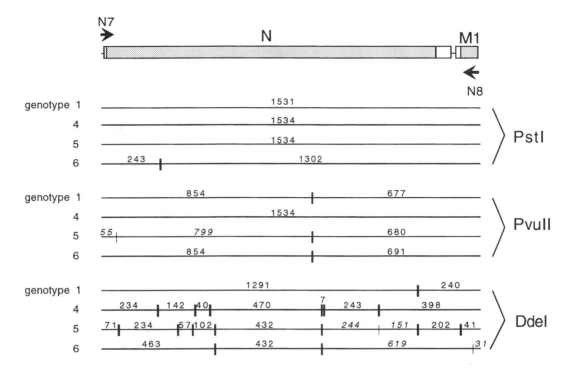

Figure 10–5 Example of typing test by RFLP to differentiate lyssaviruses from sero-genotypes 1 (rabies), 4 (Duvenhage), 5 (EBL1) and 6 (EBL2). The N_7-N_8 amplicon is digested by a panel of 3 restriction enzymes (DdeI, PstI, PvuII). PstI selectively cleaves sero-genotype 6; PvuII characterizes sero-genotype 4, which is not cleaved; DdeI distinguishes the remaining sero-genotypes 1 (only one site) and 5 (extensive cleavage). The size and location of all the restriction fragments obtained are shown. Thin bars flanked by numbers in italics indicate intrinsic variability within a sero-genotype pattern, i.e., a restriction site which is not systematically present.

primer set]; isolates of rabies 1 sero-genotype 1) and rabies-related viruses (sero-genotypes 5 and 6) spreading in Europe.

D. MOLECULAR EPIDEMIOLOGY STUDIES

The virus present in an infected animal is generally not homogeneous, but rather is composed of a population of individual molecules differing by few mutations (quasi-species). Amplification by PCR *a priori* conserves the initial distribution of the population, except for the molecules mutated in the target regions for the primers, which are negatively selected. Direct sequencing of the PCR products provides a consensus sequence, representative of the global population.[10,43,47,49] In contrast, when the PCR product is cloned before sequencing, the sequence is that of an individual and homogeneous molecular species.[47,50] When studying very closely related isolates (same biotype, same region), it must be kept in mind that the heterogeneity between isolates may be in the same range as the intrinsic heterogeneity of each isolate, which is estimated at 10^{-2} to 10^{-3} in the coding regions (Reference 50 and unpublished observation). Because of this, cloning PCR products before sequencing should be considered only with due thought.

Until recently, with the exception of taxonomy,[10,46] genetic variability studies have mostly focused on wild rabies viruses, i.e., members of sero-genotype 1. Two have focused on limited geographic areas, France[48] and Ontario,[47] comparing the Ψ and N genes, respectively. Another compared a 200 nucleotide long segment of the nucleoprotein[49] of canine rabies viruses world-wide. At present, the simple and rapid protocol described above for the direct sequencing of any double-stranded DNA purified by excision from agarose gels,[43] allows us to undertake prospective studies of the world-wide genetic variability of sero-genotype 1. Since primers internal to the initial amplicon are as efficient as PCR primers for sequencing, we are basing our comparative studies on large genomic areas such as the N (1500

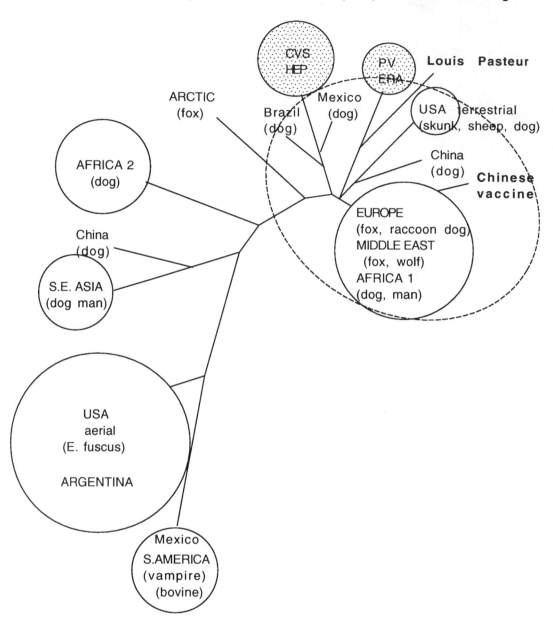

Figure 10–6 Radial phylogenetic tree illustrating the variability of rabies isolates from different countries and different hosts (sero-genotype 1). The tree was calculated by the CLUSTAL program (neighbor joining method) by comparing the sequence of the whole M2-L amplicon (2500 residues). It is unrooted. The divergence between two samples is directly proportional to length of their joining branch. Related, variable isolates are in the circles. The dotted circle outlines the Europe-Middle East-Africa 1 cluster. The vaccine strains are noted in bold characters or in grey circles.

nucleotide) or the G-Ψ-L (2500 nucleotides) regions that are amplified as single pieces and then sequenced (Figure 10–3). The fidelity of the technique was verified by repeatedly sequencing the same region (consensus sequence) amplified in different PCR reactions.

Numerous isolates of sero-genotype 1 from different hosts and countries were studied. Although the overall conservation in the ectodomain (80% at the nucleotide level -nuc, 87% at the amino acid level -aa) is important, different virus clusters are distinguishable, reflecting geographic or historical relationships, or the host species (Figure 10–6). Isolates from Europe (fox, racoon dog, and wolf) and Middle East (fox, wolf, and jackal) group close together. In Africa, two groups of canine viruses are found: first,

(Africa 1) all over the continent, joining the Europe-Middle East cluster; and second, a distinct cluster (Africa 2), mostly in Western Central countries. Viruses spreading in South East Asian dogs and in arctic foxes are also atypical. The vaccine strains form a rather heterogeneous group (90% nuc, 93% aa), more closely related to the (Europe-Middle East-Africa 1) cluster. In America, viruses circulating in bat species are genetically distinct from those circulating in terrestrial species, indicating the existence of overlapping epidemiological cycles in different vectors with no apparent interference (dog/vampire bat in North, Central, and South America; skunk-fox/insectivorous bat in North America). The terrestrial variants join the Europe-Middle East-Africa 1 cluster, except those from Ontario that are genetically related to the arctic fox virus. Among the aerial variants, vampire bat viruses form a very homogeneous group (95% nuc, 97% aa) from the North of Mexico to the South of Brazil, while the same insectivorous bat species *(Eptesicus fuscus)* found in Montana can be infected by very different viruses (86% nuc, 89% aa). Divergence from the vaccine strains is maximal with the bat viruses (80% nuc, 87% aa) in which the antigenic sites can vary. This could explain increasing cases of vaccination failures in Brazilian cattle (unpublished data).

E. PROGRESS IN TAXONOMY

The RT/PCR/sequencing technique has substantially improved understanding of the taxonomy of the Lyssaviruses. By sequence comparison of the nucleoprotein and the glycoprotein genes, six distinct genotypes were observed instead of only four serotypes[10,46] (Figure 10–7). In addition, the phylogenetical tree displays the relationships between genotypes in a more sensitive way. For example, genotypes 2 (Lagos Bat) and 3 (Mokola) are very distant from any other one. They appear more closely related to each other, as do genotypes 4 (Duvenhage) and 5 (EBL1). This last observation explained, *a posteriori,* that the first European Bat Lyssavirus isolates were initially classified into serotype 4, before being left unclassified. Despite this obvious progress in taxonomy, several lyssaviruses remain to be classified: for example Kotonkan[69] and Obodhiang[70] which were isolated from insects. The increasing number of available Lyssavirus sequences, as well as the comparison with other Rhabdo- or Paramyxoviruses,[15,56,57] should allow delineation of particularly stable genomic areas for designing primers suitable for specifically amplifying these viruses.

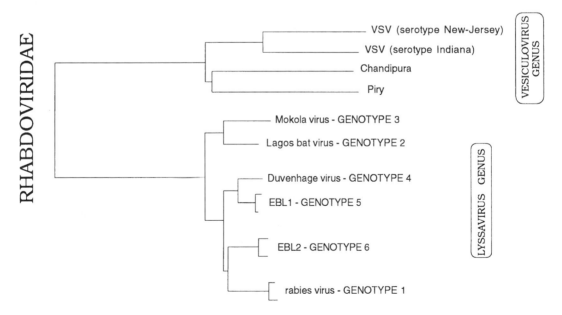

Figure 10–7 Phylogenetic tree of the *Lyssavirus* and *Vesiculovirus* genus (*Rhabdoviridae* family) by comparison of the N nucleoprotein gene using CLUSTAL program (neighbor joining method). The length of the horizontal branches are indicative of the evolutionary distance, the vertical lines are only for clarity. Six main branches corresponding to six distinct genotypes are visible. In several cases (genotypes 1, 5, 6), two isolates were studied to have an idea of the intra-genotype variability. Inter-genotype relationships are discussed in the text.

It is difficult, given the current state of knowledge, to define the limits of a genotype. Within genotype 1, which has been by far the most extensively studied, minimal conservation is observed between the insectivorous bat isolates from America and the vaccine strains (87% aa in the glycoprotein ectodomain). Between two of the more related genotypes (4 and 5), this value is 84%.

V. CONCLUSIONS

A fatal case of human rabies in New York State has been recently described by the Centers for Disease Control; this is the first case diagnosed in New York in 39 years.[71] However, death occurred before a definitive diagnosis could be made. Cerebrospinal fluid and serum were negative for rabies virus and specific antibodies; whereas brain stem, midbrain, and Purkinje cells of the cerebellum were positive when tested by the fluorescent antibody test. This led to the use of PCR on formalin-fixed brain material. Three methods of extraction of RNA were compared: standard phenol/chloroform extraction; isoquick extraction; and guanidium thiocyanate plus proteinase. The latter gave the best results. In addition, it was shown that PCR efficiency is affected by formalin. Nevertheless, RNA was amplified and the amplicon sequenced. Comparative analysis of the sequence obtained led to suspicion of the involvement of an insectivorous bat in the etiology of infection. A similar study on three immigrants who died from rabies in the U.S., had implicated variant viruses from their country of origin, identified on the basis of PCR-RFLP profiles of the amplicon.[45] Thus, as well as confirming the diagnosis of rabies, PCR was instrumental in suggesting the probable means of acquisition of the disease.

In summary, PCR appears suitable for rabies diagnosis, but still needs to be improved for routine use. It can already be considered as an excellent expert method, currently the most convenient and powerful available for the typing and molecular epidemiology of lyssaviruses, even on highly degraded samples. This invites world-wide molecular epidemiological studies, particularly in countries infected with divergent lyssaviruses suspected or demonstrated not to be reactive with current vaccines. Such studies could help, if necessary, in the selection of new vaccine strains, thus offering a greater degree of protection, or one more adapted to the local virus variants.

ACKNOWLEDGMENTS

Acknowledgments are due to P. E. Ceccaldi, P. Perrin, and Y. Rotivel for critical reading of the manuscript and to S. Boulloud for help in typing.

REFERENCES

1. **Théodoridès, J.,** *Histoire de la Rage,* Masson, Paris, 1986.
2. **Baer, G. M., Bellini, W. J., and Fishbein, D. B.,** Rhabdoviruses, in *Virology,* Fields, D. and Knipe, M., Eds., Raven Press, New York, 1990, 883–930.
3. **Pasteur, L., Chamberland, C. E., and Roux, E.,** Nouvelle communication sur la rage, *C.R. Acad. Sci. Paris,* 98, 457, 1984.
4. **Pasteur, L.,** Méthode pour prévenir la rage après morsure, *C.R. Acad. Sci. Paris,* 101, 765–774, 1885.
5. **Francki, R. I. B., Fauquet, C. M., Knudson, D. L., and Brown, F.,** Classification and nomenclature of viruses. Fifth Report of the International Committee on Taxonomy of Viruses, Springer-Verlag, Wien, 1991.
6. W.H.O., Eighth Report of the WHO Expert Committee on Rabies, number 824, Eds., W.H.O., Geneva, 1992.
7. **Dietzschold, B., Rupprecht, C. E., Tollis, M., Lafon, M., Mattei, J., Wiktor, T. J., and Koprowski, H.,** Antigenic diversity of the glycoprotein and nucleocapsid proteins of rabies and rabies-related viruses: implications for epidemiology and control of rabies, *Rev. Infect. Dis.,* 10, S785–S798, 1988.
8. **King, A. and Crick, J.,** Rabies-related viruses, in *Rabies,* Campbell, J. B. and Charlton, K. M., Eds., Kluwer Academic Publishers, Boston, 1988, 177–200.
9. **Rupprecht, C. E., Dietzschold, B., Wunner, W. H., and Koprowski, H.,** Antigenic relationships of Lyssaviruses, in *The Natural History of Rabies,* Baer, G. M., Ed., CRC Press, Boca Raton, 1991, 69–100.
10. **Bourhy, H., Kissi, B., and Tordo, N.,** Molecular diversity of the Lyssavirus genus, *Virology,* 194, 70–81, 1993.
11. **Tordo, N. and Poch, O.,** Structure of rabies virus, in *Rabies,* Campbell, J. B. and Charlton, K. M., Eds., Kluwer Academic Publishers, Boston, 1988, 25–45.

12. **Barge, A., Gaudin, Y., Coulon, P., and Ruigrok, R. W. H.,** Vesicular stomatitis virus M protein may be inside the ribonucleocapsid coil, *Virology,* 67, 7246–7253, 1993.

13. **Tordo, N., Poch, O., Ermine, A., and Keith, G.,** Primary structure of leader RNA and nucleoprotein genes of the rabies genome: segmented homology with VSV, *Nucl. Acids Res.,* 14, 2671–2683, 1986.

14. **Tordo, N., Poch, O., Ermine, A., Keith, G., and Rougeon, F.,** Walking along the rabies genome: is the large G-L intergenic region a remnant gene?, *Proc. Natl. Acad. Sci. U.S.A.,* 83, 3914–3918, 1986.

15. **Tordo, N., Poch, O., Ermine, A., Keith, G., and Rougeon, F.,** Completion of the rabies virus genome sequence determination: highly conserved domains along the L (polymerase) proteins of unsegmented negative-strand RNA viruses, *Virology,* 165, 565–576, 1988.

16. **Bourhy, N., Tordo, N., Lafon, M., and Sureau, P.,** Complete cloning and molecular organisation of a rabies-related virus: Mokola virus, *J. Gen. Virol.,* 70, 2063–2074, 1989.

17. **Conzelmann, K. K., Cox, J. H., Schneider, L. G., and Thiel, H. J.,** Molecular cloning and complete nucleotide sequence of the attenuated rabies virus SAD B19, *Virology,* 175, 485–489, 1990.

18. **Baer, G. M.,** *The Natural History of Rabies,* CRC Press, Boca Raton, 1991, 620.

19. **W.H.O.,** World Survey of Rabies 27, WHO/Rabies/93.209, W.H.O., Geneva, 1991.

20. **Plotkin, S. A.,** Vaccination in the 21th century, *J. Inf. Dis.,* 168, 29–37, 1993.

21. **Charlton, K. M.,** The pathogenesis of rabies, in *Rabies,* Campbell, J. B. and Charlton, K. M., Eds., Kluwer Academic Publishers, Boston, 1988, 101–150.

22. **Tsiang, H.,** Physiopathology of rabies virus infection of the nervous system, *Adv. Virol. Res.,* 42, 1993.

23. **Coulon, P., Derbin, C., Kucera, P., Lafay, F., Prehaud, C., and Flamand, A.,** Invasion of the peripheral nervous systems of adult mice by the CVS strain of rabies virus and its avirulent derivate AvO1, *J. Virol.,* 63, 3550–3554, 1989.

24. **Fekadu, M. and Shaddock, J. H.,** Peripheral distribution of virus in dogs inoculated with two strains of rabies virus, *Am. J. Vet. Res.,* 45, 724–729, 1984.

25. **Murphy, F. A., Bauer, S. P., Harrison, A. K., and Winn, W. C.,** Comparative pathogenesis of rabies and rabies-like viruses. Viral infection and transit from inoculation site to the central nervous system, *Lab. Invest.,* 28, 361–376, 1973.

26. **Shankar, V., Dietzschold, B., and Koprowski, H.,** Direct entry of rabies virus into the central nervous system without prior local replication, *J. Virol.,* 65, 2736–2738, 1991.

27. **Hemachudha, T., Phanuphak, B., Sriwanthana, B., Manutsathit, S., Phanthumchinda, K., Siriprasomsup, W., Ukachoke, C., Rasameechan, S., and Kaoroptham, S.,** Immunologic study of human encephalitic and paralytic rabies. Preliminary report of 16 patients, *Am. J. Med.,* 84, 673–677, 1988.

28. **Ceccaldi, P. E., Fillion, M. P., Ermine, A., Tsiang, H., and Fillion, G.,** Rabies virus selectively alters 5-HT1 receptor subtypes in rat brain, *Eur. J. Pharm.,* 245, 129–138, 1993.

29. **MacFarlan, R. I.,** Immune response to rabies virus: vaccine and natural infection, in *Rabies,* Campbell, J. B. and Charlton, K. M., Eds., Kluwer Academic Publishers, Boston, 1988, 163–176.

30. **Lafon, M., Lafage, M., Martinez-Arends, A., Vuillier, F., Lotteau, V., Charron, D., and Scott-Algara, D.,** Evidence in humans of a viral superantigen, *Nature (London),* 358, 507–509,1992.

31. **Perrin, P., Lafon, M., and Sureau, P.,** Rabies vaccines from Pasteur's time up to experimental subunit vaccines today, in *Viral Vaccines,* Mizrahi, A., Ed., Wiley-Liss, New York, 1990, 325–345.

32. **Tordo, N.,** Contribution of molecular biology to vaccine development and molecular epidemiology of rabies disease, *Mem. Inst. Butantan,* 53, 31–51, 1991.

33. **Morgeaux, S., Tordo, N., Gontier, C., and Perrin, P.,** β-propiolactone treatment impairs the biological activity of residual DNA from BHK-21 cells infected with rabies virus, *Vaccine,* 11, 82–90, 1993.

34. **Petricciani, J.,** Ongoing tragedy of rabies, *Lancet,* 342, 1067, 1993.

35. **Tignor, G. H. and Smith, A. L.,** Vaccination and challenge of mice with viruses of the rabies group, *J. Infect. Dis.,* 125, 322–324, 1972.

36. **Tignor, G. H., Murphy, F. A., Clark, H.F., Shope, R. E., Madore, P., Bauer, S. P., Buckley, S. M., and Meredith, C. D.,** Duvenhage virus: morphological, biochemical, histopathological and antigenic relationships to the rabies serogroup, *J. Gen. Virol.,* 37, 595–611, 1977.

37. **Lafon, M., Bourhy, H., and Sureau, P.,** Immunity against the European bat rabies (Duvenhage) virus induced by rabies vaccines: an experimental study in mice, *Vaccine,* 6, 362–368, 1988.

38. **Bourhy, H., Rollin, P. E., Vincent, J., and Sureau, P.,** Comparative field evaluation of the fluorescent-antibody test, virus isolation from tissue culture, and enzyme immunodiagnosis for rapid laboratory diagnosis of rabies, *J. Clin. Microbiol.,* 27, 519–523, 1989.

39. **Sureau, P., Ravisse, P., and Rollin, P.,** Rabies diagnosis by animal inoculation, identification of Negri bodies, or ELISA, in *The Natural History of Rabies,* Baer, G. M., Ed., CRC Press, Boca Raton, 1991, 203–217.

40. **Trimarchi, C. V. and Debbie, J. G.,** The fluorescent antibody in rabies, in *The Natural History of Rabies,* Baer, G. M., Ed., CRC Press, Boca Raton, 1991, 219–233.

41. **Bourhy, H. and Sureau, P.,** Laboratory methods for rabies diagnosis, *Inst. Pasteur, Paris,* edited by Commission de Laboratoire de Reference et d'Expertise de l'Institut Pasteur, Paris, 1991.

42. **Perrin, P., Gontier, C., Lecocq, E., and Bourhy, H.,** A modified rapid enzyme immunoassay for the detection of rabies and rabies-related viruses: RREID-lyssa, *Biologicals,* 20, 1992.

43. **Sacramento, D., Bourhy, H., and Tordo, N.,** PCR technique as an alternative method for diagnosis and molecular epidemiology of rabies virus, *Mol. Cell. Probes,* 6, 229–240, 1991.

44. **Kamolvarin, N., Tirawatnpong, T., Rattanasiwamoke, R., Tirawatnpong, S., Panpanich, T., and Hemachudha, T.,** Diagnosis of rabies by polymerase chain reaction with nested primers, *J. Infect. Dis.,* 167, 207–210, 1993.

45. **Smith, J. S., Fishbein, D. B., Rupprecht, C. E., and Clark, K.,** Unexplained rabies in three immigrants in the United States. A virologic investigation, *N. Engl. J. Med.,* 324, 205–211, 1991.

46. **Bourhy, H., Kissi, B., Lafon, M., Sacramento, D., and Tordo, N.,** Antigenic and molecular characterisation of bat rabies virus in Europe, *J. Clin. Microbiol.,* 30, 2419–2426, 1992.

47. **Nadin-Davies, S. A., Casey, G. A., and Wandeler, A.,** Identification of regional variants of the rabies virus within the Canadian province of Ontario, *J. Gen. Virol.,* 74, 829–837, 1993.

48. **Sacramento, D., Badrane, H., Bourhy, N., and Tordo, N.,** Molecular epidemiology of rabies in France: comparison with vaccinal strains, *J. Gen. Virol.,* 73, 1149–1158, 1992.

49. **Smith, J. S., Orciari, L. A., Yager, P. A., Seidel, H. D., and Warner, C. K.,** Epidemiologic and historical relationships among 97 rabies virus isolates as determined by limited sequence analysis, *J. Infect. Dis.,* 166, 296–307, 1992.

50. **Benmansour, A., Brahimi, M., Tuffereau, C., Coulon, P., Lafay, F., and Flamand, A.,** Rapid sequence evolution of street rabies glycoprotein is related to the heterogeneous nature of the viral population, *Virology,* 187, 33–45, 1992.

51. **Ermine, A., Larzul, D., Ceccaldi, P. E., Guesdon, J. L., and Tsiang, H.,** Polymerase chain reaction amplification of rabies virus nucleic acids from total mouse brain RNA, *Mol. Cell. Probes,* 4, 189–191, 1990.

52. **Tordo, N., Bourhy, H., and Sacramento, D.,** Polymerase chain reaction technology for rabies virus, in *Frontiers in Virology,* Becker, Y. and Darai, G., Eds., Springer-Verlag, Berlin, 1992, 389–405.

53. **Smith, J. S. and Seidel, H. D.,** Rabies: a new look at an old disease, in *Progress in Medical Virology,* Melnick, J. L., Ed., S. Karger, Basel, 1993, 82–106.

54. **Tordo, N., Sacramento, D., and Bourhy, H.,** The polymerase chain reaction (PCR) technique for diagnosis, typing and epidemiological studies of rabies, in *Laboratory Techniques in Rabies,* Meslin, F. X. and Bögel, K., Eds., World Health Organization, Geneva, 1994.

55. **Tordo, N., Bourhy, H., Sather, S., and Ollo, R.,** Structure and expression in the baculovirus of the Mokola virus glycoprotein: an efficient recombinant vaccine, *Virology,* 194, 59–69, 1993.

56. **Tordo, N., De Haan, P., Goldbach, R., and Poch, O.,** Evolution of negative-stranded RNA genomes, *Sem. Virol.,* 3, 311–417, 1992.

57. **Poch, O., Blumberg, B. M., Bougueleret, L., and Tordo, N.,** Sequence comparison of five polymerases (L proteins) of unsegmented negative-strand RNA viruses: theoretical assignments of functional domains, *J. Gen. Virol.,* 71, 1153–1162, 1990.

58. **Poch, O., Sauvaget, I., Delarue, M., and Tordo, N.,** Identification of four conserved motifs among the RNA-dependent polymerase encoding elements, *EMBO J.,* 8, 3867–3874, 1989.

59. **Delarue, M., Poch, O., Tordo, N., Moras, D., and Argos, P.,** An attempt to unify the structure of polymerases, *Protein Engineering,* 3, 461–467, 1990.

60. **Poch, O., Tordo, N., and Keith, G.,** Sequence of the 3386 3′ nucleotides of the genome of the AvO1 strain rabies virus: structural similarities of the protein regions involved in transcription, *Biochimie,* 70, 1019–1029, 1988.

61. **Smith, J.,** Rabies virus epitopic variation: use in ecologic studies, *Adv. Virus Res.,* 36, 215–253, 1989.

62. **Celis, E., Rupprecht, C. E., and Plotkin, S. A.,** New and improved vaccines against rabies, in *New Generation Vaccines,* Woodrow, G. C. and Levine, M. M., Eds., Marcel Dekker, New York, 1990, 419–439.

63. **Wunner, W. H., Larson, J. K., Dietzschold, B., and Smith, C. L.,** The molecular biology of rabies virus, *Rev. Infect. Dis.,* 10, 771–784, 1988.

64. **Barrat, J., Artois, M., and Bourhy, H.,** Two bats diagnosed rabid in France, Rabies Bulletin Europe. WHO Collaborative Center for Rabies Surveillance and Research, Tübingen, FRG, 13, 10–11, 1989.

65. **Montano Hirose, J. A., Bourhy, H., and Sureau, P.,** Retro-orbital route for the collection of brain specimens for rabies diagnosis, *Vet. Rec.,* 129, 291–292, 1991.

66. **Sambrook, J., Fritsch, E. F., and Maniatis, T.,** *Molecular Cloning: A Laboratory Manual,* Cold Spring Harbor, New York, 1989.

67. **Sanger, F., Nicklen, S., and Coulson, A. R.,** DNA sequencing with chain-terminating inhibitors, *Proc. Natl. Acad. Sci. U.S.A.,* 74, 5463–5467, 1977.

68. **Ermine, A., Tordo, N., and Tsiang, H.,** A rapid diagnosis of the rabies infection through a dot hybridization assay, *Mol. Cell. Probes,* 2, 75–82, 1988.

69. **Kemp, G. E., Lee, V. H., Moore, D. L., Shope, R. E., Causey, O. R., and Murphy, F. A.,** Kotokan, a new Rhabdovirus related to Mokola virus of the rabies serogroup, *Am. J. Epidemiol.,* 98, 43–49, 1973.

70. **Schmidt, J. R., Williams, C. C., Lule, M., Mivule, A., and Mujomba, E.,** Viruses isolated from mosquitoes collected in the Southern Sudan and Western Ethiopia, *E. Afr. Virol. Res. Inst.,* 15, 24, 1965.

71. Centers for Disease Control, Human rabies — New York, M.M.W.R., 42, 799–806, 1993.

Chapter 11

PCR Methods in the Diagnosis of HIV-1 and HIV-2 Infection

Peter Balfe

TABLE OF CONTENTS

I. INTRODUCTION

In most cases the diagnosis of human immunodeficiency virus type 1 (HIV-1) and type 2 (HIV-2) infection can be based upon serological tests,[1] however in certain cases such tests are not appropriate:

1. The immediate post-exposure/pre-seroconversion period before the development of the antibodies to HIV antigens that are the basis of almost all commercial tests
2. Infections of the very young (usually infected by parenteral transmission), where maternal antibodies may be co-circulating[2-5]
3. Screening of possibly contaminated blood products[6]
4. Confirmatory screening of sera giving ambiguous Western blot results[7,8]

In these particular cases diagnostic PCR methods can be useful for demonstrating the presence of the HIV genome. The PCR is usually performed on a DNA sample prepared from peripheral blood mononuclear cells (PBMC), detecting the presence of proviral DNA, integrated into the host genome, which is a feature of the retrovirus life cycle. Detection of free virus is also feasible, but technically more demanding since it requires first the preparation of viral RNA, followed by the synthesis of a cDNA copy. Thus both serological and PCR-based methods can be used for diagnosis, but PCR will probably be used only in the special cases outlined above.

II. DESIGN OF THE PCR SYSTEM

In the case of diagnosis the objective is merely to detect whether HIV genomic material is present in the sample. For this the most sensitive and robust PCR is the most appropriate.[9-11] Difficulties arise in determining the cutoff sensitivities for the PCR tests used and in carryover contamination, leading to false positives. For these reasons every PCR performed requires controls not only for the sensitivity, using a dilution series of known standards, but negative control reactions to demonstrate the absence of cross contamination.[12–14] Even so, like the analogous EIA results from conventional screening, no diagnosis of PCR positivity should be based on a single experiment, but should be confirmed by at least one independent repeat. PCR methods using nested sets of primers can successfully detect single molecules of DNA,[15] and are therefore the method of choice for simple diagnostic work. In common with any PCR technique the nested PCR will be fundamentally affected by the quality of the sample available, the degree of match of the primers used to the target sequence, and the length of the PCR product being generated. The degree of primer match to the target is a fundamental limitation of any PCR technique.[16] In practice the degree of mismatch is minimized by targeting the least variable of the genome under study; in HIV these are the reverse transcriptase (RT) region of *pol* and the *gag* gene.[17] For maximal sensitivity of detection, primers should be targeted to these regions. We, and others, have shown that the use of several primer pairs simultaneously in the first round of the nested PCR ("multiplexing") does not lead to any loss in sensitivity, and hence, to maximize the chance of primers matching well to the target genome, several such sets of primers can be used.[9,15,18,19]

In the case of the detection of RNA, where a necessary precursor to amplification is cDNA synthesis, the outer antisense primers of several regions to be PCR amplified can be used in the cDNA synthesis. Alternatively random hexamer priming can be used. For diagnostic use this may prove to be a better method as only one cDNA synthesis is required irrespective of the region of the genome to be subsequently PCR amplified.

III. PREPARATION OF SAMPLES

Generally whole-blood samples are available for PCR, and the majority of techniques in use are designed around either whole-blood extraction or extraction from Ficoll-prepared buffy coat cells. A second, less explored source of DNA for the PCR is dried-blood-spot specimens.[20] Direct PCR amplification of whole blood itself has generally been unsuccessful, as the reaction is rapidly inhibited even by quite low concentrations (>1% v/v) of whole blood in the PCR. However, McCusker and colleagues[21] have recently reported that a 15-min 95°C incubation of the blood before the PCR is set up enables much higher concentrations of starting material to be tolerated.

Holodniy and colleagues[22] reported that, in a comparison of DNA extracted from whole blood supplied in several anticoagulants, the commonly used anticoagulant heparin was a potent inhibitor of the PCR reaction. This inhibition could be reversed by heparinase treatment of the sample. No such problems were found with acid citrate dextrose, sodium EDTA, or potassium oxalate. Generally, several authors have reported difficulties in amplification of impure DNA, even though it is possible to obtain results from PBMCs lysed with 1% Triton-X100.[23] Several methods for removal of impurities exist. The "classical" method[24] (used in our laboratory and shown in Table 11–1), of proteinase K treatment of lysed buffy coat cells, followed by phenol-chloroform extraction, ethanol precipitation, and resuspension, is very reliable but rather cumbersome and slow. This has led to the development of a number of alternative protocols. Albert and Fenyö[25] reported that DNA from lysed buffy coat cells can be successfully PCR amplified and that the subsequent purification steps are not essential. However, we have found that samples prepared in this way can only be used fresh, since on storage DNA degradation occurs, leading to a loss of signal; for rapidity, however, the method cannot be bettered. Glass powder extraction methods, analogous to the familiar Gene-Clean™* procedure, have been used.[26] Extraction of DNA from buffy coat in a 20% Chelex suspension containing 1% Nonidet P40™,** 0.1% SDS, and 1% Tween 20 will remove much of the proteinaceous matter.[27] Differential lysis methods, where the erythrocyte fraction of the whole blood is lysed using Nonidet-NP40 and the unlysed nuclei are pelleted, washed several times, and subsequently lysed in Triton X-100, work well in our hands,[28] especially for small samples. In addition a number of commercial kits for DNA extraction from whole blood are available (e.g., the Nucleon™*** Kit and the MicroExtraction™**** Kit) and are claimed to give good PCR results. In a diagnostic setting the determinants of the DNA preparation method will be the volume of blood available and whether or not buffy coat separations will be performed. In most cases the freezing of small aliquots of intact PBMCs from buffy

---

* Bio 101, Inc., La Jolla, CA.

** Shell.

*** Scotlab.

**** Stratagene.

Table 11–1. Preparation of PBMC DNA: protocol for preparation of high-purity DNA from lymphocytes[a]

To frozen cells add an equal volume of lysis buffer and start at step 3.

1. Spin fresh buffy coat cells in a 1.5-ml Eppendorf tube (screw top) at 6500 rpm for 3 min.
2. Resuspend pellet in 600-µl "lysis buffer."
3. Add 6 µl of 10 mg/ml Proteinase K (final concentration, 100 µg/ml). Place at 65°C.
4. Check to see that pellet is dispersed; add 60 µl 10% Sarkosyl.
5. Invert mix once and allow to stand at 65°C for at least 2 h (up to 4 h has been used).
6. Add an equal volume of distilled phenol at 65°C, invert mix once, and allow to stand for 5 min.
7. Add 100 µl chloroform, spin 5 min, 12,000 rpm.
8. Remove upper aqueous phase to new tube.
9. Add an equal volume of chloroform, invert mix once, allow to stand for 30 s, spin at 12,000 rpm for 10 s.
10. Using a cut-tip pipette, remove contents to new tube and repeat chloroform extraction.
11. The DNA at this stage is of very high molecular weight and must be sheared to make it manipulable for PCR. Shear the DNA by pipetting up and down 30 times through a 1-ml (blue) Gilson tip.
12. Precipitate with 1 vol of propan-2-ol (–20°C, overnight). The amount of DNA present is large enough that salt need not be added to make the DNA precipitate.
13. Wash pellet twice with 500 µl of cold (4°C) 70% ethanol.
14. Resuspend in 100 µl of 1 × TE, pH 7.5; incubate for 1 to 2 h at 65°C to disperse pellet.

Buffers: TE: 10 mM Tris-Cl, pH 7.5, 1 mM EDTA; Lysis Buffer: 0.025 M EDTA, pH 8.0, 0.1 M Tris-Cl, pH 7.5, 50 mM NaCl.

[a] A number of techniques exist for the preparation of high-molecular-weight DNA from eukaryotes. The technique described here is based on a protocol derived in part from Diane Beatson (MRC, Edinburgh), in part from Maniatis et al.,[24] and in part from our own experience. The DNA produced is clean and will withstand repeated freeze-thaw cycles with only a modest loss in template quantity.

coats will give enough material for PCRs on simple lysates.[25] In cases where samples are small or buffy coat preparations are not available, then whole blood extraction methods will be needed.

RNA PCR methods are identical to DNA methods once the initial template has been copied by reverse transcriptase, but the lower stability of the RNA template, together with the requirement for a cDNA synthesis, make the use of RNA as a routine source of PCR template rather rare in a diagnostic setting. None of the four categories cited above requires that RNA be detected rather than DNA. However, monitoring viral levels in HIV-infected individuals during anti-retroviral therapy depends upon an accurate measurement of the amount of circulating virions, which may be possible only by RNA PCR. In this case of quantitative PCR there is a real need for the monitoring of both RNA and DNA levels.[29,30] In the traditional RNA extraction method, plasma or serum is ultracentrifuged and the pelleted virions lysed, treated with a RNase-free DNase, and finally, after heat inactivation of the DNase, cDNA is synthesized from the HIV-1 genomic RNA.[31] There are several disadvantages of this method, including the expense and inconvenience of ultracentrifugation, the slowness of the procedure, and the requirement for large volumes of starting material. This has led to the development of alternative methods. Most common among these is the "RNAzol" method of Chomczynski and Sacchi,[32] which has been used to prepare RNA from both serum and plasma. If carrier RNA (typically 2 µg of *E. coli* ribosomal RNA) is used in the extraction, then essentially all of the HIV-1 genomic RNA in a plasma or serum sample can be recovered. The method of RNA preparation and cDNA synthesis described in Table 11–2 has been successfully used for the detection of HIV-1 RNA in Factor VIII preparations.[6]

Solid-phase capture systems can be used, where the virion particles are captured onto a solid phase (usually latex or magnetic beads) coated with antibodies to the gp120 envelope protein.[33] Alternatively, the RNA can be released by lysing the virions and then captured using antisense oligonucleotides conjugated to a solid support (usually a 5′ biotinylated oligonucleotide bound to a streptavidin-coated bead, e.g., M-280 Dynabeads™*).[34] Either method allows the repeated washing of the captured material to remove proteins and other contaminants, and also acts as a concentration step. Experience has shown

Table 11–2A   **Preparation of RNA**

RNAzol Method[32] in Brief[a]

Use 800 µl RNAzol solution per 200 µl plasma or serum.

1. Add 2 µg carrier RNA (*Escherichia coli* ribosomal RNA, Boehringer Mannheim) to 200 µl plasma or serum in 1.5-ml (screw-top) Eppendorf tube.
2. Add 800 µl RNAzol solution.
3. Add 100 µl chloroform.
4. Shake vigorously for 15 s.
5. Place on ice for 15 min.
6. Microfuge for 15 min at 4°C. The RNA is in the upper phase, and the DNA and protein are at the interphase.
7. Remove the upper phase, keeping well away from the interphase.
8. Precipitate with equal volume propan-2-ol for 45 min at –20°C.
9. Microfuge for 15 min at 4°C.
10. Wash white RNA pellet twice with ice-cold 75% ethanol by vortexing and subsequently respinning (8 min at 4°C).
11. Dry under vacuum for 10 min (do not overdry as this makes the RNA insoluble).
12. Dissolve in DEPC-treated dH$_2$O; this step may require heating (at 60°C) and vortexing. Alternatively, allow to stand overnight at 4°C.

RNAzol Stock Solution A[b]

23.64 g guanidinium thiocyanate (Sigma, G-6639)
1.25 ml 1 M Na Citrate, pH 7.0
0.5 g Sarkosyl (*N*-Lauroylsarcosine, Na salt, Sigma, L-5125)

Adjust volume to 50 ml with DEPC-treated dH$_2$O. Add 0.36 ml β-mercaptoethanol before use.

RNAzol Solution[b]

1 vol stock solution A
0.1 vol 2 M Na acetate, pH 4.0
1 vol water-saturated phenol

---

[a] When working from plasma, you will find that the aqueous/phenol-chloroform interface tends to be very messy. You need to be careful and accept losses of material (you can back-extract with DEPC treated water); [b] Both solutions can be stored for up to 1 month.

that in these cases the most important factor in the likely success of the RNA detection is the quality of the sample. If the sample used has been repeatedly frozen and thawed, the probability of success is low in comparison to a fresh sample of the same material. Once the RNA is prepared, cDNA synthesis can be performed by any of the commonly available methods. Where the virion itself is captured onto the solid phase, the addition of Triton X-100 (0.1% v/v) to the synthesis buffer facilitates the entry of reverse transcriptase and allows synthesis to proceed more rapidly.[34]

IV. PCR MIXTURES

The *Taq* polymerase buffer currently in most widespread use consists of 10 mM Tris-Cl, pH 8.8, 50 mM NaCl, and 1.5 mM MgCl$_2$ (0.001 to 0.01% gelatin is sometimes also used, but usually has little effect on the reaction). To this basic buffer is added 100 µM dATP, dCTP, dGTP, and dCTP ("dNTPs"), 1 to 5 µM of each PCR primer, and 25 U/ml of *Taq* polymerase. In addition it has been shown that 0.01% Tween-20 and 0.01% Triton X-100 act to stabilize the enzyme and enhance the PCR reaction yield. Early assays for the activity of *Taq* polymerase used a 66 mM Tris-Cl, pH 8.8, 17 mM (NH$_4$)$_2$SO$_4$, 6.7 mM MgCl$_2$

---

* Dynal, Ltd.

Table 11–2B **Synthesis of cDNA**

First-Strand Synthesis from HIV-1 Bearing Total RNA

Reverse transcription (first-strand synthesis) is done in the following buffer:

10 µl RNA solution
2 µl 10 × RT buffer[a]
3 µl 3.3 mM dNTP mixture (Boehringer Mannheim, 100 mM solutions)
1 µl AMV RTase (Promega, 10 U/µl)
0.5 µl RNAsin (Promega, 30 U/µl)
1 µl antisense primer (200 ng, reverse-phase HPLC cleaned)
2.5 µl DEPC-treated dH$_2$O (ion-exchanged and then distilled)

The reaction is performed for 60 min at 42°C, the RNA is prepared as described in Table 11–2A, and subsequent PCR amplification is done in the following buffer:

2 µl RT reaction product
2 µl 10 × PCR buffer[b]
0.2 µl 10 mM dNTP mixture (Boehringer Mannheim, 100 mM solutions)
0.2 µl plus primer (40 ng)
0.2 µl minus primer (40 ng)
0.2 µl *Taq* (Cetus "Ampli*Taq*," 5 U/µl)
15 µl DEPC-treated dH$_2$O

---

[a] 10 × RT buffer is 500 mM Tris-Cl, pH 8.0, 50 mM MgCl$_2$, 50 mM DTT, 500 mM KCl, 0.5 mg/ml BSA (Pharmacia, RNase and DNase free). We have observed a definite inhibition of PCR when more than 1/10 vol of 1 × RT buffer spiked with DNA is added to the PCR reaction; this seems to be inhibition of the PCR by DTT. One solution is to dilute the RT product into 10 vol of PCR buffer or to dilute out the DTT by increasing the volume of the PCR. Alternatively the RT step could be carried out using β-mercaptoethanol, which does not show this effect; [b] 10 × PCR buffer is 100 mM Tris-Cl, pH 8.8, 500 mM NaCl, 15 mM MgCl$_2$, 0.1% Tween-20, 0.1% Triton X-100.

buffer. The newer thermostable "Vent" and *Pfu* polymerases use buffers similar to this. When using purified DNA it has been shown that the concentrations of the dNTPs can be reduced to less than 10 µM before an effect is seen on the PCR yield. Similarly, the primer concentrations can be reduced up to tenfold. However the use of the large excess of reagents gives a margin for degradation or loss of reagents, which may occur when impure DNA-containing samples, such as cell lysates, are amplified.

V. PCR CONDITIONS

Typically 25 to 35 cycles (c) of amplification are used in the PCR. The denaturing temperature is usually 94°C, though a temperature as low as 90°C also works. Times at this temperature vary with the type of machine used, but often are for 1 min. In practice once the reaction buffer reaches 90°C, strand separation is essentially instantaneous; therefore shorter times at these elevated temperatures can be used. This has the additional benefit of reducing the loss of *Taq* polymerase activity and can hence improve yields of the PCR amplicon. Annealing temperatures are largely determined by the length and GC content of the primers being used. The annealing temperature is normally defined as 5 to 10°C below the T_m (melting temperature) of the primers used. $T_{m(approx)} = 4 \times (G + C) + 2 \times (A + T)$, where G + C is the number of G or C nucleotides in the primer and A + T is the number of A or T nucleotides. Hence for a 20-base oligonucleotide containing 11 G or C, $T_{m(approx)}$ is 62°C.

Once the annealing temperature is achieved, the hybridization of the primers to the target DNA is extremely rapid as they are in molar excess. Commonly 1-min times are used at this temperature, but in practice this can be shortened. Several authors have reported that where the primers used have T_m values within 10 to 15°C of the strand-extension temperature, the anneal step can be omitted altogether. Sufficient annealing occurs at the strand-extension temperature for the reaction to proceed.[35,36] In this "two-step" PCR the extension temperature used is often 68°C, which allows both for adequate annealing and for *Taq* polymerase strand extension. Estimates of the "processivity" of *Taq* polymerase (number of bases of template copied per second) vary from 50 to 100 nucleotides per second at 72°C. Thus, the PCR product synthesis should be completed within the 1 min commonly used. A rule of thumb of 1 kb/min is often used in designing PCR reaction conditions.

In seeking to speed up the PCR reaction Perkin Elmer have increased the conductivity of their PCR tubes, by making the walls thinner, and have reduced the denaturation and anneal times to 10 s each, followed by a 30-s strand-extension reaction. This is using their 9600 thermal cycler. In seeking to copy this design with other equipment we have found that in a conventional "thick-walled" tube, setting the denature step to 97°C for 1 s is enough to separate the strands. When used with a combined anneal/extension step of 1 min at 68°C (using primers with a T_m of about 60°C), the PCR proceeds as well as when performed conventionally (35 s at 94°C, 25 s at 55°C, and 150 s at 72°C). Similar results were obtained using both a Perkin Elmer 480 and a Techne PHC-3 PCR machine (Figure 11–1). Hence, improvements in the overall speed of the PCR by reducing the time spent in each part of the PCR cycle are clearly possible. Using these faster settings allows the PCR to complete in less than an hour, making its use in a diagnostic setting more attractive.

VI. OPTIMIZATION OF THE PCR

The sensitivity and specificity of the PCR depends largely on the rapidity and accuracy of annealing of the primers used. The rapidity of annealing is greater at lower temperatures; however nonspecific hybridization is also more probable at lower temperatures, leading to the possibility of satellite bands due to the adventitious amplification of nonspecific material.[36] The concentration of Mg^{2+} in the reaction can be experimentally manipulated to reduce the intensity of satellite bands, probably by a mixture of two effects, first, as a co-factor in the activity of *Taq* polymerase, and second, via its effects on the annealing

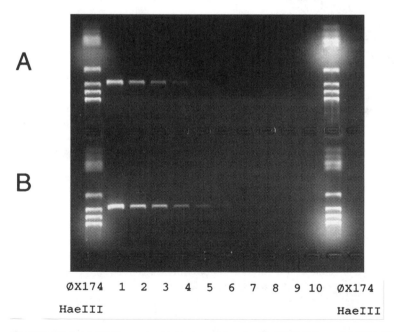

Figure 11–1 Comparison of PCR products from a "standard" PCR (35 c of 94°C, 35 s; 55°C, 25 s; 72°C, 150 s; about 2 h 45 min in total) and from a "rapid" PCR (35 c of 97°C, 1 s; 68°C, 60 s; about 55 min in total). Two tubes of 20 µl each were set up by making up 40 µl of PCR mix, adding 1 µl DNA, mixing, and transferring 20 µl to a second tube. The picture shows a 3% agarose gel of a dilution series of plasmid: (A) 1 to 8, serial dilution of plasmid from 1000 (1), 300 (2), 100 (3), 30 (4), 10 (5), 3 (6), 1 (7) and 0.3 (8) copies amplified by "standard" PCR; (B) 1 to 8, the same material as in the row above, but amplified by a "rapid" PCR. Material is visible in the 30-copy sample in both cases; fluorometry on the two 30-copy samples showed the yield in the "rapid" PCR to be 85% of that in the "standard" PCR.

temperature of the oligonucleotides. Generally these two parameters alone, Mg^{2+} and temperature, can be manipulated to optimize the PCR. An optimized single PCR can amplify DNA from as few as ten copies of starting material to a detectable threshold for either radiometric or colorimetric systems.

VII. INTERNALLY CONTROLLED PCR DESIGNS

Reliable, repeatable optimization of PCR conditions is often difficult to achieve, especially where the sample being tested is a crude lysate with unpredictable properties. In practice the only reliable quantification of a PCR relies either on a standardized high-quality extraction system, which will remove all potential inhibitors, or on some form of internal control. One of the simplest internal controls is the co-amplification of the HIV-1 target with a single-copy genomic DNA target from the infected cell. The β-globin and HLA DQ-α genes are often used.[37] More recently, internal controls based on a deleted form of the HIV-1 target itself have been developed, where a dilution series of the control DNA is titrated out in a constant volume of the sample being tested. The quantity of HIV DNA in the sample is estimated from the standard curve obtained.[38–41]

This approach, which has also been used to quantify RNA, uses the same primers for amplification of the control as the target, and hence the control and target have exactly the same reaction optima with respect to temperature and Mg^{2+} concentration, giving a theoretically ideal quantitative estimate of the number of copies of HIV in the sample. However, it is possible to get an erroneous answer even by these more elegant methods, as the variability of the HIV target may lead to primer/template mismatch in the sample material, which is absent from the synthetic control, leading to an underestimate.

VIII. DETECTION SYSTEMS FOR PCR AMPLICONS

The decision on the length of PCR amplicon is determined by two conflicting interests. It is not adequate to identify a product band merely by its size on an agarose gel; the product should be of a sufficient size to be characterized by a secondary screening method, such as hybridization or restriction enzyme digestion. However, it is better in the case of potentially degraded samples to keep the primers as close together as possible in order to maximize the chance of detecting intact molecules (the molecule need only be intact between the primers used, not an entire HIV genome). In practice it is not easy to generate high hybridization signals with very short probes, and hence a spacing of at least 50 bp between the inner primers is recommended. Furthermore, to avoid potential cross hybridization, the probe itself is best synthesized so as not to contain any of the PCR primer sequences.

Two methods of generating probes are commonly used: either labeled oligonucleotides, specific to the region between the inner primers, are synthesized or a third pair of primers, contained entirely within the inner primers (a "tertiary pair"), may be used to PCR amplify a labeled probe from cloned material. It is also possible to perform this labeling amplification with the inner primers of the nested PCR, but as the primers themselves can cross hybridize to satellite bands, care must be used to ensure high-stringency hybridization. High background signals are to be expected by this method. Several modified nucleotides have been substituted into the PCR to label DNA; α-^{32}P dATP, α-^{35}S dATP, digoxygenin dUTP, biotin dUTP, and fluorescein-dUTP have all been successfully used. It should be noted that biotin-dUTP and digoxygenin-dUTP tend to lower the efficiency of the PCR, and hence larger amounts of template should be used and the ratio of hapten-dUTP to dTTP kept low (typically 1 to 10 works well).

In a diagnostic situation the nonradioactive labeling of a synthetic oligonucleotide coupled with either biotin, digoxygenin, or dinitrophenol (DNP) will probably give an optimal probe. This is best performed by incorporating the hapten directly during oligonucleotide synthesis. One further possibility, which has not yet been exploited fully, is the direct conjugation of enzymes to an oligonucleotide via an amino link. This amino group is incorporated into the oligonucleotide during synthesis and subsequently used to cross link an enzyme (usually alkaline phosphatase) to the oligonucleotide. Hybridization of the oligonucleotide probe to the PCR product can then be measured directly, rather than via an enzyme-conjugated mAb targeted to the incorporated hapten.

It is still the case that almost all PCR-positive samples will require gel electrophoresis for their confirmation. However a large amount of time and effort has gone into replacement of the standard gel systems used in PCR with microtiter plate systems. Such systems, generically termed "PCR Enzyme ImmunoAssays" (PCR-EIAs), can be devised in a large number of ways, all of which rely on solid-phase capture of the PCR amplicon, and a plate reader compatible signal, whether that be scintillation, chemiluminescence, colorimetry, or fluorescence.

A. SOLID-PHASE CAPTURE SYSTEMS

Two or three methodologies have been applied to the capture of DNA onto solid supports. The most common is the capture of a biotinylated PCR product by immobilized streptavidin, which is either coated directly onto the microtiter plate,[28,42–44] or onto beads, either M-280 Dynabeads[34] or polystyrene spheres.[45] Immobilized antibody-based methods of capture have targeted either single-stranded DNA[46] or DNA-RNA hybrids.[42,47] In addition proteins specific for a target sequence in the amplified DNA may be coated onto the plate. An example of this is the GCN4 capture assay developed by Kemp and colleagues.[35] Vary[48] has used the triple helical structure of polypyrimidine tracts as a capture system, incorporating polypyrimidine tracts into the PCR primers and coating the microtiter plate with a capture polypyrimidine oligonucleotide. Finally it is possible to coat plates with an oligonucleotide, which will hybridize to the PCR product directly.[49]

B. SCINTILLATION COUNTING

Two companies (LKB and Canberra-Packard) have recently launched microtiter-format scintillation counters, which can measure the radioactivity incorporated in the PCR. Provided that short path length radioisotopes (^{3}H-dATP or α-^{35}S-dATP) are used, little cross talk is seen between wells. These machines therefore offer the closest microtiter-format analog of the conventional approach where densitometry is performed on autoradiographs of Southern blots or dot blots.[23,50]

C. COLORIMETRIC AND CHEMILUMINESCENT METHODS

Almost all of the colorimetric and chemiluminescent methods currently in use are adapted from methods for antibody detection used in other EIA procedures. They are therefore particularly suited to a diagnostic setting, as they will be familiar to the staff performing the tests. The principal problem with the colorimetric systems is their low dynamic range (typically 0 to 2 O.D. U), which makes them of less use as a quantitative diagnostic system. However they do offer the advantage of requiring only standard EIA reagents and standard microtiter plate readers. They are therefore very simple and cheap to introduce into a laboratory already performing EIA.

Chemiluminescent methods have nominal ranges of five orders of magnitude and hence are well-suited to quantitative detection. The reagents for the chemiluminescent assay are very similar to those used for a colorimetric system, but a specialized microtiter luminometer is required to read the completed assay. Additionally the assay is dynamic, and hence has to be read within a certain time period, unlike a colorimetric assay, which can be stopped (usually with H_2SO_4) and, if correctly stored, read at a later time.

D. ENZYMATIC COLORIMETRY AND CHEMILUMINESCENCE

Many of the tests available rely on conjugating a monoclonal antibody (mAb) to a hapten attached to the PCR amplicon, either directly to the PCR primer (e.g., digoxygenin, DNP, or biotin), or via a nucleotide triphosphate analogue incorporated during synthesis (e.g., biotin-11-dUTP, digoxygenin-11-dUTP, or fluorescein-12-dUTP). Having bound the mAb, color or luminescence reactions are performed, usually via a second mAb conjugated either to horseradish peroxidase (HRP),[35,49] to alkaline phosphatase (AP),[23,44,51] or to β-galactosidase.[42,47]

E. NONENZYMATIC CHEMILUMINESCENCE

Blackburn and colleagues[52] describe a nonenzymatic process, which they termed Electro-ChemiLuminescence (ECL), which uses an electrode to initiate chemiluminescence, and which has as good a dynamic range as enzyme-based assays. As the assay is initiated by an electrode, it would be ideally suited for unattended batch operations. Confusingly, ECL is also the acronym used by Amersham for their EnhancedChemiLuminescence system, a conventional HRP-based assay. Schmidt[53] describes an interesting nonenzymatic chemiluminescent system, in which a specific single-stranded probe for the PCR product is derivitized with an acridinium ester. On hydrolysis the chemiluminescence of this probe is rapidly lost. However, because double-stranded DNA is relatively well-protected from hydrolysis compared to single-stranded DNA, positive samples containing hybridized probe can be easily identified by their retention of signal.

F. FLUORESCENT-BASED METHODS

Two types of fluorescent labeling systems have been developed for use in PCR analysis. In the first, fluorescent probes are made and used to detect PCR amplicons by hybridization. Fluorescent probes have been made either by attaching fluorescent rare earth metals to oligonucleotides[43,54,55] or by using fluorescent dye-linked primers in the PCR. More recently it has become possible to fluorescently label PCR reactions by the direct incorporation of fluorescent nucleotide analogs.[56] Currently available analogs are the following: rhodamine-11-dUTP and fluorescein-11-dUTP (FluoroRed and FluoroGreen, Amersham); fluorescein-12-dUTP (Stratagene, Boehringer Mannheim, and others); and fluorescein-15-dATP (Boehringer Mannheim). In our hands fluorescein-11-dUTP is less well incorporated than fluorescein-12-dUTP. This second approach is attractive for diagnostic purposes as it eliminates the need for ethidium bromide (a suspected teratogen) from agarose gel analysis. In combination with microtiter-based solid-phase capture systems (see above), direct measurement of the DNA synthesized in PCR using a fluorometric microtiter plate reader becomes possible.

IX. CONCLUSION

The overview of available methodologies presented here shows that there is a range of simplified PCR assay and detection systems that can be used as the basis of HIV PCR-EIA diagnostic systems. The principal barriers to their more widespread adoption lies in their cost compared to conventional methods and in the problems of quality assurance. One potential source of savings in the method is the use of alternatives to the Perkin Elmer or Roche *Taq* polymerase, which is priced at up to ten times the cost of identical products from other suppliers. However the PCR process is patented, and especially for diagnostic use, there are potential legal pitfalls in using a licensed product. The problem of quality control, especially the possibility of false positives, has been shown to be significant in some cases.[10] However, these problems can be overcome by the incorporation of adequate controls, both for sensitivity and for contamination, into the PCR design.

ACKNOWLEDGMENTS

Thanks are due to Jane McKeating (University of Reading) for her criticisms of this manuscript and to Peter Watts (UCL) for the work on fluorescent analogs.

REFERENCES

1. **Read, S., Cassol, S., Coates, R., Major C., McLaughlin, B., Salas, T., Francis, A., Fanning, M., MacFadden, D., Shepherd, F., et al.,** Detection of incident HIV infection by PCR compared to serology, *J. Acquired Immune Deficiency Syndromes,* 5, 1075, 1992.
2. **Chadwick, E. G., Yogev, R., Kwok, S., Sninsky, J. J., Kellogg, D. E., and Wolinsky, S. M.,** Enzymatic amplification of the human immunodeficiency virus in peripheral blood mononuclear cells from pediatric patients, *J. Infect. Dis.,* 160, 954, 1989.
3. **Husson, R. N., Comeau, A. M., and Hoff, R.,** Diagnosis of human immunodeficiency virus infection in infants and children, *Pediatrics,* 86, 1, 1990.
4. **Rogers, M. F., Ou, C. Y., Kilbourne, B., and Schochetman, G.,** Early identification of human immunodeficiency virus infection in infants, *J. Pediatr.,* 118, 490, 1991.
5. **Rossi, P. and Moschese, V.,** Mother-to-child transmission of human immunodeficiency virus, *FASEB J.,* 5, 2419, 1991.
6. **Zhang, L. Q., Simmonds, P., Ludlam, C. A., and Brown, A. J.,** Detection, quantification and sequencing of HIV-1 from the plasma of seropositive individuals and from factor VIII concentrates, *Aids,* 5, 675, 1991.
7. **Jackson, J. B., MacDonald, K. L., Cadwell, J., Sullivan, C., Kline, W. E., Hanson, M., Sannerud, K. J., Stramer, S. L., Fildes, N. J., Kwok, S. Y., Sninsky, J. J., Bowman, R. J., Polesky, H. F., Balfour, H. H., and Osterholm, M. T.,** Absence of HIV infection in blood donors with indeterminate western blot tests for antibody to HIV-1, *N. Engl. J. Med.,* 322, 217, 1990.
8. **Mariotti, M., Rouger, P., Thauvin, M., Salmon, C., and Lefrere, J. J.,** Failure to detect evidence of human immunodeficiency virus type 1 (HIV-1) infection by polymerase chain reaction assay in blood donors with isolated core antibodies (anti-p24 or -p17) to HIV-1, *Transfusion,* 30, 704, 1990.
9. **Horsburgh, C. J., Ou, C. Y., Jason, J., Holmberg, S. D., Lifson, A. R., Moore, J. L., Ward, J. W., Seage, G. R., Mayer, K. H., Evatt, B. L., Schochetman, G., and Jaffe, H. W.,** Concordance of polymerase chain reaction with human immunodeficiency virus antibody detection, *J. Infect. Dis.,* 162, 542, 1990.

10. **Sheppard, H. W., Ascher, M. S., Busch, M. P., Sohmer, P. R., Stanley, M., Luce, M. C., Chimera, J. A., Madej, R., Rodgers, G. C., Lynch, C., Khayam-Bashi, H., Murphy, E. L., Eble, B., Bradford, W. Z., Royce, R. A., and Winkelstein, W.,** A multicenter proficiency trial of gene amplification (PCR) for the detection of HIV-1, *J. Acquired Immune Deficiency Syndromes,* 4, 277, 1991.

11. **Williams, P., Simmonds, P., Yap, P. L., Balfe, P., Bishop, J., Brettle, R., Hague, R., Hargreaves, D., Inglis, J., Brown, A. L., Peutherer, J., Rebus, S., and Mok, J.,** The polymerase chain reaction in the diagnosis of vertically transmitted HIV infection, *Aids,* 4, 393, 1990.

12. **Balfe, P.,** A statistical method for the detection of false positives and false negatives in microtitre format PCR-assays, *J. Virol. Meth.,* 32, 69, 1992.

13. **Fox, J. C., Ait, K. M., Webster, A., and Emery, V. C.,** Eliminating PCR contamination: is UV irradiation the answer?, *J. Virol. Meth.,* 33, 375, 1991.

14. **Isaacs, S. T., Tessman, J. W., Metchette, K. C., Hearst, J. E., and Cimino, G. D.,** Post-PCR sterilization: development and application to an HIV-1 diagnostic assay, *Nucl. Acids Res.,* 19, 109, 1991.

15. **Simmonds, P., Balfe, P., Peutherer, J. F., Ludlam, C. A., Bishop, J. O., and Leigh Brown, A. J.,** Human immunodeficiency virus-infected individuals contain provirus in small numbers of peripheral mononuclear cells and at low copy numbers, *J. Virol.,* 64, 864, 1990.

16. **Candotti, D., Jung, M., Kerouedan, D., Rosenheim, M., Gentilini, M., M'Pele, P., Huraux, J. M., and Agut, H.,** Genetic variability affects the detection of HIV by polymerase chain reaction, *Aids,* 5, 1003, 1991.

17. **Lynch, C. E., Madej, R., Louie, P. H., and Rogers, G.,** Detection of HIV-1 DNA by PCR: evaluation of primer pair concordance and sensitivity of a single primer pair, *J. Acquired Immune Deficiency Syndromes,* 5, 433, 1992.

18. **McCutchan, F. E., Sanders-Buell, E., Oster, C. W., Redfield, R. R., Hira, S. K., Perine, P. L., Ungar, B. L., and Burke, D. S.,** Genetic comparison of human immunodeficiency virus type 1 (HIV-1) isolates by polymerase chain reaction, *J. Acquired Immune Deficiency Syndromes,* 4, 124, 1991.

19. **Teglbjaerg, L. L., Nielsen, C., and Hansen, J. E.,** Sensitive non-radioactive detection of HIV-1: use of nested primers for the amplification of HIV DNA, *Mol. Cell. Probes,* 6, 175, 1992.

20. **Cassol, S. A., Lapointe, N., Salas, T., Hankins, C., Arella, M., Fauvel, M., Delage, G., Boucher, M., Samson, J., Charest, J., Montpetit, M. L., and O'Shaughnessy, M. V.,** Diagnosis of vertical HIV-1 transmission using the polymerase chain reaction and dried blood spot specimens, *J. Acquired Immune Deficiency Syndromes,* 5, 113, 1992.

21. **McCusker, J., Dawson, M. T., Noone, D., Gannon, F., and Smith, T.,** Improved method for direct PCR amplification from whole blood, *Nucl. Acids Res.,* 20, 6747, 1992.

22. **Holodniy, M., Kim, S., Katzenstein, D., Konrad, M., Groves, E., and Merigan, T. C.,** Inhibition of human immunodeficiency virus gene amplification by heparin, *J. Clin. Microbiol.,* 29, 676, 1991.

23. **Conway, B., Adler, K. E., Bechtel, L. J., Kaplan, J. C., and Hirsch, M. S.,** Detection of HIV-1 DNA in crude cell lysates of peripheral blood mononuclear cells by the polymerase chain reaction and nonradioactive oligonucleotide probes, *J. Acquired Immune Deficiency Syndromes,* 3, 1059, 1990.

24. **Maniatis, T., Fritsch, E. F., and Sambrook, J.,** *Molecular Cloning: A Laboratory Manual,* Cold Spring Harbor Laboratory Press, Cold Spring Harbor, NY, 1982.

25. **Albert, J. and Fenyö, E. M.,** Simple, sensitive, and specific detection of human immunodeficiency virus type 1 in clinical specimens by polymerase chain reaction with nested primers, *J. Clin. Microbiol.,* 28, 1560, 1990.

26. **Yamada, O., Matsumoto, T., Nakashima, M., Hagari, S., Kamahora, T., Ueyama, H., Kishi, Y., Uemura, H., and Kurimura, T.,** A new method for extracting DNA or RNA for polymerase chain reaction, *J. Virol. Meth.,* 27, 203, 1990.

27. **de Micco, P., Zandotti, C., De, L. X., Vignoli, C., and Bollet, C.,** Rapid extraction of HIV-1 DNA for PCR, *Int. Conf. Aids,* 8, 3605, 1992.

28. **Kaye, S., Loveday, C., and Tedder, R. S.,** Storage and preservation of whole blood samples for use in detection of human immunodeficiency virus type-1 by the polymerase chain reaction, *J. Virol. Meth.,* 35, 217, 1991.

29. **Eron, J. J., Gorczyca, P., Kaplan, J. C., and D'Aquila, R. T.,** Susceptibility testing by polymerase chain reaction DNA quantitation: a method to measure drug resistance of human immunodeficiency virus type 1 isolates, *Proc. Natl. Acad. Sci. U.S.A.,* 89, 3241, 1992.

30. **Lefrere, J. J., Mariotti, M., Wattel, E., Lefrere, F., Inchauspe, G., Costagliola, D., and Prince, A.,** Towards a new predictor of AIDS progression through the quantitation of HIV-1 DNA copies of PCR in HIV-infected individuals, *Br. J. Haematol.,* 82, 467, 1992.

31. **Ottmann, M., Innocenti, P., Thenadey, M., Micoud, M., Pelloquin, F., and Seigneurin, J. M.,** The polymerase chain reaction for the detection of HIV-1 genomic RNA in plasma from infected individuals, *J. Virol. Meth.,* 31, 273, 1991.

32. **Chomzynski, P. and Sacchi, N.,** *Anal. Biochem.,* 162, 156, 1987.

33. **Semple, M., Loveday, C., Weller, I., and Tedder, R.,** Direct measurement of viraemia in patients infected with HIV-1 and its relationship to disease progression and Zidovudine therapy, *J. Med. Virol.,* 35, 38, 1992.

34. **Salminen, M. and Leinikki, P.,** Solid phase direct sequencing of HIV-1 quasi-species, *Int. Conf. Aids,* 8, 2070, 1992.

35. **Kemp, D. J., Churchill, M. J., Smith, D. B., Biggs, B. A., Foote, S. J., Peterson, M. G., Samaras, N., Deacon, N. J., and Doherty, R.,** Simplified colorimetric analysis of polymerase chain reactions: detection of HIV sequences in AIDS patients, *Gene,* 94, 223, 1990.

36. **Young, B. D. and Anderson, M. L. M.,** Quantitative analysis of solution hybridisation, in *Nucleic Acid Hybridisation: A Practical Approach,* Hames, B. D. and Higgins, S. J., Eds., IRL Press, Oxford, 1985, chap. 3.

37. **Lee, T. H., Sunzeri, F. J., Tobler, L. H., Williams, B. G., and Busch, M. P.,** Quantitative assessment of HIV-1 DNA load by coamplification of HIV-1 gag and HLA-DQ-alpha genes, *Aids,* 5, 683, 1991.

38. **Jurriaans, S., Dekker, J. T., and de Ronde, R. A.,** HIV-1 viral DNA load in peripheral blood mononuclear cells from seroconverters and long-term infected individuals, *Aids,* 6, 635, 1992.

39. **Menzo, S., Bagnarelli, P., Giacca, M., Manzin, A., Varaldo, P. E., and Clementi, M.,** Absolute quantitation of viremia in human immunodeficiency virus infection by competitive reverse transcription and polymerase chain reaction, *J. Clin. Microbiol.,* 30, 1752, 1992.

40. **Pannetier, C., Delassus, S., Darche, S., Saucier, C., and Kourilsky, P.,** Quantitative titration of nucleic acids by enzymatic amplification reactions run to saturation, *Nucl. Acids Res.,* 21, 577, 1993.

41. **Telenti, A., Imboden, P., and Germann, D.,** Competitive polymerase chain reaction using an internal standard: application to the quantitation of viral DNA, *J. Virol. Meth.,* 39, 259, 1992.

42. **Coutlee, F., Saint, A. P., Olivier, C., Voyer, H., Kessous, E. A., Berrada, F., Begin, P., Giroux, L., and Viscidi, R.,** Evaluation of infection with human immunodeficiency virus type 1 by using nonisotopic solution hybridization for detection of polymerase chain reaction-amplified proviral DNA, *J. Clin. Microbiol.,* 29, 2461, 1991.

43. **Dahlen, P. O., Litia, A. J., Skagius, G., Frostell, A., Nunn, M. F., and Kwiatkowski, M.,** Detection of human immunodeficiency virus type 1 by using the polymerase chain reaction and a time-resolved fluorescence-based hybridization assay, *J. Clin. Microbiol.,* 29, 798, 1991.

44. **Suzuki, K., Okamoto, N., Watanabe, S., and Kano, T.,** Chemiluminescent microtiter method for detecting PCR amplified HIV-1 DNA, *J. Virol. Meth.,* 38, 113, 1992.

45. **Harju, L., Janne, P., Kallio, A., Laukkanen, M. L., Lautenschlager, I., Mattinen, S., Ranki, A., Ranki, M., Soares, V. R., Soderlund, H., and Syränen, A.-C.,** Affinity-based collection of amplified viral DNA: application to the detection of human immunodeficiency virus type 1, human cytomegalovirus and human papillomavirus type 16, *Mol. Cell. Probes,* 4, 223, 1990.

46. **Trenti, G., Borghi, V., Pietrosemoli, P., Codeluppi, M., and Lami, G.,** Immunodetection of PCR amplified fragments of HIV-1 DNA in children born to seropositive mothers, *Int. Conf. Aids,* 8, 2092, 1992.

47. **Coutlee, F., Yang, B. Z., Bobo, L., Mayur, K., Yolken, R., and Viscidi, R.,** Enzyme immunoassay for detection of hybrids between PCR-amplified HIV-1 DNA and a RNA probe: PCR-EIA, *Aids Res. Human Retroviruses,* 6, 775, 1990.

48. **Vary, C. P.,** Triple-helical capture assay for quantification of polymerase chain reaction products, *Clin. Chem.,* 38, 687, 1992.

49. **Keller, G. H., Huang, D. P., and Manak, M. M.,** Detection of human immunodeficiency virus type 1 DNA by polymerase chain reaction amplification and capture hybridization in microtiter wells, *J. Clin. Microbiol.,* 29, 638, 1991.

50. **Conway, B., Bechtel, L. J., Adler, K. A., D'Aquila, R. T., Kaplan, J. C., and Hirsch, M. S.,** Comparison of spot-blot and microtitre plate methods for the detection of HIV-1 PCR products, *Mol. Cell. Probes,* 6, 245, 1992.

51. **Bettens, F., Pichler, W. J., and de Weck, A.,** Incorporation of biotinylated nucleotides for the quantification of PCR-amplified HIV-1 DNA by chemiluminescence, *Eur. J. Clin. Chem. Clin. Biochem.,* 29, 685, 1991.

52. **Blackburn, G. F., Shah, H. P., Kenten, J. H., Leland, J., Kamin, R. A., Link, J., Peterman, J., Powell, M. J., Shah, A., Talley, D. B., et al.,** Electrochemiluminescence detection for development of immunoassays and DNA probe assays for clinical diagnostics, *Clin. Chem.,* 37, 1534, 1991.

53. **Schmidt, B. L.,** A rapid chemiluminescence detection method for PCR-amplified HIV-1 DNA, *J. Virol. Meth.,* 32, 233, 1991.

54. **Bush, C. E., Donovan, R. M., Peterson, W. R., Jennings, M. B., Bolton, V., Sherman, D. G., Vanden, B. K., Beninsig, L. A., and Godsey, J. H.,** Detection of human immunodeficiency virus type 1 RNA in plasma samples from high-risk pediatric patients by using the self-sustained sequence replication reaction, *J. Clin. Microbiol.,* 30, 281, 1992.

55. **Dahlen, P., Litia, A., Mukkala, V. M., Hurskainen, P., and Kwiatkowski, M.,** The use of europium (Eu3+) labelled primers in PCR amplification of specific target DNA, *Mol. Cell. Probes,* 5, 143, 1991.

56. **Woolford, A. J. and Dale, J. W.,** Simplified procedures for detection of amplified DNA using fluorescent label incorporation and reverse probing, *FEMS Microbiol. Lett.,* 99, 311, 1992.

Chapter 12

PCR for the Detection and Discrimination of Human T-Lymphotropic Virus Type I and Type II Infections

Helen H. Lee and Gregor W. Leckie

TABLE OF CONTENTS

I. INTRODUCTION

Human T-cell lymphotropic viruses type I (HTLV-I) and type II (HTLV-II) are retroviruses that infect humans. HTLV-I has been conclusively shown to be an etiologic agent of human disease, while HTLV-II has not. Due to the high degree of genetic homology between HTLV-I and HTLV-II, current HTLV-I screening enzyme immunoassays (EIAs) detect antibody to both HTLV-I and HTLV-II. Available confirmatory assays also do not reliably distinguish between anti-HTLV-I and anti-HTLV-II antibody. Consequently, there is an urgent need for confirmatory techniques that can discriminate between HTLV-I and HTLV-II infection. This chapter will review the biology and seroepidemiology of HTLV-I and HTLV-II, the current HTLV-I/II screening and confirmatory protocols, and the development of reliable discriminatory techniques, paying particular attention to the polymerase chain reaction (PCR).

II. BACKGROUND OF HTLV-I/II

HTLV-I and HTLV-II are closely related type C retroviruses belonging to the oncovirus, or cancer-causing, subfamily of the Retroviridae family. An HTLV-I/II viral particle consists of two copies of a 9000-nucleotide single-stranded RNA genome; closely associated with the genetic material is an enzyme, reverse transcriptase, that is encoded by the viral gene *pol*. Surrounding the genetic material and reverse transcriptase is an internal core, which consists of the *gag*-encoded viral proteins p15, p19, and p24. The internal core is surrounded by an external envelope consisting of a lipid bilayer and the virally encoded *env* proteins, the transmembrane protein p21e, and its external attachment protein gp46. Like human immunodeficiency virus-I (HIV-I), HTLV-I and HTLV-II preferentially infect human T-helper lymphocytes of the CD4 and CD8 lineages, although other cell types are also infected. During infection, the viral RNA genome uncoats, and it is copied by the reverse transcriptase into double-stranded DNA, which then is integrated into the host cell's DNA genome. The integrated genome, or provirus, directs the production of more HTLV-I/II virions. A more detailed description of the HTLV-I/II life cycle is given elsewhere.[1]

The nucleotide sequences of the HTLV-I and HTLV-II genomes share approximately 65% homology.[2,3] Nucleotide sequence homology is highest between the 3' halves of the viral genomes. The most highly conserved proteins are *gag* p24 (83%) and *env* p21e (84%). The least conserved is the protease (24%). Despite their homology, HTLV-I and HTLV-II are different viruses and consequently have distinct properties.

Both HTLV-I and HTLV-II can immortalize normal human peripheral blood lymphocytes (PBL) *in vitro*. This transforming ability is thought to be due to the tax gene product.[1,4] Despite the transforming ability of both viruses, only HTLV-I has been conclusively linked to human disease. HTLV-I is the etiologic agent of an aggressive adult T-cell leukemia/lymphoma (ATL) and a demyelinating neurological disease, HTLV-I-associated myelopathy or tropical spastic paraparesis (HAM-TSP). The chances of an HTLV-I-seropositive person developing either disease are relatively low. By comparison, HTLV-II has not been conclusively linked to human disease even though HTLV-II has been isolated from patients with a wide variety of medical conditions, including two individuals (MO and NRA) with hairy cell leukemia (HCL).[1,4]

Seroepidemiologic studies have demonstrated that HTLV-I infection is found worldwide, with high rates being observed in Japan, the Caribbean, and some parts of Africa; it has also been detected in the Americas, the Pacific, and Europe. Cases of ATL and HAM/TSP have also been observed worldwide.[4] The seroepidemiology of HTLV-II is much less well understood. In large part this is because the HTLV-I-screening EIAs detect antibody to both HTLV-I and HTLV-II.[5-7] Workers who have used techniques capable of discriminating HTLV-II infection from that of HTLV-I have demonstrated that both viruses infect U.S. blood donors at an equal rate, and that HTLV-I/II-seropositive intravenous drug users (IVDU) from the U.S. and Italy are usually infected with HTLV-II, not HTLV-I.[6,8,9] HTLV-II has also been detected in the ancient populations of North America, Central America, and Africa.[10-13] HTLV-I and HTLV-II are transmitted by intravenous drug use, by the transfusion of infected blood units containing cellular components, by sexual contact, and postnatally by breast feeding.[1,4]

III. CURRENT TESTING PROTOCOLS FOR HTLV-I/II

To reduce the transmission of HTLV-I/II by blood transfusion, the U.S., Canada, and France screen all donated blood units for antibody to these viruses using HTLV-I viral lysate-based EIAs. Plasma is not tested because it does not transmit HTLV-I/II infection.[14] Blood units that are EIA repeat reactive are discarded. However, the repeat reactive blood units are also tested by a confirmatory pathway that includes two different HTLV-I viral lysate-based assays, Western blot (WB) and radioimmunoprecipitation assay (RIPA). Blood units are considered confirmed if either WB or RIPA, or both, detect antibody to *gag* p24 and *env* gp46 (WB) or *env* gp61/68 (RIPA). Samples are considered indeterminate if antibody against *gag* alone or *env* alone is detected and nonconfirmed if no antibody is detected.[5] Neither the HTLV-I screening EIAs nor the confirmatory assays can reliably discriminate between antibody to HTLV-I and to HTLV-II.[5-7] Detection of antibody to *gag* p24 alone, by WB, is indicative of HTLV-II infection but is not conclusive.[1,15] Consequently, reliable discriminatory assays are required.

IV. DEVELOPMENT OF DISCRIMINATORY TECHNIQUES

The development of techniques to reliably distinguish HTLV-I from HTLV-II infection is important for the following reasons: (1) the epidemiology of HTLV-II is poorly understood; (2) well-defined epidemiology is required so that possible disease associations for HTLV-II can be recognized; and (3) HTLV-I/II-seropositive persons need to know if they are infected by the disease-causing virus, HTLV-I, or by the presumably harmless virus, HTLV-II. Researchers have developed a number of methods to discriminate HTLV-I from HTLV-II infection. All the methods will be discussed, but particular attention will be paid to the polymerase chain reaction (PCR).

A. DISCRIMINATORY IMMUNOASSAYS

Numerous research and manufacturing groups have identified short (approximately 18 to 30 amino acid residues), synthetic peptides from the *gag* and *env* regions of HTLV-I and HTLV-II for immunologic differentiation of the two viruses.[13,16-19] Discriminatory recombinant proteins have also been identified from the *env* region of HTLV-I and HTLV-II.[20] Using synthetic-peptide-based immunoassays, we analyzed 135 HTLV-I/II seropositive blood samples that had been typed by PCR as being either HTLV-I or HTLV-II infected. An HTLV-I gp46 synthetic peptide detected all 59 HTLV-I-confirmed blood samples, while a *gag* p19 peptide detected only 29 of 59 samples. Neither peptide cross reacted with any of the HTLV-II blood samples. The HTLV-II peptides were not as effective as the HTLV-I peptides. The HTLV-II gp46 peptide detected 63 of 76 HTLV-II blood samples, while the HTLV-II p19 peptide detected 45 of 76 HTLV-II samples. However, the HTLV-II p19 peptide cross reacted with 12 of the 59 HTLV-I-seropositive blood samples. Our experience suggests that with careful selection, synthetic-peptide-based immunoassays will be able to reliably discriminate more than 90% of confirmed HTLV-I/II infections. Such results have been reported in the literature.[17–19] It is therefore likely that these reliable, quick immunoassays will replace PCR for routine differentiation of HTLV-I/II seropositives.

B. CO-CULTURE

Co-culture of presumed infected HTLV-I/II peripheral blood lymphocytes (PBLs) with uninfected PBLs takes 10 to 20 days to produce a result, is labor intensive and potentially biohazardous, and has a sensitivity of 63%.[21] HTLV-I/II replication is measured by immunological and molecular biological techniques, which can be made discriminatory if required.

C. DNA:DNA HYBRIDIZATION

Southern blot hybridization performed on DNA extracted from the PBLs of HTLV-I/II-seropositive persons is faster and safer than co-culture and directly differentiates HTLV-I from HTLV-II. However, Southern blot is labor intensive and has a sensitivity of only 29%.[22] This lack of sensitivity is because (1) Southern blot requires that between 1 and 10% of PBLs be infected by HTLV-I/II before it can produce a positive signal and (2) quantitative PCR assays have shown that between 0.067 and 6.25% of the PBLs from asymptomatic HTLV-I seropositives contain integrated provirus, while 3.7 to 20% of PBLs from HTLV-I-seropositive patients with HAM/TSP are infected.[23,24]

V. POLYMERASE CHAIN REACTION (PCR)

Numerous groups have developed PCR assays for detection of HTLV-I/II because it is the only direct detection technique with enough sensitivity to be able to consistently detect integrated HTLV-I/II provirus from the PBLs of asymptomatic seropositive persons. PCR can either detect both HTLV-I and HTLV-II or discriminate between the two viruses. However, PCR is labor intensive and can produce false positives if performed with poor technique.

A. SELECTION OF PCR PRIMERS

Many of the groups that have developed PCR assays for detection of HTLV-I/II have used PCR primer pairs and probes originally identified by Kwok and colleagues.[25–27] The primer pairs were selected to be either specific for HTLV-I or HTLV-II or to detect both viruses. The nucleotide sequences of the Kwok primers and probes have been published, and they are commercially available from Perkin-Elmer (Norwalk, CT) and from Synthetic Genetics (San Diego, CA).[28] Other groups have also developed HTLV-I/II PCR primer pairs and probes. Table 1 describes primers routinely used in our laboratory.

We used Kwok's *tax/rex* SK43/44 primer pairs as our screening tool for routine detection of both HTLV-I and HTLV-II.[28] The 20-nucleotide SK43 primer is identical to the coding sequence of the *tax/rex* region of HTLV-I, but has two mismatches with the equivalent region of the HTLV-II sequence. The 21-nucleotide SK44 primer is exactly complementary to the HTLV-I nucleotide sequence, but has two mismatches with the HTLV-II sequence. Despite the mismatches with HTLV-II, the *tax/rex* SK43/44 primer pair successfully amplifies a 159-nucleotide fragment from the vast majority of HTLV-I/II isolates. The ability of one PCR primer pair to detect nearly all HTLV-I/II isolates is yet more proof of the genetic conservation of both HTLV-I and HTLV-II.[29,30]

Despite the effectiveness of the *tax/rex* SK43/44 primer pair, we also used five other primer pairs for confirmation of PCR results. Using standard protocols, we designed primer pair 57/58 to amplify a 116-nucleotide region from the *env* p21e region of both HTLV-I and HTLV-II. Upstream primer 57 is 20 nucleotides long, is identical to the HTLV-I sequence, and has one mismatch with HTLV-II. Downstream primer 58 is 23 nucleotides long, is perfectly complementary to the HTLV-I sequence, but has three mismatches with the HTLV-II sequence. We also designed primer pairs against

Table 12–1   Primer pairs used for the amplification and detection of HTLV-I and HTLV-II

Gene, virus	Primer ID	Position (nt)	Amplified Product Size (nt)	Sequence (5′ → 3′)
tax, HTLV-I	SK 43	7359-7378	159	CGGATACCCAGTCTACGTGT
	SK 44	7517-7497		GAGCCGATAACGCGTCCATCG
tax, HTLV-II	SK 43	7248-7267	159	tGGATACCCcGTCTACGTGT
	SK 44	7406-7386		GAGCtGAcAACGCGTCCATCG
env p21e, HTLV-I	57	6154-6173	116	GTCTGGCTTGTCTCCGCCCT
	58	6269-6212		TGGGAAATATCTTTGTCCACCTC
env p21e, HTLV-II	57	6119-6138	116	GTgTGGCTTGTCTCCGCCCT
	58	6234-6212		TGGGAgATgTCTTTGTCaACCTC
gag p19, HTLV-I	1	834-853	124	TCTTTTCCCGTAGCGCTAGC
	4	957-938		TGGTGGAAATCGTAACTGGA
gag p19, HTLV-II	3	897-816	147	GCTTACCGCTTGCAGCCTAG
	7	1043-1023		CCTTCCTGGATATCCCTTGGG
pol, HTLV-I	SK 54	3366-3385	119	CTTCACAGTCTCTACTGTGC
	SK 55	3484-3466		CGGCAGTTCTGTGACAGGG
pol, HTLV-II	SK 58	4198-4217	103	ATCTACCTCCACCATGTCCG
	SK 59	4300-4281		TCAGGGGAACAAGGGGAGCT
env gp46, HTLV-II	82	5323-5342	539	CGACCTTAATTCCCTAACAA
	88	5861-5842		ATCTATCCACGCAAACCATG
env gp46, HTLV-II	85	5618-5637	434	CGACTACACTTCTCTAAGTG
	86	6051-6032		AGGATAATGGAGTTGTTGCA

GenBank accession numbers are J02029 for HTLV-I and M10060 for HTLV-II, MacVector software version 3.5. Mismatches between HTLV-I and HTLV-II for primers SK43, SK44, 57 and 58 are represented by lower case letters.

the *gag* p19 region that would only detect either HTLV-I (primers 1 and 4, Table 1) or HTLV-II (primers 3 and 7, Table 1). The specific *pol* primer pairs were designed by Kwok and colleagues and were designated SK54/55 (HTLV-I) and SK58/59 (HTLV-II).

B. SAMPLE PREPARATION

All of the HTLV-I/II PCR protocols detect HTLV-I/II provirus that is integrated into infected PBLs. We normally collect 50 ml of heparinized blood (other groups collect as little as 5 ml) and layer a portion of it onto a Ficoll-Hypaque density gradient. During a low-speed centrifugation, the PBLs are separated from the other blood components on the basis of their density. The PBL band is readily visualized and harvested. After a wash step, the collected PBLs are counted before being processed immediately or cryopreserved (–70°C) until they are analyzed by PCR or culture. The isolated PBLs are lysed by proteinase K-SDS treatment to release genomic DNA. Cellular protein is removed by phenol-chloroform treatment, and the DNA is collected by ethanol precipitation. The DNA is reconstituted in Tris-EDTA (TE) buffer before its concentration and purity are determined by optical density analysis. The processed DNA samples are either stored at 2 to 8°C or frozen (–70°C) until analysis.

C. PCR AMPLIFICATION

We add 5 μg of purified DNA, or the equivalent of 750,000 PBLs, from each sample to a tube that already contains the PCR reaction components. Other groups typically add 1 to 2 μg of DNA. The reaction tubes are capped, and placed into a thermocycler, where they are first heated to 95°C to denature double-stranded DNA into its single-stranded counterpart. The heating step is followed by a defined thermocycling program, which varies with the primer pair used. After thermocycling, the complementary PCR product molecules hybridize to one another to form double-stranded DNA.

D. DETECTION OF PCR PRODUCT

HTLV-I and HTLV-II DNA can be differentiated by using primer sets that are specific for each virus or by using PCR primers that amplify a region from both viruses and differentiating after amplification. We end label the 5′ terminus of the upstream primer of each probe set with [γ-^{32}P] so that the PCR product is radiolabeled. After PCR amplification, we load a portion of the reaction onto an 8% PAGE gel. After electrophoresis the gel is dried, and the radiolabeled PCR product is detected by autoradiography. Figure 12–1 illustrates the 159-nucleotide *tax/rex* PCR product and the 116 nucleotide *env* p21e PCR product while Figure 12–2 illustrates the HTLV-I and HTLV-II *gag* p19 and *pol* PCR products. It is evident that DNA from both HTLV-I and HTLV-II infected cells is amplified by both the *tax/rex* and *env* p21e primer pairs, while the *gag* p19 and *pol* primer pairs are specific for either HTLV-I or HTLV-II proviral DNA. Dilution analysis has demonstrated that the *tax/rex* PCR assay has a sensitivity of 10 HTLV-I/II proviral genomes in a background of 750,000 uninfected PBLs.

Differentiation of the *tax/rex* and *env* p21e products, or their equivalents, can be achieved by probing with a specific, radiolabeled probe in a dot blot, Southern blot, or solution hybridization format. Most researchers use one of these options. We use a different option in which we subject the radiolabeled PCR product to diagnostic restriction endonuclease analysis prior to electrophoresis. Both the *tax/rex* and *env* p21e PCR products contain different restriction endonuclease sites, depending on whether they are derived from HTLV-I or from HTLV-II. The 159-nucleotide HTLV-1 *tax/rex* PCR product has one *Taq* I site and one *Sau* 3A1 site, while the HTLV-II *tax/rex* PCR product has two *Taq* I sites and no *Sau* 3A1 site. Figure 12–1 illustrates that if an HTLV-I *tax/rex* PCR product is digested with *Taq* I, it produces a labeled fragment of 138 nucleotides and an unlabeled fragment of 21 nucleotides. Autoradiography will detect only the radiolabeled fragment. By comparison, a *Taq* I digest of a *tax/rex* HTLV-II PCR product will produce a radiolabeled fragment of 85 nucleotides and an unlabeled fragment of 74 nucleotides. If *Sau* 3A1 is used for digestion, the radiolabeled HTLV-I PCR product will be 120 nucleotides long, while the HTLV-II equivalent will be 159 nucleotides long.

Figure 12–1 also shows that similar results can be produced with the 116-nucleotide *env* p21e PCR product. Digestion with *Taq* I does not change the size of an HTLV-I PCR product, but cleaves an HTLV-II PCR product into a radiolabeled fragment of 91 nucleotides and an unlabeled fragment of 25 nucleotides. Other detection methodologies, including capture EIA and a semiquantitative slot blot, have also been used for HTLV-I/II PCR.[18,24]

VI. INFORMATION OBTAINED BY HTLV-I/II PCR ASSAYS

A. EPIDEMIOLOGICAL STUDIES

In the mid to late 1980s, apart from the identification of a few clinical isolates of HTLV-II and reports of HTLV-II-infected IVDUs from England and New York City, little was known about the epidemiology of HTLV-II.[4,31,32] We therefore developed a discriminatory PCR assay to study the epidemiology of HTLV-I and HTLV-II. Using the *tax/rex* PCR primers and diagnostic restriction endonuclease analysis, we determined that 21/23 HTLV-I/II-seropositive IVDUs from New Orleans were actually infected by HTLV-II and not HTLV-I.[6] This finding was of considerable interest since it had been previously shown that IV drug use was a major risk factor for HTLV-I/II seropositivity.[31–33] Other workers who used discriminatory PCR assays confirmed our findings when they found that a large proportion of HTLV-I/II-seropositive IVDUs from California, New Jersey, New York, and Italy were infected by HTLV-II, not HTLV-I.[9,34–36]

We extended our study to 480,000 volunteer U.S. blood donors and determined that 207 of them, or 0.043%, were HTLV-I/II seropositive.[8] Of the 65 HTLV-I/II-seropositive blood donors that we further examined, 28 (43%) were infected with HTLV-I and 34 (52%) with HTLV-II. Donor interviews established a strong link between HTLV-I infection and origin, or sexual contact with a person, from

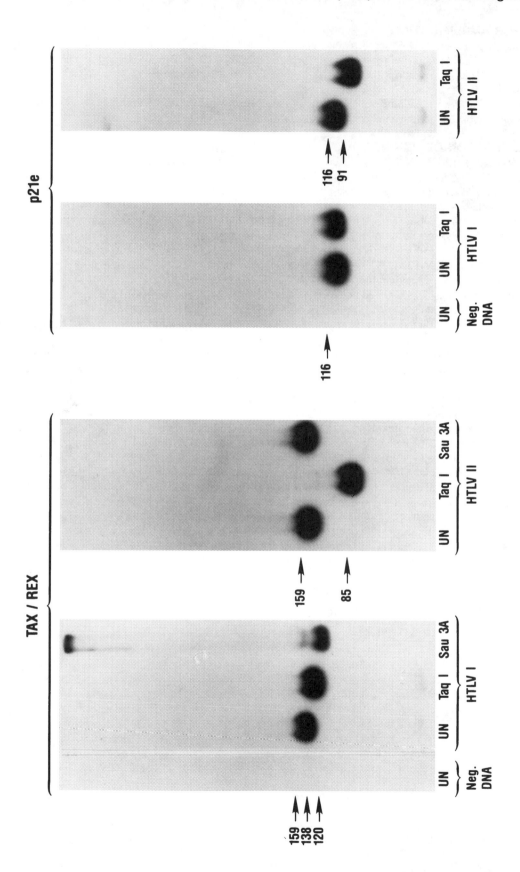

Figure 12–1 Amplification and diagnostic restriction endonuclease analysis of the *tax/rex* and *env* p21e HTLV-I/II PCR products. The sizes of the PCR products (in base pairs) are shown on the side of the figure. UN, uncut; *Taq* I and *Sau* 3AI, restriction endonucleases.

an HTLV-I endemic region. HTLV-II infection, by comparison, was strongly associated with intravenous drug use or sexual contact with an IVDU. Transfusion appeared to be an equivalent risk factor for both HTLV-I and HTLV-II. Consequently, there appear to be differences in the risk factors for HTLV-I and HTLV-II infection. Given these results, it is interesting to note that PCR studies have identified pockets of endemic HTLV-II infection in Guaymi Indians of Panama, in native Americans of New Mexico, and in Africans.[10–12] The HTLV-II-infected persons from these regions did not have the normal HTLV-II risk factors, suggesting that HTLV-II infection has been present in these populations for many years. Further epidemiologic studies are required to study whether currently defined HTLV-I-endemic regions also harbor HTLV-II, to find other HTLV-II-endemic populations, to determine if HTLV-II infection has any disease associations and to find out why HTLV-II, not HTLV-I, is so prevalent among IVDUs.

B. STUDY OF SEROLOGICAL INDETERMINATES

As previously mentioned, blood units that are repeat reactive by the HTLV-I-screening EIA, but have antibody to *gag* p19, *gag* p19 and *gag* p24, or *gag* p19 and *env* gp61/68, are currently defined as indeterminate.[5] Of 2851 EIA repeat-reactive blood samples that we tested using the current confirmatory protocol, 285 (10%) were confirmed, 1796 (63%) were indeterminate, and 770 (27%) were nonconfirmed. Because it is hard to counsel persons with indeterminate HTLV-I/II serology, we decided to analyze such blood samples by PCR. The *tax/rex* PCR assay confirmed HTLV-I/II infection in 9 of 49 (18.4%) samples with indeterminate serology. All 12 samples with reactivity to *gag* p19 only were negative by PCR, while 4 of 25 samples with *gag* p24 only were infected by HTLV-II. Of the 12 samples that had *gag* p19 and *gag* p24 reactivity or *gag* p19 and *env* gp61/68 reactivity, 5 were determined to be infected by HTLV-I.

Other workers have also studied indeterminate samples with varying results. Three groups did not find any PCR-confirmed indeterminate samples, while two groups found that 3 and 21% of indeterminate blood samples had PCR-confirmed HTLV-I/II infections.[37–41] The group that detected HTLV-I/II in 21% (4 of 19) of their indeterminate blood samples confirmed HTLV-I/II infection in all 4 by co-culture.[41] The variation in results could be due to methodological differences or to variations in the populations of indeterminate blood samples.

C. TRANSMISSION OF HTLV-II

The routes of transmission have already been established for HTLV-I by numerous groups in Japan, an HTLV-I-endemic area.[1,4] Since HTLV-I and HTLV-II infection are equally common in the U.S., all the groups that have studied transmission of HTLV-II have used PCR assays, and some have used discriminatory peptide immunoassays as well. Transmission of HTLV-II has been documented by blood transfusion and by sexual contact.[42–45] Considering the high rate of HTLV-II infection in IVDUs, it is highly likely, if not yet proven, that intravenous drug use is another HTLV-II transmission route. As with HTLV-I, transplacental transmission of HTLV-II either never occurs or is very rare.[46] Transmission of HTLV-II by breast feeding has not been documented.

D. GENETIC VARIATION OF HTLV-I/II ISOLATES

We sequenced the entire genome of HTLV-II strain NRA and found that it had 95.2% homology with the HTLV-II prototype strain, Mo.[30] In the same report we described PCR differentiation of HTLV-II isolates into Mo-like and NRA-like subtypes. We used two different PCR primer pairs against the HTLV-II *env* gene: primer pair 82/88, which defined a 539-bp fragment from nucleotides 5323 to 5861, and primer pair 85/86, which defined a 434-bp fragment from nucleotides 5618 to 6051. Each primer is 20 nucleotides long and primer positions correspond to nucleotide sequence in an entire proviral genome of HTLV II Mo isolate. After PCR amplification, the radiolabeled double-stranded PCR products were subjected to restriction endonuclease analysis before being analyzed by PAGE and autoradiography. Figure 12–3 illustrates that the restriction enzyme patterns of the 82/88 PCR product vary, depending on whether the isolate is Mo or NRA, or is from an Italian IVDU. The diagnostic enzyme is *Rsa* I; it cleaves at different positions within the PCR product of the Mo isolate and the Italian IVDU; however, it does

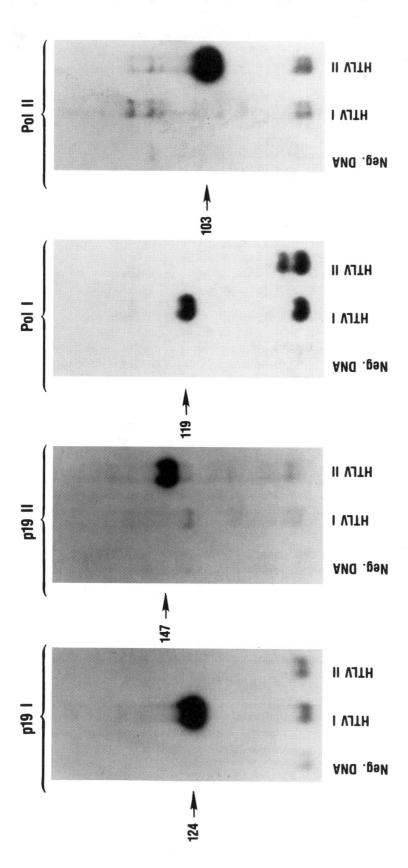

Figure 12–2 Specific HTLV-I and HTLV-II PCR products from the *gag* p19 and *pol* genes. The sizes of the PCR products (in base pairs) are shown on the side of the figure.

not cleave the NRA PCR product. By comparison, Figure 12–4 illustrates that when we used the PCR primers 85/86, we determined that the Mo and NRA restriction patterns were still clearly distinguishable but that the Italian HTLV-II isolate now resembled the NRA pattern.

When we analyzed other HTLV-II isolates, we found that 5 of 8 isolates from U.S. blood donors had the Mo restriction enzyme pattern, as did 6 of 10 isolates from U.S. IVDUs. The NRA-like pattern was observed in the remaining HTLV-II-infected U.S. blood donors and IVDUs. However, we observed the Italian restriction digest pattern in all 7 of the HTLV-II-infected Italian IVDUs that we studied.

Nucleotide sequence analysis of HTLV-II PCR products resulted in the determination of sequence from nucleotides 5291 to 5610. This analysis revealed that the Mo and NRA isolates were different at 4.1% of their nucleotide positions. The Italian isolates were NRA-like with the exception that they all had a G instead of an A at nucleotide 5413. This substitution created the additional *Rsa* I site that resulted in the unique restriction pattern of PCR product 82/88. Other workers have also noticed the division of HTLV-II isolates into Mo or NRA-like subtypes.[47,48]

Similar work has been performed on isolates of HTLV-I. Like HTLV-II, HTLV-I isolates are highly conserved but they do not appear to be divided into two distinct subtypes.[29] The nucleotide sequence varies with the geographical origin of the isolate; variation has not been tied to either of the HTLV-I-associated disease states.[29,49] Consequently, PCR analysis is a useful tool with which to study the molecular epidemiology of HTLV-I and HTLV-II.

VII. SUMMARY

As applied to diagnosis and discrimination of HTLV-I/II infection, PCR is an extremely powerful tool. Currently it is the standard against which other discriminatory techniques such as synthetic-peptide-based immunoassays are measured. PCR has helped to define the seroepidemiology of HTLV-I and HTLV-II as well as the routes of transmission for HTLV-II; it has also been used to study the molecular variation of HTLV-I and HTLV-II isolates. However, due to the high skill level needed to successfully perform PCR, it will not become part of the normal confirmatory protocol. That role will be played by discriminatory immunoassays. But PCR will still be used for research applications and for differentiation of blood samples that give serologically confusing results.

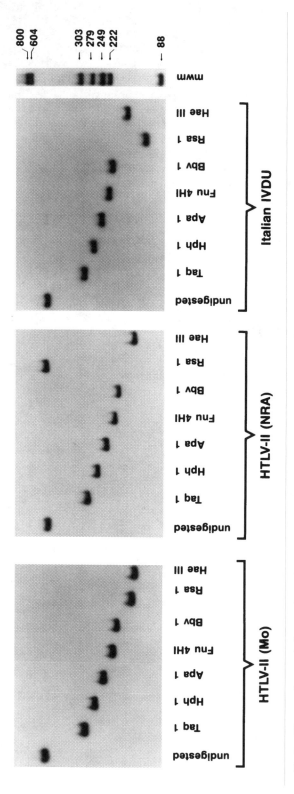

Figure 12–3   Diagnostic restriction endonuclease analysis of the *env* 82/88 PCR product amplified from different isolates of HTLV-II. The sizes of the PCR products in (base pairs) are established in relation to a molecular weight marker (mwm). *Taq* I, *Hph* I *Apa* I, *Fnu* 4HI, *Bbv* I *Rsa* I, *Hae* III are restriction endonucleases used.

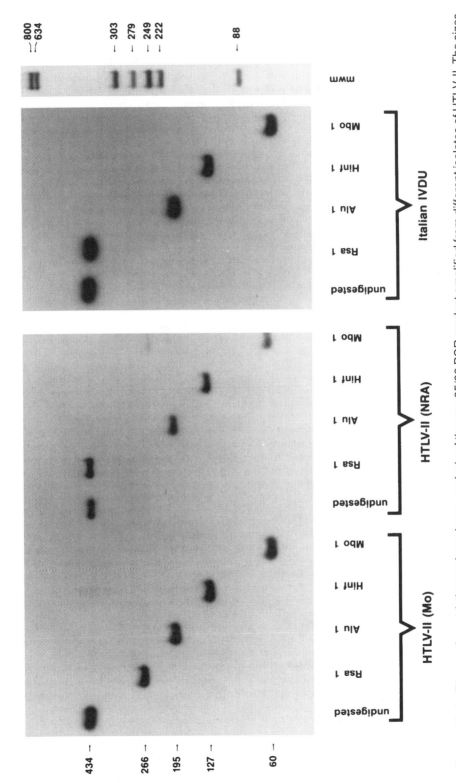

Figure 12–4 Diagnostic restriction endonuclease analysis of the *env* 85/86 PCR product amplified from different isolates of HTLV-II. The sizes of the PCR products (in base pairs) are established in relation to a molecular weight marker (mwm). *Rsa* I, *Alu* I, *Hinf* I, *Mbo* I are restriction endonucleases used.

REFERENCES

1. **Hjelle, B.,** Human T-cell leukemia/lymphoma viruses: life cycle, pathogenicity, epidemiology, and diagnosis, *Arch. Pathol. Lab. Med.,* 115, 440, 1991.
2. **Seiki, M., Hattori, S., Hirayama, Y., and Yoshida, M.,** Human adult T-cell leukemia virus: complete nucleotide sequence of the provirus genome integrated in leukemia cell DNA, *Proc. Natl. Acad. Sci. U.S.A.,* 80, 3618, 1983.
3. **Shimotohno, K., Takahashi, Y., Shimizu, N., Gojobori, T., Golde, D. W., Chen, I. S. Y., Miwa, M., and Sugimura, T.,** Complete nucleotide sequence of an infectious clone of human T-cell leukemia virus type II: an open reading frame for the protease gene, *Proc. Natl. Acad. Sci. U.S.A.,* 82, 3101–3105, 1986.
4. **Sandler, S., Fang, C. T., and Williams, A. E.,** Human T-cell lymphotropic virus type I and II in transfusion medicine, *Transfusion Med. Rev.,* 5, 93, 1991.
5. **Anderson, D. W., Epstein, J. S., Lee, T.-H., Lairmore, M. D., Saxinger, C., Kalyanaraman, V. S., Slamon, D., Parks, W., Poiesz, B. J., Pierik, L. T., Lee, H., Montagna, R., Roche, P. A., Williams, A., and Blattner, W.,** Serological confirmation of human T-lymphotropic virus type I infection in healthy blood and plasma donors, *Blood,* 74, 2585, 1989.
6. **Lee, H., Swanson, P., Shorty, V. S., Zack, J. A., Rosenblatt, J. D., and Chen, I. S. Y.,** High rate of HTLV-II infection in seropositive drug abusers in New Orleans, *Science,* 244, 471, 1989.
7. **Wiktor, S. Z., Pate, E. J., Weiss, S. H., Gohd, R. S., Correa, P., Fontham, E. T., Hanchard, B., Biggar, R. J., and Blattner, W. A.,** Sensitivity of HTLV-I antibody assays for HTLV-II, *Lancet,* 338, 512, 1991.
8. **Lee, H. H., Swanson, P., Rosenblatt, J. D., Chen, I. S. Y., Sherwood, W. C., Smith, D. E., Tegtmeier, G. E., Fernando, L. P., Fang, C. T., Osame, M. and Kleinman, S. H.,** Relative prevalence and risk factors of HTLV-I and HTLV-II infection in U.S. blood donors, *Lancet,* 337, 1435, 1991.
9. **Zella, D., Mori, L., Sala, M., Ferrante, P., Casoli, C., Magnani, G., Achilli, G., Cattaneo, E., Lori, F., and Bertazzoni, U.,** HTLV-II infection in Italian drug abusers, *Lancet,* 336, 575, 1990.
10. **Hjelle, B., Scalf, R., and Swenson, S.,** High frequency of human T-cell leukemia-lymphoma virus type II infection in New Mexico blood donors: determination by sequence-specific oligonucleotide hybridization, *Blood,* 76, 450, 1990.
11. **Lairmore, M. D., Jacobson, S., Gracia, F., De, B. K., Castillo, L., Larreategui, M., Roberts, B. D., Levine, P. H., Blattner, W. A., and Kaplan, J. E.,** Isolation of human T-cell lymphotropic virus type 2 from Guaymi Indians in Panama, *Proc. Natl. Acad. Sci. U.S.A.,* 87, 8840, 1990.
12. **Delaporte, E., Monplaisir, N., Louwagie, J., Peeters, M., Martin-Prevel, Y., Louis, J.-P., Trebucq, A., Bedjabaga, L., Ossari, S., Honore, C., Larouze, B., d'Auriol, L., Van Der Groen, G., and Piot, P.,** Prevalence of HTLV-I and HTLV-II infections in Gabon, Africa: comparison of serological and PCR results, *Int. J. Cancer,* 49, 373, 1991.
13. **Goubau, P., Desmyter, J., Swanson, P., Reynders, M., Shih, J., Kazadi, K., and Lee, H.,** Detection of HTLV-I and HTLV-II infection in Africans using type-specific envelope peptides, *J. Med. Virol.,* 39, 28, 1993.
14. **Okochi, K., Sato, H., and Hinuma, Y.,** A retrospective study on transmission of adult T-cell leukemia virus by blood transfusion: seroconversion in recipients, *Vox Sanguinis,* 46, 245, 1984.
15. **Wiktor, S. Z., Alexander, S. S., Shaw, G. M., Weiss, S. H., Murphy, T. L., Wilks, R. J., Shortly, V. J., Hanchard, B., Blattner, W. A.,** Distinguishing between HTLV-I and HTLV-II by western blot, *Lancet,* 335, 1533, 1990.
16. **Jelle, B., Cyrus, S., Swenson, S., and Mills, R.,** Serologic distinction between human T-lymphotropic virus (HTLV) type I and HTLV type II, *Transfusion,* 31, 731, 1991.
17. **Horal, P., Hall, W. W., Svennerholm, B., Lycke, J., Jeansson, S., Rymo, L., Kaplan, M. H., and Vahlne, A.,** Identification of type-specific linear epitopes in the glycoproteins gp46 and gp21 of human T-cell leukemia viruses type I and type II using synthetic peptides, *Proc. Natl. Acad. Sci. U.S.A.,* 88, 5754, 1991.
18. **Viscidi, R. P., Hill, P. M., Li, S., Cerny, E. H., Vlahov, D., Farzadegan, H., Halsey, N., Kelen, G. D., and Quinn, T. C.,** Diagnosis and differentiation of HTLV-I and HTLV-II infection by enzyme immunoassays using synthetic peptides, *J. Acquired Immune Deficiency Syndromes,* 4, 1190, 1991.

19. **Roberts, C. R., Mitra, R., Hyams, K., Brodine, S. K., and Lal, R. B.,** Serologic differentiation of human T lymphotropic virus type I from type II infection by synthetic peptide immunoassays, *J. Med. Virol.,* 36, 298, 1992.

20. **Lipka, J. J., Miyoshi, I., Hadlock, K. G., Reyes, G. R., Chow, T. P., Blattner, W. A., Shaw, G. M., Hanson, C. V., Gallo, D., Chan, L., and Foung, S. K. H.,** Segregation of human T cell lymphotropic virus type I and II infections by antibody reactivity to unique viral epitopes, *J. Infect. Dis.,* 165, 268, 1992.

21. **Hjelle, B., Mills, R., Goldsmith, C., Swenson, S. G., and Cyrus, S.,** Primary isolation of human T-cell leukemia-lymphoma virus types I and II: use for confirming infection in seroindeterminate blood donors, *J. Clin. Microbiol.,* 30, 2195, 1992.

22. **Gessain, A., Saal, F., and Gout, O.,** High human T-cell lymphotropic virus type I proviral DNA load with polyclonal integration in peripheral blood mononuclear cells of French, West Indian, Guianese, and African patients with tropical spastic paraparesis, *Blood,* 15, 428, 1990.

23. **Shaw, G. M., Hahn, B. H., Arya, S. K., Groopman, J. E., Gallo, R. C., and Wong-Staal, F.,** Molecular characterization of human T-cell leukemia (lymphotropic) virus type III in the acquired immune deficiency syndrome, *Science,* 226, 1165, 1984.

24. **Wattel, E., Mariotti, M., Agis, F., Gordien, E., Le Coeur, F. F., Prin, L., Rouger, P., Chen, I. S. Y., Wain-Hobson, S., and Lefrere, J.-J.,** Quantification of HTLV-I proviral copy number in peripheral blood of symptomless carriers from the French West Indies, *J. Acquired Immune Deficiency Syndromes,* 5, 943, 1992.

25. **Kwok, S., Ehrlich, G., Poiesz, B., Kalish, R., and Sninsky, J.,** Enzymatic amplification of HTLV-I viral sequences from peripheral blood mononuclear cells and infected tissues, *Blood,* 72, 1117, 1988.

26. **Kwok, S., Kellogg, D., Ehrlich, G., Poiesz, B., Bhagavati, S., and Sninsky, J.,** Characterization of a sequence of human T cell leukemia virus type I from a patient with chronic progressive myelopathy, *J. Infect. Dis.,* 158, 1193, 1988.

27. **Greenberg, S., Ehrlich, G., Abbott, M., Hurwitz, B., Waldmann, T., and Poiesz, B.,** Detection of exogenous sequences homologous to human retroviral DNA in multiple sclerosis by gene amplification, *Proc. Natl. Acad. Sci. U.S.A.,* 86, 2878, 1989.

28. **Ehrlich, G. D., Greenberg, S., and Abbott, M. A.,** Detection of human T-cell lymphoma/leukemia viruses, in *PCR Protocols: A Guide to Methods and Applications,* Innis, M. A., Gelfand, D. H., Sninsky, J. J., and White, T. J., Eds., Academic Press, San Diego, CA, 1989, chap. 39.

29. **Ratner, L., Philpott, T., and Trowbridge, D. B.,** Nucleotide sequence analysis of isolates of human T-lymphotropic virus type I of diverse geographical origins, *AIDS Res. Human Retroviruses,* 7, 923, 1991.

30. **Lee, H., Idler, K. B., Swanson, P., Aparicio, J., Chen, K. K., Lax, J. P., Nguyen, M., Mann, T., Leckie, G., Zanetti, A., Marinucci, G., Chen, I. S. Y., and Rosenblatt, J. D.,** Complete nucleotide sequence of HTLV-II isolate NRA; comparison of envelope sequence variation of HTLV-II isolates from U.S. blood donors and U.S. and Italian IV drug users, *Virology,* in press.

31. **Tedder, R. S., Shanson, D. C., Jeffries, D. J., Chiengsong-Popov, R., Clapham, P., Dalgleish, A., Nagy, K., and Weiss, R. A.,** Low prevalence in the U.K. of HTLV-I and HTLV-II infection in subjects with AIDS, with extended lymphadenopathy, and at risk for AIDS, *Lancet,* 2, 125, 1984.

32. **Robert-Guroff, M., Weiss, S. H., Giron, J. A., Jennings, A. M., Ginzburg, H. M., Margolis, I. B., Blattner, W. A., and Gallo, R. C.,** Prevalence of antibodies to HTLV-I, -II and -III in intravenous drug abusers from an AIDS endemic region, *JAMA,* 255, 3133, 1986.

33. **Williams, A. E., Fang, C. T., Slamon, D. J., Poiesz, B. J., Sander, S. G., Darr, W. F., Schulman, G., McGowan, E. I., Douglas, D. K., and Bowman, R. J.,** Seroprevalence and epidemiological correlates of HTLV-I infection in U.S. blood donors, *Science,* 240, 643, 1988.

34. **Ehrlich, G. D., Glaser, J. B., LaVigne, K., Quan, D., Mildvan, D., Sninsky, J. J., Kwok, S., Papsidero, L., and Poiesz, B. J.,** Prevalence of human T-cell leukemia/lymphoma virus (HTLV) type II infection among high-risk individuals: type-specific identification of HTLVs by polymerase chain reaction, *Blood,* 74, 1658, 1989.

35. **Kwok, S., Gallo, D., Hanson, C., McKinney, N., Poiesz, B., and Sninsky, J. J.,** High prevalence of HTLV-II among intravenous drug abusers: PCR confirmation and typing, *AIDS Res. Human Retroviruses,* 6, 561, 1990.

36. **Palumbo, P. E., Weiss, S. H., McCreedy, B. J., Alexander, S. S., Denny, T. N., Klein, C. W., and Altman, R.,** Evaluation of human T cell lymphotropic virus infection in a cohort of injecting drug users, *J. Infect. Dis.,* 166, 896, 1992.

37. **Lipka, J. J., Young, K. K. Y., Kwok, S. Y., Reyes, G. R., Sninsky, J. J., and Foung, S. K. H.,** Significance of human T-lymphotropic virus type I indeterminant serological findings among healthy individuals, *Vox Sanguinis,* 61, 171, 1991.

38. **Khabbaz, R. F., Heneine, W., Grindon, A., Hartley, T. M., Shulman, G., and Kaplan, J.,** Indeterminate HTLV serologic results in U.S. blood donors: are they due to HTLV-I or HTLV-II? *J. Acquired Immune Deficiency Syndromes,* 5, 400, 1992.

39. **Lal, R. B., Rudolph, D. L., Coligan, J. E., Brodine, S. K., and Roberts, C. R.,** Failure to detect evidence of human T-lymphotropic virus (HTLV) type I and type II in blood donors with isolated gag antibodies to HTLV-I/II, *Blood,* 80, 544, 1992.

40. **Kwok, S., Lipka, J. J., McKinney, N., Kellogg, D. E., Poiesz, B., Foung, S. K. H., and Sninsky, J. J.,** Low incidence of HTLV infections in random blood donors with indeterminate western blot patterns, *Transfusion,* 30, 491, 1990.

41. **Hjelle, B., Mills, R., Goldsmith, C., Swenson, S. G., and Cyrus, S.,** Primary isolation of human T-cell leukemia-lymphoma virus types I and II: use for confirming infection in seroindeterminate blood donors, *J. Clin. Microbiol.,* 30, 2195, 1992.

42. **Donegan, E., Busch, M. P., Galleshaw, J. A., Shaw, G. M., and Mosley, J. W.,** Transfusion Safety Study Group, Transfusion of blood components from a donor with human T-lymphotropic virus type II (HTLV-II) infection, *Ann. Internal Med.,* 113, 7, 555, 1990.

43. **Hjelle, B., Mills, R., Mertz, G., and Swenson, S.,** Transmission of HTLV-II via blood transfusion, *Vox Sanguinis,* 59, 119, 1990.

44. **Kleinman, S., Swanson, P., Allain, J. P., and Lee, H.,** Transfusion transmission of human T-lymphotropic virus types I and II: serologic and polymerase chain reaction results in recipients identified through look-back investigations, *Transfusion,* 33, 14, 1993.

45. **Hjelle, B., Cyrus, S., and Swenson, S. G.,** Evidence for sexual transmission of human T lymphotropic virus type II, *Ann. Internal Med.,* 116, 90, 1992.

46. **Kaplan, J. E., Abrams, E., Shaffer, N., Cannon, R. O., Kaul, A., Krasinski, K., Bamji, M., Hartley, T. M., Roberts, B., Kilbourne, B., Thomas, P., Rogers, M., and Heneine, W.,** NYC Perinatal HIV Transmission Collaborative Study, Low risk of mother-to-child transmission of human T lymphotropic virus type II in non-breast-fed infants, *J. Infect. Dis.,* 166, 892, 1992.

47. **Hall, W. W., Takahashi, H., Liu, C., Kaplan, M. H., Scheewind, O., Ijichi, S., Nagashima, K., and Gallo, R. C.,** Multiple isolates and characteristics of human T-cell leukemia virus type II, *J. Virol.,* 66, 2465, 1992.

48. **Dube, D. K., Sherman, M. P., Saksena, N. K., Bryz-Gornia, V., Mendelson, J., Love, J., Arnold, C. B., Spicer, T., Dube, S., Glaser, J. B., Williams, A. E., Nishimura, M., Jacobsen, S., Ferrer, J. F., Del Pino, N., Quiruelas, S., and Poiesz, B. J.,** Genetic heterogeneity in human T-cell leukemia/lymphoma virus type II, *J. Virol.,* 67, 1175, 1993.

49. **Daenke, S., Nightingale, S., Cruickshank, J. K., and Bangham, C. R. M.,** Sequence variants of human T-cell lymphotropic virus type I from patients with tropical spastic paraparesis and adult T-cell leukemia do not distinguish neurologic from leukemic isolates, *J. Virol.,* 64, 1278, 1990.

Chapter 13

Herpesvirus PCR

Elisabeth Puchhammer-Stöckl

TABLE OF CONTENTS

I. INTRODUCTION

The Herpesviridae family includes a number of different pathogens. Some of the herpesviruses frequently cause human disease; these include the herpes simplex viruses (HSV1 and HSV2), varicella zoster virus (VZV), cytomegalovirus (CMV), Epstein-Barr virus (EBV), and human herpesvirus type 6 (HHV-6). These viruses have in common a linear, double-stranded DNA genome encoding a large number of polypeptides.[1] Some DNA sequences of the herpesviruses cross hybridize with each other, and there also exist homologies to parts of the human genome. After primary infection, all of the herpesviruses mentioned above remain in a latent state in their human hosts for their lifetime. In situations of immunosuppression, herpesviruses may reactivate and sometimes again cause disease. Within the last few years, PCR technology has been used in different ways for the detection of herpesvirus nucleic acids in clinical samples. The major problem in the application of PCR for diagnostic purposes is the presence of latent herpesviruses, with genomes that may also serve as target for PCR, in a high percentage of the healthy population. Therefore a critical evaluation of positive PCR results is necessary and clinical specimens that are likely to show an acute herpesvirus infection or a reactivation must be carefully chosen. Genomic variation between single strains of the same herpesvirus and sequence homology between the different herpesviruses provide further difficulties in the establishment of a PCR assay, requiring special care in primer choice. Considering these aspects as well as the contamination danger inherent to PCR (necessitating special laboratory equipment), it is obvious that the establishment of reliable PCR assays for diagnostic purposes may be difficult to achieve. In the following sections an overview will be given about the various applications of PCR for the detection of Herpesvirus-DNA.

0-8493-4833-1/95/$0.00+$.50

II. HERPES SIMPLEX VIRUSES

A. CLINICAL MANIFESTATIONS[2,3]

The herpes simplex virus group includes two different genotypes (HSV1 and HSV2), which in parts of the genome, show extensive DNA sequence homology. Primary infection with HSV1 usually occurs by contact with the mucosal surface of mouth or throat and may be asymptomatic or present as vesicular gingivostomatitis, pharyngitis, or tonsillitis. HSV1 is then transported by neurons primarily to the trigeminal ganglia, where it remains in a latent state and from where reactivation may occur. Reactivation may be asymptomatic or, in uncomplicated cases, may cause herpes labialis. Infection with HSV2 is usually acquired by sexual transmission. The clinical picture of both primary and reactivated infection shows vesicular genital lesions, sometimes together with fever and malaise. The location of HSV2 latency is primarily the sacral ganglion. Primary infection as well as recurrence may also occur without symptoms. In rare cases HSV1 may be the cause of herpes genitalis and HSV2 the cause of herpes labialis. The most severe complication associated with herpes simplex viruses is the infection of the central nervous system (CNS). HSV1 is the most common cause of endemic encephalitis in many industrial countries. The infection causes a severe focal inflammation of the brain, typically including the temporal lobes, and is lethal in about 70% of untreated patients. Early treatment with acyclovir which inhibits efficiently HSV replication, clearly reduces the mortality rate and also the rate of sequelae. HSV2 may also infect the CNS but usually causes a more benign meningitis or meningoencephalitis. However, in newborns infected with HSV2 during delivery by a maternal herpes genitalis, a severe generalized infection may occur. This "herpes neonatorum" may primarily affect the brain tissue or may be spread to the CNS in the course of a disseminated hematogenous infection. Both courses of disease are associated with a high mortality rate. Another severe complication is the HSV1 keratoconjunctivitis, which is often followed by corneal blindness.

B. CONVENTIONAL DIAGNOSIS[4]

The typical clinical picture of HSV primary infection or reactivation does not usually require virological confirmation. If necessary, virus can be isolated directly from clinical materials such as vesicular fluid, throat washings, or corneal swabs. A differentiation between the two HSV types may be made by using monoclonal antibodies (MABs) specific for HSV1 or HSV2. Primary infection with HSV can be serologically identified by the detection of HSV-specific IgM-antibodies (ABs). Reactivation can sometimes cause an increase of complement fixing (CF) and/or IgG ABs. The major diagnostic problem is the early detection of herpes simplex encephalitis (HSE). The detection of HSV-specific intrathecal AB production is not possible early in the course of disease, and recovery of virus from cerebrospinal fluid (CSF) by conventional means has most been unsuccessful. Thus the early diagnosis of HSE has been limited to the identification of HSV in brain biopsy tissue, a procedure that is usually avoided due to its invasive nature.

C. PCR FOR HSV DIAGNOSIS

Because of tremendous diagnostic problems with HSE, PCR was applied early for the detection of HSV in the CSF of HSE patients. It has been shown independently by different groups by retrospective testing of confirmed HSE cases that HSV DNA can be detected in CSF within the first days of disease in a high percentage of HSE patients[5–11] (Figure 13–1). The PCR assays could be completed within 1 to 2 days, thus allowing a rapid detection of HSE. It was demonstrated, using a semiquantitative PCR assay, that only background signals are derived from CSF of latently infected persons or of individuals undergoing HSV reaction in the absence of HSE (Figure 13–2). Thus positive PCR results were shown to be highly specific for HSE cases. A recently completed prospective study, using the PCR for routine investigation of patients with clinically suspected HSE, showed that this method not only allowed an earlier diagnosis of HSE than serological tests but also identified, in agreement with clinical and neurodiagnostic parameters, a number of additional HSE cases that had not been detected by serological methods.[12] While positive PCR results are considered positive proof of HSE, false-negative test results may be obtained. Failure of PCR seems to be a rare event, however, and may occur especially with CSF samples drawn very early in the course of disease or with CSFs obtained from patients with a more benign course of HSE.[12] The PCR has also been shown to be an excellent tool for monitoring the efficacy of acyclovir treatment during HSE. Initiation of therapy before sampling of CSF does not prevent the detection of HSV DNA in CSF,

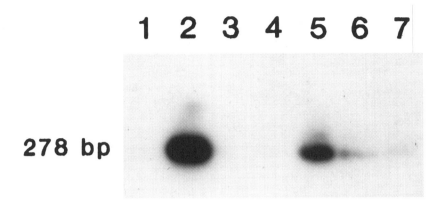

Figure 13–1 Southern blot hybridization of the 278-bp DNA fragments amplified by HSV PCR. Lane 1, CSF of a negative control patient; lanes 2 to 4, CSF samples obtained from a patient with confirmed HSE on the 5th, 15th, and 26th day of disease; lanes 5 to 7, HSV1-infected fibroblasts diluted in a PCR-negative CSF sample to concentrations of 1000, 100, and 10 infectious units. (From Puchhammer-Stöckl, E., Popow-Kraupp, Th., Heinz, F. X., Mandl, C. W., and Kunz, C., *J. Med. Virol.*, 32, 77, 1990. With permission.)

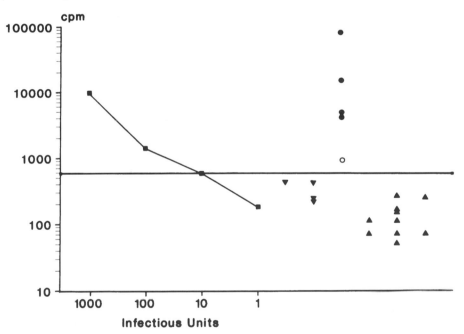

Figure 13–2 Evaluation of the radioactive counts obtained after HSV PCR and hybridization of CSF samples (●) from patients with confirmed HSE in the first week of disease; (○) from a newborn infected with HSV2; and ( ) from negative control patients. (∇) concentrated blood lymphocytes of four seropositive patients; (■) infected fibroblasts at different concentration. (From Puchhammer-Stöckl, E., Popow-Kraupp, Th., Heinz, F. X., Mandl, C. W., and Kunz, C., *J. Med. Virol.*, 32, 77, 1990. With permission.)

and positive results are usually obtained up to the eighth to tenth day of therapy.[8,12] Relapses of HSE, which sometimes occur after termination of therapy, were also detected immediately by PCR.[10] It was also shown that the HSV DNA identified by PCR in CSF is primarily free DNA,[6] a finding that would explain the inability to isolate virus from CSF. As expected, HSV DNA was also detectable in the brain tissue of HSE patients.[13,14] PCR assays established for HSE diagnosis are indicated in Table 13–1.

Table 13–1　**PCR assays established for the detection of HSV**

Genomic Region	HSV1/2	Amplification Conditions	Detection System	Clinical Material	Ref.
UL42 278 bp[a]	HSV1 > 2	1′, 94; 40; 68°C 35 c	Slot-blot Semiquantitative	CSF	7
gpD 221/138 bp	HSV1 > 2	30″, 95; 55; 1′ 72°C 20/30 c	Nested PCR	CSF	8
gpB 488 bp	HSV1 + 2	1′, 97; 53; 74°C 30 c	Southern blot Semiquantitative	Swabs of genital lesions	17
DNA polymerase gene	HSV1 or 2	1′, 96; 2′, 67; 3′, 72°C 35 c	Gel analysis Southern blot	Swabs of skin and mucosal lesions	37

[a] Length of the amplicon with the primers shown below:

Puchhammer-Stöck et al.[7]:

Primers:　HS13: 5′ACGACGACGTCCGACGGCGA 3′
　　　　　HS14: 5′GTGCTGGTGCTGGACGACAC 3′
Probe:　　HS15: 5′ATAGTGCCACGCCCACCACGTTCGA 3′

Aurelius et al.[8]:

Outer primers:　BJHSV1.1: 5′ATCACGGTAGCCCGGCCGTGTGACA 3′
　　　　　　　　BJHSV1.2: 5′CATACCGGAACGCACCACACAA 3′
Inner primers:　BJHSV1.3: 5′CCATACCGACCACACCGACGA 3′
　　　　　　　　BJHSV1.4: 5′GGTAGTTGGTCGTTCGCGCTGAA 3′

Cone et al.[17]:

Primer:　5′CAGAACTACACGGAGGGCATC 3′
　　　　　5′GTAAAACGGGGACATGTAC 3′
Probe:　　5′GTCTCGTGGTCGTCCCGGTGAAACGCGGTG 3′

Kimura et al.[37]:

Universal primer:　DNAP5: 5′ATGGTGAACATCGACATGTACGG 3′
HSV-1 primer:　　　DNAP3–1: 5′CCTCGCGTTCGTCCTCGTCCTCC 3′
HSV-2 primer:　　　DNAP3–2: 5′CCTCCTTGTCGAGGCCCCGAAAC 3′

HSV PCR has been shown to be an important tool for the detection of neonatal HSV infection. Application of PCR proved to be especially important for the diagnosis of neurological forms of neonatal herpes. It was shown that HSV DNA can be amplified from CSF as well as from serum samples of infected newborns,[7,15] whereas virus isolation from both materials is only rarely successful. In contrast, PCR on skin lesions during generalized fetal infection has been reported to be only slightly more successful than virus isolation from the same material.[15] The PCR results also made possible a better understanding of the course of neonatal HSV infection and indicated that CNS infections of the newborn may often be due to virus distribution by the hematogenous route.[15] HSV DNA was detectable in infected children up to 14 days after onset of therapy.

Uncomplicated HSV infection or reactivation certainly does not require the application of PCR. However, it was described as being of value for the identification of atypical or partially resolved dermal lesions or for differential diagnosis of HSV and VZV cutaneous eruptions.[16] PCR assays have also been applied to the detection of symptomatic or asymptomatic herpes genitalis.[17–20] HSV DNA was detected several days longer than virus could be isolated, thus facilitating a retrospective diagnosis of unclear genital lesions.[17] In another study, with pregnant women, it was pointed out that PCR allowed the identification of HSV DNA in genital specimens within 1 day, in contrast to virus isolation, which takes up to a few days.[20] Detection of HSV DNA is, however, not comparable to detection of replicative virus in genital secretions and may also reflect the presence of nonviable virus. Therefore the consequences of these PCR-positive results are still unclear.

Various studies have described the detection of HSV DNA in corneal tissue.[21–25] However, since it was shown that HSV DNA may be latently present in corneas without recent HSV-caused disease,[21–24] the value of PCR for diagnosis of herpetic keratoconjunctivitis seems to be limited. Several authors have used PCR to investigate the association of individual diseases to HSV infection. HSV has been believed to be one cause for erythema multiforme (EM), and amplification of HSV DNA from EM lesions, further supports this theory.[26–30] Thus PCR may possibly be of value in providing an indication for the initiation of acyclovir therapy. PCR was also one method used to show that peptic ulcers may be associated with HSV infection of gastric cells.[31] PCR findings also suggested that HSV may be associated with Mollarets meningitis[32] and that it may perhaps play a role in progressive stages of anal cancer[33] or in the development of idiopathic corneal endotheliopathy.[34] The presence of HSV DNA in brain tissue of Alzheimer's disease patients[35] did not seem to be significantly correlated with this disease.

In addition to the PCR studies summarized above, two studies have described test systems enabling the differentiation of HSV1 and HSV2, which may be of importance for diagnostic purposes as well as for epidemiological studies.[36,37]

III. VARICELLA ZOSTER VIRUS

A. CLINICAL MANIFESTATIONS[38,39]

VZV is the causative agent of two different clinical conditions. Primary infection with VZV causes varicella, which presents in uncomplicated cases as a vesicular rash. Complications are rare in immuno-competent persons, principally including varicella pneumonia, postinfectious encephalitis, and possibly Guillain-Barré or Reye's syndrome. In immunocompromised patients, severe and generally disseminated infections can be observed. Primary infection with VZV especially during the first third of pregnancy may affect the fetus. In rare cases, this leads to abortion or to the development of a "congenital varicella syndrome," which is characterized by cutaneous scarifications, eye abnormalities, or hypoplasia of the limbs. Maternal varicella immediately before or after delivery can be the cause of severe disseminated and sometimes lethal VZV infection of the newborn. After primary infection, VZV remains latent in dorsal root ganglia. Reactivation leads to herpes zoster, characterized by often painful vesicular eruptions in single dermatomes. The most common complication observed is postherpetic neuralgia. However, VZV reactivation may also cause other neurological symptoms such as aseptic meningitis, facial palsy, encephalitis, or myelitis. VZV-caused neurological disease may also occur in the absence of dermal lesions as "zoster sine herpete". In immunocompromised persons, a severe disseminated herpes zoster, including numerous dermatomes, may be observed.

B. CONVENTIONAL DIAGNOSIS[40]

Due to the distinctive clinical picture of varicella and herpes zoster, virological diagnosis is only rarely required in uncomplicated cases. VZV can be isolated from vesicular fluid and, in rare cases, from CSF. Primary infection can be confirmed by the detection of VZV-specific IgM-ABs. Herpes reactivation may sometimes be detected serologically by an increase in CF-ABs and/or IgG-ABs. VZV-caused neurological disease can be serologically confirmed by detection of intrathecal AB production only late in the course of disease. Cases of zoster sine herpete, where no dermal eruptions are observed, often remain undetected.

C. PCR FOR VZV DIAGNOSIS

PCR was shown to be of value and superior to conventional methods in the early detection of such VZV-caused neurological diseases as meningoencephalitis, myelitis, or facial palsy, especially when no cutaneous lesions are simultaneously observed.[41,42] A semiquantitative PCR assay was described, which enabled the sensitive detection of VZV DNA in CSF samples[41] (Figure 13–3). PCR signals above an established test-cutoff value proved to be highly specific for VZV-caused neurological disease. VZV DNA was shown to be detectable within the first days after onset of neurological symptoms, thus allowing

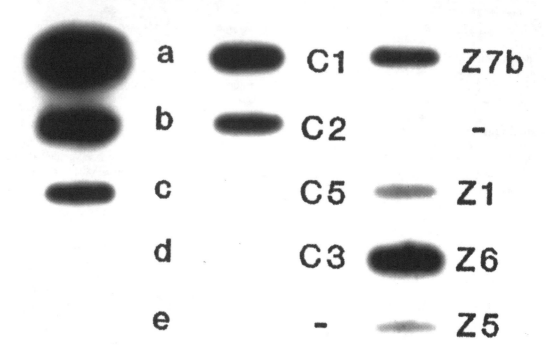

Figure 13–3   Varicella zoster virus PCR results after slot blot hybridization from cells diluted to (a) 10, (b) 1, (c) 0.1, (d) 0.05, and (e) 0.01 infectious units as well as from CSF samples. C1, C2, C3, and C5, CSFs from children with post-chicken-pox encephalitis; Z1, Z5, Z6, and Z7b, CSFs from patients with herpes zoster associated neurological symptoms; –, CSFs of negative control patients. (From Puchhammer-Stöckl, E., Popow-Kraupp, Th., Heinz, F. X., Mandl, C. W., and Kunz, C., *J. Clin. Microbiol.*, 29, 1513, 1991. With permission.)

the early initiation of acyclovir treatment. PCR also enabled the detection of VZV DNA in the CSF of children with post-chicken-pox cerebellitis (Figure 13–3). These findings have no diagnostic significance, but suggest that this clinical picture may not only be caused by immunological processes and could in some cases be treated with acyclovir.

PCR has also been applied for the detection of VZV in other clinical samples.[43–45] It was shown that the PCR is clearly superior to virus isolation in detecting virus from vesicular fluid samples in sensitivity as well as rapidity,[45] and it was suggested that PCR could generally replace the routine culture method in clinically ambiguous cases where virological confirmation seems necessary. PCR even enabled the detection of VZV from healing crusts in the course of varicella.[44,45] PCR was further applied as a possible diagnostic tool for the detection of prenatally acquired VZV embryopathy. Prenatal diagnosis by PCR, however, was described as unsuccessful as it was shown that presence of VZV DNA in chorionic villus samples is not associated with fetal disease.[46] In contrast, retrospective confirmation of fetal VZV infection seems to be possible by detection of VZV DNA after fetal death in various tissue samples (Puchhammer-Stöckl, unpublished data).

Besides application for diagnostic purposes, PCR was used for the detection of virus in oropharyngeal secretions during chicken pox.[43,44,47] However, contradictory results were obtained by different authors. Two groups have described the successful and frequent detection of VZV DNA in throat swabs, especially after onset of clinical symptoms.[44,47] In contrast, other authors detected VZV DNA in oropharyngeal secretions in only 1 of 30 patients[43] and concluded that these data, which are in agreement with the low virus-isolation rate from oropharyngeal secretions, reflect the low prevalence of VZV in these specimens during the symptomatic stage of disease. The reasons for these discrepancies remain to be elucidated. Investigation of PBMCs during primary VZV infection by PCR showed that within the first 24 h of chicken pox, viremia can be detected in 67% of immunocompetent subjects.[43] VZV viremia was also demonstrated in bone marrow transplant recipients during subclinical or symptomatic VZV reactivation.[48] In addition, VZV DNA was also detected in PBMCs of healthy individuals.[49] The presence of viral DNA in these cells was shown to be associated with higher age but not with postherpetic neuralgia.

Table 13–2 **PCR assays established for the detection of VZV**

Genomic Region	Amplification Conditions	Detection System	Clinical Material	Ref.
Gene 13, 36, 62, 68	1', 90; 2', 60; 5', 72°C 30 c	Gel analysis Southern blot	Throat swabs Vesicular fluid	44
gp11 249 bp[a]	1', 95; 30", 60; 1', 72°C 35 c	Gel analysis Dot blot	PBMCs; oropharyngeal secretions; cutaneous lesions	43
XbalM 276 bp	1', 94; 40; 68°C 35 c	Slot blot Semiquantitative	CSF	41
IE63 326 bp	1', 94; 1.6', 72°C 35 c	Gel analysis Southern blot	Vesicular fluid	45

[a] Length of the amplicon with the primers shown below:

Kido et al.[44]:

Primers: gpI-1: 5′CCGTATATGAGCCTTACTACCATTC 3′
    gpI-2: 5′GAGTTCATCAAACAGTGTGCTCGTG 3′

Koropchak et al.[43]:

Primers: TK4: 5′GAGGAAGTTGAAGCCACATC 3′
    TK5: 5′GAAGGTCAAGGTTGGTTGGC 3′
Probe: CK1: 5′GCCTACGGGATGGTGCATAC 3′

Puchhammer-Stöckl et al.[41]:

Primers: VZ7: 5′ATGTCCGTACAACATCAACT 3′
VZ8: 5′CGATTTTCCAAGAGAGACGC 3′
Probe: VZ9: 5′GGTGGAGACGACTTCAATAGC 3′

Dlugosch et al.[45]:

Primers: 5′GCTCGTTGAGGACATCAACCGTGTT 3′
    5′CATCGTCGCTATCGTCTTCACCAC 3′

It was concluded that these findings may be due to subclinical VZV reaction or to clinically evident viral reactivation in the absence of dermal lesions. Investigation of human ganglia showed that VZV DNA is latently present in thoracic as well as in trigeminal ganglia of the same patients.[45,50] Epidemiological studies were also enriched by PCR. *In vitro* amplification combined with cloning and sequencing allowed the detection of thymidine kinase gene mutations in drug-resistant VZV populations.[51] Table 13–2 summarizes some VZV PCR assays established so far.

IV. CYTOMEGALOVIRUS

A. CLINICAL MANIFESTATIONS[52,53]

CMV infection is transmitted by shedding of virus in such body fluids as saliva or genital secretions, by transfusion of blood products, or by organ transplantation. The primary infection usually presents subclinically or with mild symptoms such as hepatitis, fever and malaise, or lymphocytosis. CMV then remains latent in PBMCs or epithelial cells of various organs. CMV reactivation or reinfection with another CMV strain occurs without symptoms in immunocompetent persons. In immunocompromised patients such as transplant recipients or HIV-infected individuals, however, CMV infection or reactivation may be a dangerous event. Almost every organ may then be infected by the virus, and in particular CMV interstitial pneumonia, as well as CMV encephalitis, are associated with a high mortality rate. In

HIV-infected patients a CMV chorioretinitis is commonly observed. CMV infections or reactivations in immunocompromised patients can be treated with gancyclovir, a potent inhibitor of CMV replication. CMV infection may also be dangerous when acquired during pregnancy. Primary CMV infection of pregnant women is one of the major viral causes of congenital malformations in the newborn. Congenital CMV infection may result in mental retardation or palsies as well as in eye abnormalities and splenomegalie. In contrast, CMV reactivation during pregnancy almost never leads to congenital malformations.

B. CONVENTIONAL DIAGNOSIS[54]

The early diagnosis of CMV infection or reactivation in immunocompromised patients is of paramount importance. CMV may be isolated from various clinical samples such as urine, throat washings, or PBMCs. Because virus replicates slowly in cell cultures, an immunofluorescence assay using MABs against a CMV immediate early protein is usually applied, and this allows the detection of CMV replication in cell cultures within a few hours after inoculation.[55] In addition, the detection of a CMV lower matrix protein in PBMCs using MABs has recently been shown to be useful for the rapid identification and quantification of CMV viremia. Primary CMV infection as well as reactivation may also be detected serologically. A congenitally acquired CMV infection may be diagnosed with certainty only by detection of virus in the urine within the first weeks of life.

C. PCR FOR CMV DIAGNOSIS

The first study applying CMV PCR for diagnostic purposes showed the successful detection of CMV DNA in the urine of newborns who were prenatally infected with CMV.[56] It was demonstrated that PCR using two different pairs of primers allowed the detection of CMV from the urine of newborns with a sensitivity and specificity equal to that of routine virus isolation. Both primer pairs were shown to identify a number of different CMV strains, as shown in Table 13–3. The advantages of PCR were its applicability to small amounts of clinical material and its rapidity. DNA amplification from urine may be hampered by the presence of enzyme inhibitors. These inhibitors were shown to be either urea or metabolites derived from parenteral nutrition.[56] Inhibitory effects may be overcome by the application of smaller amounts of urine sample to PCR,[56] by dialysis or ultrafiltration,[57] or by purification of DNA with glass beads.[58]

One of the most controversial subjects in herpesvirus PCR is the application of this extremely sensitive method for the diagnosis of CMV infection or reactivation in immunocompromised patients. While detection of CMV DNA in PBMCs by PCR may be a valuable tool for the early detection of CMV primary infection, the situation is different in cases of CMV reactivation or reinfection. It has been shown previously by *in situ* hybridization that CMV is present in PBMCs of latently infected healthy persons.[59,60] In accordance with these findings, the investigation of PBMCs of healthy blood donors by PCR allowed the detection of CMV DNA.[61–65] PCR was even shown to identify CMV DNA in PBMCs from some healthy seronegative blood donors.[61] This indicates that a higher sensitivity may be achieved by the PCR in identifying latent CMV infection than by serological tests. Thus serological screening for CMV in blood products, which are to be transfused to seronegative transplant recipients, could miss a number of infected blood products. In contrast to these results, several authors who have used the PCR for the investigation of CMV viremia in transplant or HIV patients found no positive results with healthy seropositive or -negative persons and described the PCR as being specifically positive only with PBMCs of patients undergoing active viral reactivation.[66–70] The contradictory results from the various PCR studies may be due to different test sensitivities. However, only one of the groups using PCR for diagnostic purposes has described their test sensitivity as arbitrarily restricted to exclude the detection of latent infection.[66] A general sensitivity limit required for excluding the detection of latent infection cannot be determined so far because some authors have not given a detailed description of their test systems and the test sensitivities indicated by individual authors could often not be compared. Several of the PCR test systems mentioned above are included in Table 13–3, and their sensitivity and ability to detect latent CMV infection in PBMCs is indicated.

Another problem arises with the finding that both asymptomatic and clinically relevant CMV reactions are detected by PCR[67,71–73] and cannot be differentiated. Thus the clinical significance of a positive PCR result from PBMCs is unclear. Similar results were observed by testing granulocyte preparations.[74,75] Therefore at the moment, PCR from blood cells seems to provide no alternative to conventional diagnostic test systems, which allow not only the detection but also the exact quantification of viremia and antigenemia[68] and can thus predict the clinical significance of a CMV reactivation event. However,

Table 13–3 **PCR assays for the detection of CMV**

Genomic Region	Amplification Conditions	Detection Systems	Clinical Material	Detection of CMV Strains	Detection Limit (PBMCs)	Ref.
Major IE gene	1', 95; 3', 65°C 32 c	Gel analysis	PBMCs Lip:[a] neg		>5 genomes/ 10^4 cells	66
DNA polymerase gene	1', 94; 55; 72°C 35 c	Solution hybridization	PBMCs Lip: neg		30 copies of plasmid DNA	67
Major IE gene	1', 85; 2', 50; 3', 70°C 40 c	Southern blot	PBMCs urine, saliva	8/8	1 genome 4×10^4 cells	64
Major IE gene/gp64	30", 95; 15", 14; 1', 72°C 50 c	Dot blot	PBMCs Lip: neg	IE: 27/28 gp64: 27/28	1 pfu	70
Major IE gene/LA	2', 94; 90", 65; 1', 72°C[b] 40 c	Dot blot	Urine of newborns	LA: 38/46 IE:43/46		56
EcoRID	2', 94; 2', 37; 3', 72°C 20–25 c	Southern or slot blot	Urine	5/5		80
a-seq L-S junction	1', 94; 55; 72°C 25 c	Gel analysis Southern blot				105

[a] Lip, latently infected persons; [b] increasing length of annealing time.

Demmler et al.[56]:

Primers: MIE-4: 5'CCAAGCGGCCTCTGATAACCAAGCC 3'
         MIE-5: 5'CAGCACCATCCTCCTCTTCCTCTGG 3'
Probe: MIE: 5'GAGGCTATTGTAGCCTACACTTTGG 3'

Jiwa et al.[66]:

Primers: IE:A: 5'AGCTGCATGATGTGAGCAAG 3'
         B: 5'GAAGGCTGAGTTCTTGGTAA 3'
Probe: 5'GGGTGCACTGCAGGCTAAGGCCCGTGCTAAAAAGGATGAA 3'

Shibata et al.[70]:

Primers: IE1: 5'CCACCCGTGGTGCCAGCTCC 3'
         IE2: 5'CCCGCTCCTCCTGAGCACCC 3'
Probe: IE3: 5'CTGGTGTCACCCCCAGAGTCCCCTGTACCCGCGACTATCC 3'

Zala et al.[105]:

Primers: a-seq: 5'TTCCCCGGGGAATCAA(C)ACAG 3' and 3'AAAGTGGGGGGGCGATTTTT 5'

PCR was described as being of possible value for the control of gancyclovir therapy. After onset of treatment, the presence of CMV was detectable by PCR in PBMCs for a much longer period than by conventional diagnostic methods such as virus isolation or IE antigen detection.[68,69,76] It was shown that patients from whom positive PCR results were still obtained after termination of treatment might have a greater risk of a relapse. Thus it was recommended that disappearance of viral DNA from PBMCs should be achieved by therapy and that remaining PCR-positive persons should be monitored more often by other tests for the reappearance of clinically relevant viremia.[68]

    The detection of CMV DNA by PCR was also described for other clinical specimens. One study showed that CMV DNA can be successfully identified in the serum of transplant patients during the active stage of viral infection.[77] The assay was shown to be an early marker for CMV pneumonia and to overcome the diagnostic problems with latent CMV. These findings, however, will have to be further evaluated by more extensive studies. Several authors have applied PCR for the detection of CMV DNA in urine or saliva samples,[64,67,71,76,78–81] and some of them describe this method as a possible diagnostic tool for detection of CMV infection or reactivation. However, since it has been shown by the less sensitive means of dot blot hybridization[82,83] as well as by PCR[64] that viral DNA is also detectable in these specimens in healthy persons and by PCR even in seronegative persons[64] it is questionable whether a positive PCR result from these specimens can be of diagnostic significance. CMV DNA was also amplified from nephrectomy samples and renal biopsies[84,85] as well as from paraffin-embedded or fresh lung biopsy materials.[71,86–88] A semiquantitative PCR assay was applied to the detection of CMV in bronchoalveolar lavage specimens and was described to differentiate latency from active viral replication.[89] CMV DNA was further detected by PCR in aqueous subretinal fluid and in vitreous specimens of patients with CMV retinitis;[90,91] in the adrenal gland of autopsied HIV patients;[92] in myocardial biopsy specimens of a heart transplant patient undergoing CMV myocarditis;[93] as well as in cervicovaginal cells.[94] PCR was also used to investigate pancreatic cells of type 2 diabetes patients for CMV[95,96] and to investigate the role of CMV in the pathogenesis of atherosclerosis.[97,98] PCR showed no association between CMV infection and Kaposi's sarcomas,[99] IgA nephropathy,[100] idiopathic pneumonitis,[86] or spontaneous abortion in the first trimester of pregnancy.[101]

PCR has been shown to be a valuable tool for investigating the epidemiology of CMV infection. The differentiation of CMV strains was shown to be facilitated by it.[102–104] Two authors have described the identification of different CMV strains using PCR primers from the L-S junction region.[105,106] Because this region is hypervariable and contains various inserts and deletions, fragments of different length are generated from unrelated strains (Figure 13–4). In addition to the studies mentioned above, numerous authors have published the establishment of individual CMV PCR assays.[107–112] The amplification of viral mRNA was described[67,113] as well as application of different detection systems for the identification of amplified products[81,111,112] and the analysis of fresh and fixed tissue samples by PCR.[110,114]

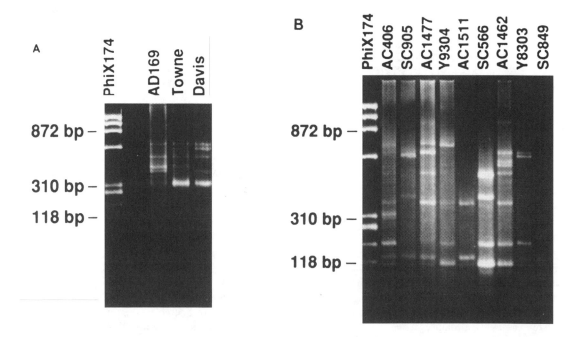

Figure 13–4 PCR products from epidemiologically unrelated CMV strains amplified with L-S junction primers. (A) Results obtained with laboratory reference strains AD 169, Towne and Davis. (B) PCR products from clinical CMV isolates. PhiX174, molecular weight marker. (From Sokol, D. M., Demmler, G. J., and Buffone, G. J., *J. Clin. Microbiol.*, 30, 839, 1992. With permission.)

V. EPSTEIN-BARR VIRUS

A. CLINICAL MANIFESTATIONS[115]

Infection with EBV is initiated in the oropharynx by saliva exchange and occurs primarily in the epithelial cells followed by an infection and polyclonal transformation of the B lymphocytes. EBV infection may present without clinical symptoms or with the typical clinical picture of infectious mononucleosis (IM), which includes high fever for about 10 days, a sore throat, generalized lymphadenopathy, and tonsillitis. Atypical T lymphocytes, which are clearing infected B lymphocytes, are present in the peripheral blood. Sometimes in addition rash, pneumonitis, or hepatosplenomegaly are observed. As a complication of primary EBV infection several CNS syndromes including aseptic meningitis, encephalitis, or transverse myelitis as well as a number of hematological syndromes such as thrombocytopenia, neutropenia, hemolytic, or aplastic anemia have been described. Mononucleosis may end fatally in immunodeficient individuals either by progressive infiltration of different organs with infected B cells or by the development of agammaglobulinaemia and pancytopenia. EBV is a virus known to be associated with the development of cancer. It was identified in Burkitt's lymphoma and nasopharyngeal carcinomas, and EBV-associated lymphomas have been described in immunosuppressed patients.

B. CONVENTIONAL DIAGNOSIS[116]

The virological confirmation of EBV infection is usually based on the detection of EBV-specific ABs by immunofluorescence. A recent EBV infection can be detected either by identification of IgM ABs against the viral capsid antigen (VCA) or, especially in younger persons, by the detection of IgG ABs against early antigen (EA) and VCA in the absence of ABs against the nuclear antigen (EBNA).

C. PCR FOR EBV DIAGNOSIS

The PCR has been used successfully for the specific detection of EBV DNA in clotted blood samples of patients with IM[117] (Figure 13–5). However, general application of PCR is probably not required for the routine diagnosis of IM and was also not intended by the authors in view of the effectiveness of conventional diagnostic tests. Based on the finding that in latent EBV infection only about one infected

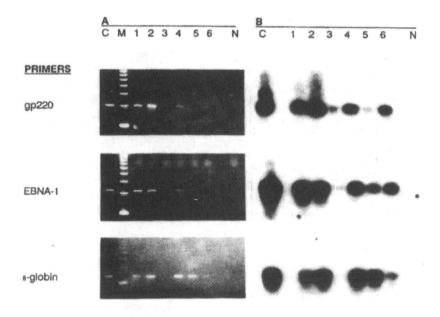

Figure 13–5 (A) Electrophoresis and (B) Southern blot of amplified EBV sequences and the human β-globin gene from blood and tissue samples. (C) Positive control (Daudi cells). M, molecular weight marker; N, negative control. Lane 1, transplant lymphoma; lane 2, AIDS lymphoma (fresh frozen tissue), lane 3, transplant lymphoma (paraffin embedded); lanes 4 and 5, blood from transplant patients; lane 6, clotted blood from IM patient. (From Telenti, A., Marshall, W. F., and Smith, T. F., *J. Clin. Microbiol.*, 28, 2187, 1990. With permission.)

PBMC is present in 10^6 to 10^8 uninfected cells several authors showed that the use of a limited number of PBMCs as templates for PCR excluded the detection of latent EBV infection in healthy persons.[117–119] Only in one study was EBV detected in healthy adults when 10^5 lymphocytes were used for PCR.[120] However, EBV DNA was detected in the peripheral blood of a high proportion of immunosuppressed transplant patients when testing only as few as 15 to 75×10^3 leukocytes[117] (Figure 13–5), which certainly reflects EBV reactivation. The significance of these PCR-positive results for the clinical management of immunosuppressed patients remains to be evaluated. The application of PCR for the monitoring of EBV primary infection in two liver transplant patients[121] showed a correlation of the course of disease with the presence of EBV DNA in liver and lymph node biopsy samples.

PCR was also applied for the detection of EBV infections in patients with X-linked lymphoproliferative disease.[122] Because an EBV infection may be lethal in these patients and they are therefore often prophylactically treated with immunoglobulins, serological diagnosis may be difficult and the PCR may provide in these cases a valuable diagnostic tool.

In addition to diagnostic applications, PCR has been used to elucidate the still-unclear relation between EBV and Sjögren's syndrome (SS), an autoimmune disease involving lacrimal and salivary glands and showing an increased frequency of lymphomas. However, discrepant PCR results were obtained from different groups. Some authors[123,124] found a significantly increased level of EBV DNA in salivary gland biopsies of SS patients compared to control subjects and concluded that EBV may play a role in the pathogenesis of SS. Other authors showed, however, that EBV DNA is also present in salivary glands of non-SS patients, and they doubt that a correlation exists between SS and EBV.[125,126] Also the observation that EBV DNA is present in lacrimal glands in normal persons[119] and at a higher frequency in SS patients[127] provides no proof of an association between EBV infection and SS. EBV PCR was also applied to the investigation of genital secretions.[118,128] EBV DNA was detected in urethral discharges from 48% of men with genital infections and from 13% of healthy men as well as in genital secretions of 28% of healthy women. Possible viral transmission cannot, however, be concluded from these results as the positive PCR results could also be derived from nonviable virus.

PCR was further used to investigate the correlation between EBV and different forms of cancer. EBV DNA was detected specifically in fine-needle aspiration of nasopharynx carcinoma (NPC) tissue and from NPC metastatic lesions but not from other head and neck tumors.[129] Therefore it was supposed that PCR-detectable EBV DNA may be a marker for NPC, especially in cases where the primary tumor is occult and only unclear metastatic lesions are observed.[129] EBV DNA was also amplified from NPC tissue of Japanese[130] and Taiwanese[131] patients. The amount of EBV in NPCs was shown to be especially high in undifferentiated cells. In addition, PCR was used for screening for EBV in non-Hodgkin's lymphomas (NHL) of HIV-infected patients. In one study, detection of EBV DNA was described in lymph node biopsy samples of HIV-infected persons with persistent generalized lymphadenopathy (PGL) but in none of the biopsies from normal individuals.[132] The presence of EBV DNA in PGL seemed to be a prognostic marker, being associated with an increased incidence of concurrent lymphomas at another site or with subsequent development of lymphomas. EBV DNA was detected by others in about 50% of already-developed NHL of HIV-infected persons,[133] suggesting a possible role of EBV in the pathogenesis of this disease. It remains to be elucidated, however, whether EBV is present in lymph node tissues as a pathogenetic factor or perhaps only due to the lack of T-cell suppression of EBV-transformed lymphocytes. Numerous authors have also used PCR to investigate the relationship between EBV and Hodgkin's disease (HD) and have detected EBV DNA in lymph node tissues in some HD patients.[134–144] Contradictory conclusions were, however, drawn from these studies: some authors suggesting, others doubting a correlation of HD to EBV. EBV was also detected in lung biopsy material of a patient with interstitial lung disease;[145] in hairy leukoplakia lesions and normal oral mucosa of HIV patients;[146] in epidermal skin lesions of an immunosuppressed patient;[147] in normal corneal epithelium;[27] and in different tissues in a number of other diseases of unclear origin.[148–155]

Some authors have used PCR for the investigation of EBV epidemiology and have established test systems for the differentiation of EBV strains.[133,156–159] Investigation of healthy EBV carriers showed that usually only one single virus strain is predominant in oropharyngeal secretions.[160] Another study has shown that PCR used in addition to RFLP analysis was valuable for the identification of EBV strains transmitted by organ transplantation.[161] Besides the studies mentioned above, other PCR assays for the detection of EBV DNA have been published.[162,163] Some PCR assays for EBV DNA detection are summarized in Table 13–4.

Table 13–4 **PCR assays for the detection of EBV**

Genomic Region	Amplification Conditions	Detection System	Clinical Samples	Ref.
gp220 EBNA 1	1.5′, 94; 45″, 60; 2′, 70°C 30 c	Southern blot	Clotted blood PBMCs Tissue samples	117
BamHI Nhet Y,E	1′, 95; 2′, 55; 2′, 72°C 32 c 50 c	Southern blot	Nasopharynx carcinoma biopsies	131
EBNA 2 A/B	30″, 94; 1′, 55; 3′, 72°C 50 c	Gel analysis	EBV transformed cell lines	156

Telenti et al.[117]:

Primers: gp220: 5′GGCTGGTGTCACCTGTGTTA 3′
    5′CCTTAGGAGGAACAAGTCCC 3′
Probe: 5′GGTGGAGGGGCTGAGTGTCTCTGGGTTTGAACTGGG 3′

Chang et al.[131]:

Primers: E1: 5′GAATTCAGACCCACCATGGAATCATTTG 3′
    E2: 5′CTTAAGCGCTTGCAGGTGCGATTGCTAA 3′

Jilg et al.[156]:

Primers: Type A: 2A-FS: 5′TCTTGATAGGGATCCGCTAGGATA 3′
    2A-CS: 5′ACCGTGGTTCTGGACTATCTGGATC 3′
  Type B: 2B-FS: 5′ACTGGATATGAATCCCCTGGGCAG 3′
    2B-CS: 5′GAGTCCTGTACTATCAGAACTACAATG 3′

VI. HUMAN HERPESVIRUS TYPE 6

A. CLINICAL MANIFESTATIONS[164]

HHV-6 was first isolated in 1986 from peripheral blood mononuclear cells of AIDS patients and of patients with other lymphoproliferative disorders.[165] Since then it has been established that primary infection with HHV-6 is the cause of exanthema subitum.[166] Infection usually occurs within the first 3 years of life as shown by the high presence of HHV-6-specific ABs in older children and adults. Virus then remains latent in the host, possibly in cells such as T lymphocytes or glial cells. Reactivation of HHV-6 has not yet been significantly associated with a particular clinical picture and is thus supposed

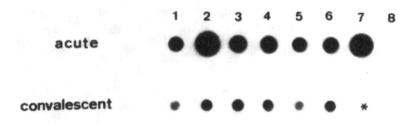

Figure 13–6 Detection of HHV-6 DNA in PBMCs of patients with exanthema subitum collected during the febrile phase (patients 1, 2, 7, and 8) and during the exanthem phase (patients 3, 4, and 5). *, no sample collected. (From Kondo, K., Hayakawa, Y., Mori, H., Sato, S., Kondo, T., Takahashi, K., Minamishima, Y., Takahashi, M., and Yamanishi, K., *J. Clin. Microbiol.,* 28, 970, 1990. With permission.)

to be generally symptomless. The presumed correlation of the chronic fatigue syndrome to HHV-6 proved to be no more significant than to other viral infections. Presence of HHV-6 in immune cells was discussed as a co-factor for the progression of HIV infections or for the development of malignancies. Until now, however, such associations have not been proven.

B. CONVENTIONAL DIAGNOSIS

Conventional diagnosis is based primarily on the detection of HHV-6-specific ABs by an indirect immunofluorescence assay using infected cord blood lymphocytes as the antigen.[164] Isolation of virus from PBMCs may be successful, especially during acute disease, but is usually not applied for routine diagnosis.

C. PCR FOR HHV-6 DIAGNOSIS

Kondo et al. have used PCR for the detection of HHV-6 from PBMCs of children with exanthema subitum.[167] It was shown that HHV-6 DNA was detectable in these cells during the febrile or exanthem period when HHV-6 specific ABs were not yet detectable and also during the convalescent phase of disease (Figure 13–6). Some authors[168,169] have described the amplification of HHV-6 DNA from PBMCs of healthy volunteers. This finding certainly reduces the value of HHV-6 PCR for diagnostic purposes. However, the detection rate of the PCR was shown to be dependent on the number of PBMCs used in the PCR assay.[168] Considering that during latent HHV-6 infection <1 infected cell is present in 10^5 lymphocytes,[169] while during exanthema subitum up to 20,000 virus-carrying cells may be present within 10^5 PBMCs,[164] a selective detection of acute HHV-6 infection seems to be possible, provided that the number of PBMCs used for PCR is limited. Limitation of the test sensitivity to establish an assay specific for acute disease may also be achieved by reduction of the number of PCR cycles or by establishment of a semiquantitative assay.

The transmission of HHV-6 has been supposed to occur via oropharyngeal secretions from acutely as well as from latently infected persons. However, although some authors describe the successful isolation of HHV-6 from the saliva of healthy persons, others have failed to recover infectious virus from such individuals.[164] Similarly contradictory results were obtained by screening for HHV-6 DNA in saliva by PCR.[120,168,170] The detection rates of HHV-6 DNA in healthy persons varied between 3^{170} and 90%.[168] Thus the prevalence of viral DNA as well as of infectious virus in saliva of healthy persons needs to be further investigated.

Some authors have used the PCR to detect HHV-6 DNA in different tissues. The examination of corneas of HIV-infected persons showed that HHV-6 may be capable of invading corneal tissue but that this occurs infrequently.[171] HHV-6 DNA was also detected in retinal tissue of one patient with AIDS-associated retinitis together with CMV.[172] Investigation of lymphoid tissue showed that HHV-6 DNA was detected in none of 41 non-Hodgkin's lymphomas but in 3 of 25 Hodgkin's lymphomas.[173] In another study[174] HHV-6 DNA was detected by PCR in bone marrow cells of patients with acute T-cell leukemia. Further investigation will, however, be necessary to elucidate the significance of these findings. Other studies investigating PBMCs of patients with different lymphoproliferative disorders showed no significant association of HHV-6 to these diseases.[169] Detection of HHV-6 DNA was also described in PBMCs of patients with chronic fatigue syndrome;[175] this, however, does not itself confirm a connection of this disease to HHV-6. In addition, PCR was shown to be valuable in facilitating the identification of different HHV-6 strains.[176] The use of PCR for HHV-6 DNA detection is summarized in Table 13–5.

VII. CONCLUSIONS

Considering the herpesvirus PCR studies summarized here, it is obvious that in specific situations the use of PCR represents a significant improvement over conventional herpesvirus diagnostics, making the routine application of certain well-established PCR assays clearly desirable. In contrast, in the majority of the applications of herpesvirus PCR described it is not yet clear whether this technology can or should be used for routine diagnostic purposes. One major advantage of PCR over some conventional tests has been shown to be its rapidity, and this makes it the method of choice in some cases. In other cases it is either not yet clear or already unlikely that PCR could ever be an alternative to the excellent conventional tests that already exist. As was to be expected, problems arose in various PCR studies, especially with the discrimination between herpesvirus latency and active infection. In some cases arbitrary limitation of the test sensitivity and/or semiquantitative evaluation of the PCR results have allowed or perhaps would

Table 13–5 **PCR assays for the detection of HHV-6**

Genomic Region	Amplification Conditions	Detection System	Clinical Material	Ref.
SalI (Hashimoto) 776 bp[a]	1', 90; 2', 62; 5', 72°C 30 c	Dot blot	PBMCs	167
pZH14 clone 161/90 bp	0.1', 94; 50; 90", 72°C 40/40 c	Nested PCR	PBMCs saliva	168
Clone pHC-5	1', 90; 55; 72°C	Southern blot	Tissue culture	177

[a] Length of the amplicon with the primers shown below:

Kondo et al.[167]:

Primers: 5'GTGTTTCCATTGTACTGAAACCGGT 3'
          5'TAAACATCAATGCGTTGCATACAGT 3'

Jarrett et al.[168]:

Primers: 5'TCTCACAGCCCAGGACAATGGATTATATAT 3'
          5'TGAGATCATTCTCCCGTTCTTGAGGG 3'

Collandre et al.[177]:

Primers: A: 5'GATCCGACGCCTACAAACAC 3'
          B: 5'TACCGACATCCTTGACATATTAC 3'
Probe: S: 5'GGCTGATTAGGATTAATAGGAGA 3'

Table 13–6 **Detection and discrimination of different herpesviruses with one PCR assay**

Virus Detected	Genomic Region	Amplification Conditions	Detection System	Clinical Material	Ref.
HSV1,2 EBV, CMV 518/524/589 bp	DNA polymerase	1', 94; 60; 72°C 40 c	Restriction enzyme analysis	CSF	10

Primers: P1: 5'CGACTTTGCCAGCCTGTACC 3'
          P2: 5'AGTCCGTGTCCCCGTAGATG 3'

allow a differentiation between latent infection, asymptomatic or clinically relevant reactivation. Contradictory PCR results, sometimes obtained by different authors, are certainly often due to the variety of PCR conditions used as well as to differences in the test sensitivities.

In addition to its value for diagnostic purposes, the PCR has been found to be especially important for epidemiological studies. PCR was further shown to be a possible tool for investigating the pathogenesis of viral infections and the association between herpesviruses and different diseases of unclear genesis.

As shown above, a number of excellent PCR assays have been established in the last few years, using primers highly specific for individual herpesviruses as well as carefully developed sample preparation and PCR product-detection methods. Also the use of generic primers for amplification of all herpesviruses, and subsequent discrimination between the amplicons based on their DNA sequence, may prove to be useful[10] (see Table 13–6). It can be expected that based on the progress made so far, a number of the problems inherent to herpesvirus PCR will be overcome and that the PCR will continue to increase in value as a diagnostic tool as well as for other applications.

REFERENCES

1. **Roizman, B.,** Herpesviridae: A brief introduction, in *Virology,* Fields, B. N. and Knipe, D. M., Eds., Raven Press, New York, 1990, chap. 64.
2. **Whitley, R. J.,** Herpes simplex viruses, in *Virology,* Fields, B. N. and Knipe, D. M., Eds., Raven Press, New York, 1990, chap. 66.
3. **Adam, E.,** Herpes simplex virus infections, in *Human Herpesvirus Infection,* Glaser, G. and Gotlieb-Stematsky, T., Eds., Marcel Dekker, New York, 1982.
4. **Lycke, E. and Jeansson, S.,** Herpes viridae: Herpes simplex virus, in *Laboratory Diagnosis of Infectious Diseases: Principles and Practice,* Lenette, E. H., Halonen, P., and Murphy, F. A., Eds., Springer-Verlag, New York, 1988, chap. 11.
5. **Powell, K. F., Anderson, N. E., Frith, R. W., and Croxon, M. C.,** Noninvasive diagnosis of herpes simplex encephalitis, *Lancet,* i, 357, 1990.
6. **Rowley, A. H., Whitley, R. J., Lakeman, F. D., and Wolinsky, S. M.,** Rapid detection of herpes-simplex-virus DNA in cerebrospinal fluid of patients with herpes simplex encephalitis, *Lancet,* i, 440, 1990.
7. **Puchhammer-Stöckl, E., Popow-Kraupp, Th., Heinz, F. X., Mandl, C. W., and Kunz, C.,** Establishment of PCR for the early diagnosis of herpes simplex encephalitis, *J. Med. Virol.,* 32, 77, 1990.
8. **Aurelius, E., Johansson, B., Sköldenberg, B., Staland, A., and Forsgren, M.,** Rapid diagnosis of herpes simplex encephalitis by nested polymerase chain reaction assay of cerebrospinal fluid, *Lancet,* 337, 189, 1991.
9. **Klapper, P. E., Cleator, G. M., Dennett, C., and Lewis, A. G.,** Diagnosis of herpes simplex encephalitis via southern blotting of cerebrospinal fluid DNA amplified by polymerase chain reaction, *J. Med. Virol.,* 32, 261, 1990.
10. **Rozenberg, F. and Lebon, P.,** Amplification and characterization of herpesvirus DNA in cerebrospinal fluid from patients with acute encephalitis, *J. Clin. Microbiol.,* 29, 2412, 1991.
11. **Pohl-Koppe, A., Dahm, C., Elgas, M., Kühn, J. E., Braun, R. W., and ter Meulen, V.,** The diagnostic significance of the polymerase chain reaction and isoelectric focusing in herpes simplex virus encephalitis, *J. Med. Virol.,* 36, 147, 1992.
12. **Puchhammer-Stöckl, E., Heinz, F. X., Kundi, M., Popow-Kraupp, T., Grimm, G., Millner, M., and Kunz, C.,** Evaluation of the polymerase chain reaction for the diagnosis of herpes simplex encephalitis, *J. Clin. Microbiol.,* 31, 196, 1993.
13. **Nicoll, J. A., Maitland, N. J., and Love, S.,** Use of the polymerase chain reaction to detect herpes simplex virus DNA in paraffin sections of human brain at necropsy, *J. Neurol. Neurosurg. Psychiatry,* 54, 167, 1991.
14. **Nicoll, J. A., Maitland, N. J., and Love, S.,** Autopsy neuropathological findings in burnt out herpes simplex encephalitis and use of the polymerase chain reaction to detect viral DNA, *Neuropathol. Appl. Neurobiol.,* 17, 375, 1991.
15. **Kimura, H., Futamura, M., Kito, H., Ando, T., Goto, M., Kuzushima, K., Shibata, M., and Morishima, T.,** Detection of viral DNA in neonatal herpes simplex virus infections, frequent and prolonged presence in serum and cerebrospinal fluid, *J. Infect. Dis.,* 164, 289, 1991.
16. **Penneys, N. S., Goldstein, B., Nahass, G. T., Leonardi, C., and Zhu, W. Y.,** Herpes simplex virus DNA in occult lesions, demonstration by the polymerase chain reaction, *J. Am. Acad. Dermatol.,* 24, 689, 1991.
17. **Cone, R. W., Hobson, A. C., Palmer, J., Remington, M., and Corey, L.,** Extended duration of herpes simplex virus DNA in genital lesions detected by the polymerase chain reaction, *J. Infect. Dis.,* 164, 757, 1991.
18. **Hardy, D. A., Arvin, A. M., Yasukawa, L. L., Bronzan, R. N., Lewinsohn, D. M., Hensleigh, P. A., and Prober, C. G.,** Use of polymerase chain reaction for successful identification of asymptomatic genital infection with herpes simplex virus in pregnant women at delivery, *J. Infect. Dis.,* 162, 1031, 1990.
19. **Rogers, B. B., Josephson, S. L., and Mak, S. K.,** Detection of herpes simplex virus using the polymerase chain reaction followed by endonuclease cleavage, *Am. J. Pathol.,* 139, 1, 1991.
20. **Rogers, B. B., Josephson, S. L., Mak, S. K., and Sweeney, P. J.,** Polymerase chain reaction amplification of herpes simplex virus DNA from clinical samples, *Obstet. Gynecol.,* 79, 464, 1992.
21. **Cantin, E. M., Chen, J., McNeill, J., Willey, D. E., and Openshaw, H.,** Detection of herpes simplex virus DNA sequences in corneal transplant recipients by polymerase chain reaction assays, *Curr. Eye Res.,* 10, 15, 1991.

22. **Crouse, C. A., Pflugfelder, S. C., Pereira, I., Cleary, T., Rabinowitz, S., and Atherton, S. S.,** Detection of herpes viral genomes in normal and diseased corneal epithelium, *Curr. Eye Res.,* 9, 569, 1991.

23. **Kaye, S. B., Lynas, C., Patterson, A., Risk, J. M., McCarthy, K., and Hart, C. A.,** Evidence for herpes simplex viral latency in the human cornea, *Br. J. Ophthalmol.,* 75, 195, 1991.

24. **Rong, B. L., Pavan Langston, D., Weng, Q. P., Martinez, R., Cherry, J. M., and Dunkel, E. C.,** Detection of herpes simplex virus thymidine kinase and latency-associated transcript gene sequences in human herpetic corneas by polymerase chain reaction amplification, *Invest. Ophthalmol. Visual Sci.,* 32, 1808, 1991.

25. **Holbach, L. M., Font, R. L., Baehr, W., and Pittler, S. J.,** HSV antigens and HSV DNA in avascular and vascularized lesions of human herpes simplex keratitis, *Curr. Eye Res.,* 10, 63, 1991.

26. **Aslanzadeh, J., Helm, K. F., Espy, M. J., Muller, S. A., and Smith, T. F.,** Detection of HSV-specific DNA in biopsy tissue of patients with erythema multiforme by polymerase chain reaction, *Br. J. Dermatol.,* 126, 19, 1992.

27. **Darragh, T. M., Egbert, B. M., Berger, T. G., and Yen, T. S. B.,** Identification of herpes simplex virus DNA in lesions of erythema multiforme by the polymerase chain reaction, *J. Am. Acad. Dermatol.,* 24, 23, 1991.

28. **Weston, W. L., Brice, S. J., Jester, J. D., Lane, A. T., Stockert, S., and Huff, J. C.,** Herpes simplex virus in childhood erythema multiforme, *Pediatrics,* 89, 32, 1992.

29. **Brice, S. L., Krzemien, D., Weston, W. L., and Huff, J. C.,** Detection of herpes simplex virus DNA in cutaneous lesions of erythema multiforme, *J. Invest. Dermatol.,* 93, 183, 1989.

30. **Miura, S., Smith, C. C., Burnett, J. W., and Aurelian, L.,** Detection of viral DNA within skin of healed recurrent herpes simplex infection and erythema multiforme lesions, *J. Invest. Dermatol.,* 98, 68, 1992.

31. **Löhr, J. M., Nelson, J. A., and Oldstone, M. B.,** Is herpes simplex virus associated with peptic ulcer disease?, *J. Virol.,* 64, 2168, 1990.

32. **Yamamoto, L. J., Tedder, D. G., Ashley, R., and Levin, M. J.,** Herpes simplex virus type 1 DNA in cerebrospinal fluid of a patient with Mollaret's meningitis, *N. Engl. J. Med.,* 325, 1082, 1991.

33. **Palefsky, J. M., Holly, E. A., Gonzales, J., Berline, J., Ahn, D. K., and Greenspan, J. S.,** Detection of human papillomavirus DNA in anal intraepithelial neoplasia and anal cancer, *Cancer Res.,* 51, 1014, 1991.

34. **Ohashi, Y., Yamamoto, S., Nishida, K., Okamoto, S., Kinoshita, S., Hayashi, K., and Manabe, R.,** Demonstration of herpes simplex virus DNA in idiopathic corneal endotheliopathy, *Am. J. Ophthalmol.,* 112, 419, 1991.

35. **Jamieson, G. A., Maitland, N. J., Wilcock, G. K., Craske, J., and Itzhaki, R. F.,** Latent herpes simplex virus type 1 in normal and Alzheimer's disease brains, *J. Med. Virol.,* 33, 224, 1991.

36. **Piiparinen, H. and Vaheri, A.,** Genotyping of herpes simplex viruses by polymerase chain reaction, *Arch. Virol.,* 119, 275, 1991.

37. **Kimura, H., Shibata, M., Kuzushima, K., Nishikawa, K., Nishiyama, Y., and Morishima, T.,** Detection and direct typing of herpes simplex virus by polymerase chain reaction, *Med. Microbiol. Immunol. (Berlin),* 179, 177, 1990.

38. **Gelb, L. D.,** Varicella-zoster-virus, in *Virology,* 2nd ed., Fields, B. N. and Knipe, D. M., Eds., Raven Press, New York, 1990, 2011.

39. **Grose, C.,** Varicella zoster virus infections, chickenpox (varicella) and shingles (zoster), in *Human Herpesvirus Infections,* Glaser, R. and Gotlieb-Stematsky, T., Eds., Dekker, New York, 1982, 85.

40. **Takahashi, M.,** Herpesviridae, varicella-zoster virus, in *Laboratory Diagnosis of Infectious Diseases: Principles and Practice,* Vol. 2, Lennete, E. H., Halonen, P., and Murphy, F. A., Eds., Springer-Verlag, New York, 1988, 261.

41. **Puchhammer-Stöckl, E., Popow-Kraupp, Th., Heinz, F. X., Mandl, C. W., and Kunz, C.,** Detection of varicella-zoster virus DNA by polymerase chain reaction in the cerebrospinal fluid of patients suffering from neurological complications associated with chicken pox or herpes zoster, *J. Clin. Microbiol.,* 29, 1513, 1991.

42. **Shoji, H., Honda, Y., Murai, I., Sato, Y., Oizumi, K., and Hondo, R.,** Detection of varicella-zoster virus DNA by polymerase chain reaction in cerebrospinal fluid of patients with herpes zoster meningitis, *J. Neurol.,* 239, 69, 1992.

43. **Koropchak, C. M., Graham, G., Palmer, J., Winsberg, M., Ting, S. F., Wallace, M., Prober, C. G., and Arvin, A. M.,** Investigation of varicella-zoster virus infection by polymerase chain reaction in the immunocompetent host with acute varicella, *J. Infect. Dis.,* 163, 1016, 1991.

44. **Kido, S., Ozaki, T., Asada, H., Higashi, K., Kondo, K., Hayakawa, Y., Morishima, T., Takahashi, M., and Yamanishi, K.,** Detection of varicella-zoster virus (VZV DNA in clinical samples from patients with VZV by the polymerase chain reaction, *J. Clin. Microbiol.,* 29, 76, 1991.

45. **Dlugosch, D., Eis-Hübinger, A. M., Kleim, J. P., Kaiser, R., Bierhoff, E., and Schneeweis, K. E.,** Diagnosis of acute and latent varicella-zoster virus infections using the polymerase chain reaction, *J. Med. Virol.,* 35, 136, 1991.

46. **Isada, N. B., Paar, D. P., Johnson, M. P., Evans, M. I., Holzgreve, W., Qureshi, F., and Straus, S. E.,** *In utero* diagnosis of congenital varicella zoster virus infection by chorionic villus sampling and polymerase chain reaction, *Am. J. Obstet. Gynecol.,* 165, 1727, 1991.

47. **Ozaki, T., Miwata, H., Matsui, Y., Kido, S., and Yamanishi, K.,** Varicella zoster virus DNA in throat swabs, *Arch. Dis. Childhood,* 65, 333, 1990.

48. **Wilson, A., Sharp, M., Koropchak, C. M., Ting, S. F., and Arvin, A. M.,** Subclinical varicella-zoster virus viremia, herpes zoster, and T lymphocyte immunity to varicella-zoster viral antigens after bone marrow transplantation, *J. Infect. Dis.,* 165, 119, 1992.

49. **Devlin, M. E., Gilden, D. H., Mahalingam, R., Dueland, A. N., and Cohrs, R.,** Peripheral blood mononuclear cells of the elderly contain varicella-zoster virus DNA, *J. Infect. Dis.,* 165, 619, 1992.

50. **Mahalingam, R., Wellish, M., Wolf, W., Dueland, A. N., Cohrs, R., Vafai, A., and Gilden, D.,** Latent varicella-zoster viral DNA in human trigeminal and thoracic ganglia, *N. Engl. J. Med.,* 323, 627, 1990.

51. **Lacey, S. F., Suzutani, T., Powell, K. L., Purifoy, D. J., and Honess, R. W.,** Analysis of mutations in the thymidine kinase genes of drug-resistant varicella-zoster virus populations using the polymerase chain reaction, *J. Gen. Virol.,* 72, 623, 1991.

52. **Alford, C. A. and Britt, W. J.,** Cytomegalovirus, in *Virology,* 2nd ed., Fields, B. N. and Knipe, D. M., Eds., Raven Press, New York, 1990, 1981.

53. **Sullivan, J. L. and Hanshaw, J. B.,** Human cytomegalovirus infection, in *Human Herpesvirus Infections,* Glaser, R. and Gotlieb-Stematsky, T., Eds., Marcel Dekker, New York, 1982, 57.

54. **Drew, W. L.,** Herpesviridae, cytomegalovirus, in *Laboratory Diagnosis of Infectious Diseases: Principles and Practice,* Lenette, E. H., Halonen, P., and Murphy, F. A., Eds., Springer-Verlag, New York, 1988, chap. 13.

55. **Gleaves, C. A., Smith, T. F., Shuster, E. A., and Pearson, G. R.,** Rapid detection of cytomegalovirus in MRC-5 cells inoculated with urine specimens by using low-speed centrifugation and monoclonal antibody to an early antigen, *J. Clin. Microbiol.,* 19, 917, 1984.

56. **Demmler, G. J., Buffone, G. J., Schimbor, C. M., and May, R. A.,** Detection of cytomegalovirus in urine from newborns by using polymerase chain reaction DNA amplification, *J. Infect. Dis.,* 185, 1177, 1988.

57. **Khan, G., Kangro, H. O., Coates, P. J., and Heath, R. B.,** Inhibitory effects of urine on the polymerase chain reaction for cytomegalovirus DNA, *J. Clin. Pathol.,* 44, 360, 1991.

58. **Buffone, G. J., Demmler, G. J., Schimbor, C. M., and Greer, J.,** Improved amplification of cytomegalovirus DNA from urine after purification of DNA with glass beads, *Clin. Chem.,* 37, 1945, 1991.

59. **Schrier, R., Nelson, J., and Oldstone, M.,** Detection of human cytomegalovirus in peripheral blood lymphocytes in a natural infection, *Science,* 230, 1048, 1985.

60. **Stöckl, E., Popow-Kraupp, Th., Heinz, F. X., Mühlbacher, F., Balcke, P., and Kunz, C.,** Potential of *in situ* hybridization for early diagnosis of productive cytomegalovirus infection, *J. Clin. Microbiol.,* 26, 2536, 1988.

61. **Stanier, P., Taylor, D. L., Kitchen, A. D., Wales, N., Tryhorn, Y., and Tyms, A. S.,** Persistence of cytomegalovirus in mononuclear cells in peripheral blood from blood donors, *Br. Med. J.,* 299, 897, 1989.

62. **Morris, D. J., Kimpton, C. P., and Corbitt, C.,** Persistence of cytomegalovirus in peripheral blood from blood donors, *Br. Med. J.,* 299, 1164, 1989.

63. **Bevan, I. S., Daw, R. A., Day, P. J., Ala, F. A., and Walker, M. R.,** Polymerase chain reaction for detection of human cytomegalovirus infection in a blood donor population, *Br. J. Hematol.,* 78, 94, 1991.

64. **Cassol, S. A., Poon, M. C., Pal, R., Naylor, M. J., Culver, J. J., Bowen, T. J., Russel, J. A., Krawetz, S. A., Pon, R. T., and Hoar, D. I.,** Primer-mediated enzymatic amplification of cytomegalovirus (CMV) DNA. Application to the early diagnosis of CMV infection in marrow transplant recipients, *J. Clin. Invest.,* 83, 1109, 1989.

65. **Stanier, P., Kitchen, A. D., Taylor, D. L., and Tyms, A. S.,** Detection of human cytomegalovirus in peripheral mononuclear cells and urine samples using PCR, *Mol. Cell. Probes,* 6, 51, 1992.
66. **Jiwa, N. M., Van Gemert, G. W., Raap, A. K., Van de Rijke, F. M., Mulder, A., Lens, P. F., Salimans, M. M., Zwaan, F. E., Van Dorp, W., and Van der Ploeg, M.,** Rapid detection of human cytomegalovirus DNA in peripheral blood leukocytes of viremic transplant recipients by the polymerase chain reaction, *Transplantation,* 48, 72, 1989.
67. **Rowley, A. H., Wolinsky, S. M., Sambol, S. P., Barkholt, L., Ehrnst, A., and Andersson, J. P.,** Rapid detection of cytomegalovirus DNA and RNA in blood of renal transplant patients by in vitro enzymatic amplification, *Transplantation,* 51(1), 1028, 1991.
68. **Gerna, G., Zipeto, D., Parea, M., Revello, M. G., Silini, E., Percivalle, E., Zavattoni, M., Grossi, P., and Milanesi, G.,** Monitoring of human cytomegalovirus infections and ganciclovir treatment in heart transplant recipients by determination of viremia, antigenemia, and DNAemia, *J. Infect. Dis.,* 164, 488, 1991.
69. **Gerna, G., Parea, M., Percivalle, E., Zipeto, D., Silini, E., Barbarini, G., and Milanesi, G.,** Human cytomegalovirus viraemia in HIV-1-seropositive patients at various clinical stages of infection, *AIDS,* 4, 1027, 1990.
70. **Shibata, D., Martin, W. J., Appleman, M. D., Causey, D. M., Leedom, J. M., and Arnheim, N.,** Detection of cytomegalovirus DNA in peripheral blood of patients infected with human immunodeficiency virus, *J. Infect. Dis.,* 158, 1185, 1988.
71. **Einsele, H., Steidle, M., Vallbracht, A., Saal, J. G., Ehninger, G., and Müller, C. A.,** Early occurrence of human cytomegalovirus infection after bone marrow transplantation as demonstrated by the polymerase chain reaction technique, *Blood,* 77, 1104, 1991.
72. **Delgado, R., Lumbreras, C., Concepcion, A., Pedraza, M. A., Otero, J. R., Gomez, R., Moreno, E., Noriega, A. R., and Paya, C. V.,** Low predictive value of polymerase chain reaction for diagnosis of cytomegalovirus disease in liver transplant recipients, *J. Clin. Microbiol.,* 30, 1876, 1992.
73. **Zipeto, D., Revello, M. G., Silini, E., Parea, M., Percivalle, E., Zavattoni, M., Milanesi, G., and Gerna, G.,** Development and clinical significance of a diagnostic assay based on the polymerase chain reaction for detection of human cytomegalovirus DNA in blood samples from immunocompromised patients, *J. Clin. Microbiol.,* 30, 527, 1992.
74. **Boland, G. J., De Weger, R. A., Tilanus, M. G. J., Ververs, C., Bosboom-Kalsbeek, K., and De Gast, G. C.,** Detection of cytomegalovirus (CMV) in granulocytes by polymerase chain reaction compared with the CMV antigen test, *J. Clin. Microbiol.,* 30, 1763, 1992.
75. **Gerna, G., Zipeto, D., Parea, M., Percivalle, E., Zavattoni, M., Gaballo, A., and Milanesi, G.,** Early virus isolation, early structural antigen detection and DNA amplification by the polymerase chain reaction in polymorphonuclear leukocytes from AIDS patients with human cytomegalovirus viraemia, *Mol. Cell. Probes,* 5, 365, 1991.
76. **Einsele, H., Ehninger, G., Steidle, M., Vallbracht, A., Müller, M., Schmidt, H., Saal, J. G., Waller, H. D., and Müller, C. A.,** Polymerase chain reaction to evaluate antiviral therapy for cytomegalovirus disease, *Lancet,* ii, 1170, 1991.
77. **Ishigaki, S., Takeda, M., Kura, T., Ban, N., Saitoh, T., Sakamaki, S., Watanabe, N., Kohgo, Y., and Niitsu, Y.,** Cytomegalovirus DNA in the sera of patients with cytomegalovirus pneumonia, *Br. J. Hematol.,* 79, 198, 1991.
78. **Olive, D. M., Al Mufti, S., Simsek, M., Fayez, H., and Al Nakib, W.,** Direct detection of human cytomegalovirus in urine specimens from renal transplant patients following polymerase chain reaction amplification, *J. Med. Virol.,* 29, 232, 1989.
79. **Olive, D. M. Simsek, M., and Al Mufti, S.,** Polymerase chain reaction assay for detection of human cytomegalovirus, *J. Clin. Microbiol.,* 27, 1238, 1989.
80. **Hsia, K., Spector, D. H., Lawrie, J., and Spector, S. A.,** Enzymatic amplification of human cytomegalovirus sequences by polymerase chain reaction, *J. Clin. Microbiol.,* 27, 1802, 1989.
81. **Brytting, M., Sundqvist, V. A., Stalhandske, P., Linde, A., and Wahren, B.,** Cytomegalovirus DNA detection of an immediate early protein gene with nested primer oligonucleotides, *J. Virol. Meth.,* 32, 127, 1991.
82. **Spector, S. A., Rua, J. A., Spector, D. H., and McMillan, R.,** Detection of human cytomegalovirus in clinical specimens by DNA-DNA hybridization, *J. Infect. Dis.,* 150, 121, 1984.
83. **Augustin, S., Popow-Kraupp, Th., Heinz, F. X., and Kunz, C.,** Problems in detection of cytomegalovirus in urine samples by dot blot hybridization, *J. Clin. Microbiol.,* 25, 1973, 1987.

84. **Chen, Y. T., Mercer, G. O., Cheigh, J. S., and Mouradian, J. A.,** Cytomegalovirus infection of renal allografts: detection by polymerase chain reaction, *Transplantation,* 53, 99, 1992.

85. **Loning, T., Stilo, K., Riviere, A., and Helmchen, U.,** Cytomegalovirus detection in kidney transplants, results obtained from the polymerase chain reaction, *Clin. Nephrol.,* 37, 78, 1992.

86. **Jiwa, M., Steenbergen, R. D., Zwaan, F. E., Kluin, P. M., Raap, A. K., and Van-der-Ploeg, M.,** Three sensitive methods for the detection of cytomegalovirus in lung tissue of patients with interstitial pneumonitis, *Am. J. Clin. Pathol.,* 93, 491, 1990.

87. **Persons, D. L., Moore, J. A., and Fishback, J. L.,** Comparison of polymerase chain reaction, DNA hybridization, and histology with viral culture to detect cytomegalovirus in immunosuppressed patients, *Mod. Pathol.,* 4, 149, 1991.

88. **Burgart, L. J., Heller, M. J., Reznicek, M. J., Greiner, T. C., Teneyck, C. J., and Robinson, R. A.,** Cytomegalovirus detection in bone marrow transplant patients with idiopathic pneumonitis. A clinicopathologic study of the clinical utility of the polymerase chain reaction on open lung biopsy specimen tissue, *Am. J. Clin. Pathol.,* 96, 572, 1991.

89. **Cagle, P. T., Buffone, G., Holland, V. A., Samo, T., Demmler, G. J., Noon, G. P., and Lawrence, E. C.,** Semiquantitative measurement of cytomegalovirus DNA in lung and heart-lung transplant patients by in vitro DNA amplification, *Chest,* 101, 93, 1992.

90. **Fox, G. M., Crouse, C. A., Chuang, E. L., Pflugfelder, S. C., Cleary, T. J., Nelson, S. J., and Atherton, S. S.,** Detection of herpesvirus DNA in vitreous and aqueous specimens by the polymerase chain reaction, *Arch. Ophthalmol.,* 109, 266, 1991.

91. **Fenner, T. E., Garweg, J., Hufert, F. T., Boehnke, M., Schmitz, H.,** Diagnosis of human cytomegalovirus-induced retinitis in human immunodeficiency virus type 1-infected subjects by using the polymerase chain reaction, *J. Clin. Microbiol.,* 29, 2621, 1991.

92. **Shibata, D. and Klatt, E. C.,** Analysis of human immunodeficiency virus and cytomegalovirus infection by polymerase chain reaction in the acquired immunodeficiency syndrome: an autopsy study, *Arch. Pathol. Lab. Med.,* 113, 1239, 1989.

93. **Powell, K. F., Bellamy, A. R., Catton, M. G., Cooper, I. P., Philip, B. A., Rainer, S. P., Coverdale, H. A., and Croxson, M. C.,** Cytomegalovirus myocarditis in a heart transplant recipient, sensitive monitoring of viral DNA by the polymerase chain reaction, *J. Heart Transplant.,* 8, 465, 1989.

94. **Yuan, C. F., Kao, S. M., Wang, D. C., Ng, H. T., and Pao, C. C.,** Detection of human cytomegalovirus in cervicovaginal cells by culture, *in situ* DNA hybridization and DNA amplification methods, *Mol. Cell. Probes,* 4, 475, 1990.

95. **Lohr, J. M. and Oldstone, M. B.,** Detection of cytomegalovirus nucleic acid sequences in pancreas in type 2 diabetes, *Lancet,* 336, 644, 1990.

96. **Hattersley, A. T., Lo, Y. M., Read, S. J., Eglin, R. P., Wainscoat, J. S., and Clark, A.,** Failure to detect cytomegalovirus DNA in pancreas in type-2 diabetes, *Lancet,* 339, 335, 1992.

97. **Hendrix, M. G., Salimans, M. M., Van-Boven, C. P., and Bruggeman, C. A.,** High prevalence of latently present cytomegalovirus in arterial walls of patients suffering from grade III atherosclerosis, *Am. J. Pathol.,* 136, 23, 1990.

98. **Hendrix, M. G., Daemen, M., and Bruggeman, C. A.,** Cytomegalovirus nucleic acid distribution within the human vascular tree, *Am. J. Pathol.,* 138, 563, 1991.

99. **Van-den-Berg, F., Schipper, M., Jiwa, M., Rook, R., Van-de-Rijke, F., and Tigges, B.,** Implausibility of an aetiological association between cytomegalovirus and Kaposi's sarcoma shown by four techniques, *J. Clin. Pathol.,* 42, 128, 1989.

100. **Bene, M. C., Tang, J. Q., and Faure, G. C.,** Absence of cytomegalovirus DNA in kidneys in IgA nephropathy, *Lancet,* 335, 868, 1990.

101. **Putland, R. A., Ford, J., Korban, G., Evdokiou, A., and Tremaine, M.,** Investigation of spontaneously aborted concepti for microbial DNA, investigation for cytomegalovirus DNA using polymerase chain reaction, *Aust. N.Z. J. Obstet. Gynaecol.,* 30, 248, 1990.

102. **Chou, S. W.,** Differentiation of cytomegalovirus strains by restriction analysis of DNA sequences amplified from clinical specimens, *J. Infect. Dis.,* 162, 738, 1990.

103. **Brytting, M., Wahlbert, J., Lundberg, J., Wahren, B., Uhlén, M., and Sundqvist, V. A.,** Variations in the cytomegalovirus major immediate-early gene found by direct genomic sequencing, *J. Clin. Microbiol.,* 30, 955, 1992.

104. **Lehner, R., Stamminger, T., and Mach, M.,** Comparative sequence analysis of human cytomegalovirus strains, *J. Clin. Microbiol.,* 29, 2494, 1991.

105. **Zaia, J. A., Gallez-Hawkins, G., Churchill, M. A., Morton-Blackshere, A., Pande, H., Adler, S. P., Schmidt, G. M., and Forman, S. J.,** Comparative analysis of human cytomegalovirus a-sequence in multiple clinical isolates by using polymerase chain reaction and a restriction fragment length polymorphism assays, *J. Clin. Microbiol.,* 28, 2602, 1990.

106. **Sokol, D. M., Demmler, G. J., and Buffone, G. J.,** Rapid epidemiologic analysis of cytomegalovirus by using polymerase chain reaction amplification of the L-S junction region, *J. Clin. Microbiol.,* 30, 839, 1992.

107. **Kimpton, C. P., Morris, D. J., and Corbitt, G.,** Sensitive non-isotopic DNA hybridisation assay or immediate-early antigen detection for rapid identification of human cytomegalovirus in urine, *J. Virol. Meth.,* 32, 89, 1991.

108. **Sandin, R. L., Rodriguez, E. R., Rosenberg, E., Porter-Jordan, K., Caparas, M., Nasim, S., Rockis, M., Keiser, J. F., and Garrett, C. T.,** Comparison of sensitivity for human cytomegalovirus of the polymerase chain reaction, traditional tube culture and shell vial assay by sequential dilutions of infected cell lines, *J. Virol. Meth.,* 32, 181, 1991.

109. **Shibata, M., Morishima, T., Terashima, M., Kimura, H., Kuzushima, K., Hanada, N., Nishikawa, K., and Watanabe, K.,** Human cytomegalovirus infection during childhood, detection of viral DNA in peripheral blood by means of polymerase chain reaction, *Med. Microbiol. Immunol. (Berlin),* 179, 245, 1990.

110. **Chehab, F. F., Xiao, X., Kan, Y. W., and Yen, T. S.,** Detection of cytomegalovirus infection in paraffin-embedded tissue specimens with the polymerase chain reaction, *Mod. Pathol.,* 2, 75, 1989.

111. **Harju, L., Janne, P., Kallio, A., Laukkanen, M. L., Lautenschlager, I., Mattinen, S., Ranki, A., Ranki, M., Soares, V. R., Soderlund, H., and Syvänen, A.-C.,** Affinity-based collection of amplified viral DNA, application to the detection of human immunodeficiency virus type 1, human cytomegalovirus and human papillomavirus type 16, *Mol. Cell. Probes.,* 4, 223, 1990.

112. **Porter-Jordan, K., Rosenberg, E. I., Keiser, J. F., Gross, J. D., Ross, A. M., Nasim, S., and Garrett, C. T.,** Nested polymerase chain reaction assay for the detection of cytomegalovirus overcomes false positives caused by contamination with fragmented DNA, *J. Med. Virol.,* 30, 85, 1990.

113. **Buffone, G. J., Hine, E., and Demmler, G. J.,** Detection of mRNA from the immediate early gene of human cytomegalovirus in infected cells by *in vitro* amplification, *Mol. Cell. Probes,* 4, 143, 1990.

114. **Rogers, B. B., Alpert, L. C., Hine, E. A., and Buffone, G. J.,** Analysis of DNA in fresh and fixed tissue by the polymerase chain reaction, *Am. J. Pathol.,* 136, 541, 1990.

115. **Miller, G.,** Epstein-Barr virus, in *Virology,* Fields, B. N. and Knipe, D. M., Eds., Raven Press, New York, 1990, chap. 68.

116. **Lenette, E. T.,** Herpesviridae, Epstein-Barr virus, in *Laboratory Diagnosis of Infectious Diseases: Principles and Practice,* Lenette, E. H., Halonen, P., and Murphy, F. A., Eds., Springer-Verlag, New York, 1988, chap. 12.

117. **Telenti, A., Marshall, W. F., and Smith, T. F.,** Detection of Epstein-Barr virus by polymerase chain reaction, *J. Clin. Microbiol.,* 28, 2187, 1990.

118. **Israele, V., Shirley, P., and Sixbey, J. W.,** Excretion of the Epstein-Barr virus from the genital tract of men, *J. Infect. Dis.,* 163, 1341, 1991.

119. **Crouse, C. A., Pflugfelder, S. C., Cleary, T., Demick, S. M., and Atherton, S. S.,** Detection of Epstein-Barr virus genomes in normal human lacrimal glands, *J. Clin. Microbiol.,* 28, 1026, 1990.

120. **Gopal, M. R., Thomson, B. J., Fox, J., Tedder, R. S., and Honess, R. W.,** Detection by PCR of HHV-6 and EBV DNA in blood and oropharynx of healthy adults and HIV-seropositives, *Lancet,* 335, 1598, 1990.

121. **Telenti, A., Smith, T. F., Ludwig, J., Keating, M. R., Krom, R. A., and Wiesner, R. H.,** Epstein-Barr virus and persistent graft dysfunction after liver transplantation, *Hematology,* 14, 282, 1991.

122. **Okano, M., Bashir, R. M., Davis, J. R., and Purtilo, D. T.,** Detection of primary Epstein-Barr virus infection in a patient with x-linked lymphoproliferative disease receiving immunoglobulin prophylaxis, *Am. J. Hematol.,* 36, 294, 1991.

123. **Saito, I., Servenius, B., Compton, T., and Fox, R.,** Detection of Epstein-Barr virus DNA by polymerase chain reaction in blood and tissue biopsies from patients with Sjögren's syndrome, *J. Exp. Med.,* 169, 2191, 1989.

124. **Mariette, X., Gozlan, J., Clerc, D., Bisson, M., and Morinet, F.,** Detection of Epstein-Barr virus DNA by *in situ* hybridization and polymerase chain reaction in salivary gland biopsy specimens from patients with Sjögren's syndrome, *Am. J. Med.,* 90, 286, 1991.

125. **Deacon, E. M., Matthews, J. B., Potts, A. J. C., Hamburger, J., Bevan, I. S., and Young, L. S.,** Detection of Epstein-Barr virus antigens and DNA in major and minor salivary glands using immunocytochemistry and polymerase chain reaction, possible relationship with Sjögren's syndrome, *J. Pathol.,* 163, 351, 1991.

126. **Deacon, L. M., Shattles, W. G., Mathews, J. B., Young, L. S., and Venables, P. J. W.,** Frequency of EBV DNA detection in Sjögren's syndrome, *Am. J. Med.,* 92, 453, 1992.

127. **Pflugfelder, S. C., Crouse, C., Pereira, I., and Atherton, S.,** Amplification of Epstein-Barr virus genomic sequences in blood cells, lacrimal glands and tears from primary Sjögren's syndrome patients, *Ophthalmology,* 97, 976, 1990.

128. **Naher, H., Gissmann, L., Freese, U. K., Petzold, D., and Helfrich, S.,** Subclinical Epstein-Barr virus infection of both the male and female genital tract: indication for sexual transmission, *J. Invest. Dermatol.,* 98, 791, 1992.

129. **Feinmesser, R., Miyazaki, I., Cheung, R., Freeman, J. L., Noyek, A. M., and Dosch, H. M.,** Diagnosis of nasopharyngeal carcinoma by DNA amplification of tissue obtained by fine-needle aspiration, *N. Engl. J. Med.,* 326, 17, 1992.

130. **Akao, I., Sato, Y., Mukai, K., Uhara, H., Furuya, S., Hoshikawa, T., Shimosato, Y., and Takeyama, I.,** Detection of Epstein-Barr virus DNA in formalin-fixed paraffin-embedded tissue of nasopharyngeal carcinoma using polymerase chain reaction and *in situ* hybridization, *Laryngoscope,* 101, 279, 1991.

131. **Chang, Y. S., Tyan, Y. S., Liu, S. T., Tsai, M. S., and Pao, C. C.,** Detection of Epstein-Barr virus DNA sequences in nasopharyngeal carcinoma cells by enzymatic DNA amplification, *J. Clin. Microbiol.,* 28, 2398, 1990.

132. **Shibata, D., Weiss, L. M., Nathwani, B. N., Brynes, R. K., and Levine, A. M.,** Epstein-Barr virus in benign lymph node biopsies from individuals infected with the human immunodeficiency virus is associated with concurrent or subsequent development of non-Hodgkin's lymphoma, *Blood,* 77, 1527, 1991.

133. **Boyle, M. J., Sewell, W. A., Sculley, T. B., Apolloni, A., Turner, J. J., Swanson, C. E., Penny, R., and Cooper, D. A.,** Subtypes of Epstein-Barr virus in human immunodeficiency virus-associated non-Hodgkin lymphoma, *Blood,* 78, 3004, 1991.

134. **Shibata, D., Hansmann, M. L., Weiss, L. M., and Nathwani, B. N.,** Epstein-Barr virus infections and Hodgkin's disease: a study of fixed tissues using the polymerase chain reaction, *Human Pathol.,* 22, 1262, 1991.

135. **Knecht, H., Odermatt, B. F., Bachmann, E., Teixeira, S., Sahli, R., Hayoz, D., Heitz, P., and Bachmann, F.,** Frequent detection of Epstein-Barr virus DNA by the polymerase chain reaction in lymph node biopsies from patients with Hodgkin's disease without genomic evidence of B- or T-cell clonality, *Blood,* 78, 760, 1991.

136. **Masih, A., Weisenburger, D., Duggan, M., Mrmitage, J., Bashir, R., Mitchell, D., Wickert, R., and Purtilo, D. T.,** Epstein-Barr virus genome in lymph nodes from patients with Hodgkin's disease may not be specific to Reed-Sternberg cells, *Am. J. Pathol.,* 139, 37, 1991.

137. **Wright, C. F., Reid, A. H., Tsai, M. M., Ventre, K. M., Murai, P. J., Frizzera, G., and O'Leary, T. J.,** Detection of Epstein-Barr virus sequences in Hodgkin's disease by the polymerase chain reaction, *Am. J. Pathol.,* 139, 393, 1991.

138. **Bignon, Y. J., Bernard, D., Cure, H., Fonck, Y., Pauchard, J., Travade, P., Legros, M., Dastugue, B., and Plagne, R.,** Detection of Epstein-Barr viral genomes in lymph nodes of Hodgkin's disease patients, *Mol. Carcinogenesis,* 3, 9, 1990.

139. **Brocksmith, D., Angel, C. A., Pringle, J. H., and Lauder, I.,** Epstein-Barr viral DNA in Hodgkin's disease, amplification and detection using the polymerase chain reaction, *J. Pathol.,* 165, 11, 1991.

140. **Herbst, H., Midobitek, G., Kneba, M., Hummel, M., Finn, T., Anagnostopoulos, I., Bergholz, M., Krieger, G., and Stein, H.,** High incidence of Epstein-Barr virus genomes in Hodgkin's disease, *Am. J. Pathol.,* 137, 13, 1990.

141. **Weiss, L. M., Chen, Y. Y., Liu, X. F., and Shibata, D.,** Epstein-Barr virus and Hodgkin's disease. A correlative *in situ* hybridization and polymerase chain reaction study, *Am. J. Pathol.,* 139, 1259, 1991.

142. **Khan, G., Coates, P. J., Gupta, R. K., Kangro, H. O., and Slavin, G.,** Presence of Epstein-Barr virus in Hodgkin's disease is not exclusive to Reed-Sternberg cells, *Am. J. Pathol.,* 140, 757, 1992.

143. **Samoszuk, M. and Ravel, J.,** Frequent detection of Epstein-Barr viral deoxyribonucleic acid and absence of cytomegalovirus deoxyribonucleic acid in Hodgkin's disease and acquired immunodeficiency syndrome-related Hodgkin's disease, *Lab. Invest.,* 65, 631, 1991.

144. **Moran, C. A., Tuur, S., Angritt, P., Reid, A. H., and O'Leary, T. J.,** Epstein Barr virus in Hodgkin's disease from patients with human immunodeficiency virus infection, *Mod. Pathol.,* 5, 85, 1992.

145. **Myers, J. L., Peiper, S. C., and Katzenstein, A. L.,** Pulmonary involvement in infectious mono-nucleosis, histopathologic features and detection of Epstein-Barr virus-related DNA sequences, *Mod. Pathol.,* 2, 444, 1989.

146. **Snjiders, P. J., Schulten, E. A., Mullink, H., Ten-Kate, R. W., Jiwa, M., van der Waal, I., Meijer, C. J., and Walboomers, J. M.,** Detection of human papillomavirus and Epstein-Barr virus DNA sequences in oral mucosa of HIV-infected patients by the polymerase chain reaction, *Am. J. Pathol.,* 137, 659, 1990.

147. **Fermand, J. P., Gozlan, J., Bendelac, A., Deleuche-Cavallier, M. C., Brouet, J. C., and Morinet, F.,** Detection of Epstein-Barr virus in epidermal skin lesions of an immunocompromised patient, *Ann. Int. Med.,* 112, 511, 1990.

148. **Katzenstein, A. L. and Peiper, S. C.,** Detection of Epstein-Barr virus genomes in lymphomatoid granulomatosis, analysis of 29 cases by the polymerase chain reaction technique, *Mod. Pathol.,* 3, 435, 1990.

149. **Knecht, H., Sahli, R., Shaw, P., Meyer, C., Bachmann, E., Odermatt, B. F., and Bachmann, F.,** Detection of Epstein-Barr virus DNA by polymerase chain reaction in lymph node biopsies from patients with angioimmunoblastic lymphadenopathy, *Br. J. Haematol.,* 75, 610, 1990.

150. **Gal, A. A., Unger, E. R., Koss, M. N., and Yen, T. S.,** Detection of Epstein-Barr virus in lymphoepithelioma-like carcinoma of the lung, *Mod. Pathol.,* 4, 264, 1991.

151. **Bignon, Y. J., Clavelou, P., Ramos, F., Jouvet, A., Tommasi, M., Tournilhac, M., Dastugue, B., and Plagne, R.,** Detection of Epstein-Barr virus sequences in primary brain lymphoma without immunodeficiency, *Neurology,* 41, 1152, 1991.

152. **Rouah, E., Rogers, B. B., Wilson, D. R., Kirkpatrick, J. B., and Buffone, G. J.,** Demonstration of Epstein-Barr virus in primary central nervous system lymphomas by the polymerase chain reaction and *in situ* hybridization, *Hum. Pathol.,* 21, 545, 1990.

153. **Burke, A. P., Yen, T. S., Shekitka, K. M., and Sobin, L. H.,** Lymphoepithelial carcinoma of the stomach with Epstein-Barr virus demonstrated by polymerase chain reaction, *Mod. Pathol.,* 3, 377, 1990.

154. **Kikuta, H., Nakanishi, M., Ishikawa, N., Konno, M., and Matsumoto, S.,** Detection of Epstein-Barr virus sequences in patients with Kawasaki disease by means of the polymerase chain reaction, *InterVirology,* 33, 1, 1992.

155. **Min, K. W., Holmquist, S., Peiper, S. C., and O'Leary, T. J.,** Poorly differentiated adenocarcinoma with lymphoid stroma (lymphoepithelioma-like-carcinomas of the stomach). Report of three cases with Epstein-Barr virus genome demonstrated by the polymerase chain reaction, *Am. J. Clin. Pathol.,* 96, 219, 1991.

156. **Jilg, W., Sieger, E., Alliger, P., and Wolf, H.,** Identification of type A and B isolates of Epstein-Barr virus by polymerase chain reaction, *J. Virol. Meth.,* 30, 319, 1990.

157. **Miyashita, T., Kawaguchi, H., Asada, M., Mizutani, S., and Ibuka, T.,** Epstein-Barr virus type B in patient with T-cell lymphoma, *Lancet,* 337, 1045, 1991.

158. **Kunimoto, M., Tamura, S., Tabata, T., and Yoshie, O.,** One-step typing of Epstein Barr virus by polymerase chain reaction, predominance of type 1 virus in Japan, *J. Gen. Virol.,* 73, 455, 1992.

159. **Chen, X. Y., Pepper, S. D., and Arrand, J. R.,** Prevalence of the A and B types of Epstein-Barr virus DNA in nasopharyngeal carcinoma biopsies from southern China, *J. Gen. Virol.,* 73, 463, 1992.

160. **Yao, Q. Y., Rowe, M., Martin, B., Young, L. S., and Rickinson, A. B.,** The Epstein-Barr virus carrier state, dominance of a single growth-transforming isolate in the blood and in the oropharynx of healthy virus carriers, *J. Gen. Virol.,* 72, 1579, 1991.

161. **Cen, H., Breinig, M. C., Atchison, R. W., Ho, M., and McKnight, J. L.,** Epstein-Barr virus transmission via the donor organs in solid organ transplantation, polymerase chain reaction and restriction fragment length polymorphism analysis of IR2, IR3, and IR4, *J. Virol.,* 65, 976, 1991.

162. **Ambinder, R. F., Lambe, B. C., Mann, R. B., Hayward, S. D., Zehnbauer, B. A., Burns, W. S., and Charache, P.,** Oligonucleotides for polymerase chain reaction amplification and hybridization detection of Epstein-Barr virus DNA in clinical specimens, *Mol. Cell. Probes,* 4, 397, 1990.

163. **Glukhov, A. L., Gordeev, S. A., Vinogradov, S. V., Kiselev, V. I., Kramarov, V. M., Kiselev, O. I., and Severin, E. S.,** Amplification of DNA sequences of Epstein-Barr and human immunode-ficiency viruses using DNA-polymerase from *Thermus thermophilus, Mol. Cell. Probes,* 4, 435, 1990.

164. **Lopez, C. and Honess, R. W.,** Human herpesvirus-6, in *Virology,* Fields, B. N. and Knipe, D. M., Eds., Raven Press, New York, 1990, chap. 72.

165. **Salahuddin, S. Z., Ablashi, D. V., Markham, P. D., Josephs, S. F., Stutzenegger, S., Kaplan, M., Halligan, G., Biberfeld, P., Wong-Staal, F., Kramarsky, B., and Gallo, R. C.,** Isolation of a new virus, HBLV, in patients with lymphoproliferative disorders, *Science,* 234, 596, 1986.

166. **Yamanishi, K., Kondo, T., Kondo, K., Hayakawa, Y., Kido, S., Takahashi, K., and Takahasi, M.,** Exanthem subitum and human herpesvirus 6 (HHV-6) infection, *Adv. Exp. Med. Biol.,* 278, 29, 1990.

167. **Kondo, K., Hayakawa, Y., Mori, H., Sato, S., Kondo, T., Takahashi, K., Minamishima, Y., Takahashi, M., and Yamanishi, K.,** Detection by polymerase chain reaction amplification of human herpesvirus-6 DNA in peripheral blood of patients with exanthem-subitum, *J. Clin. Microbiol.,* 28, 970, 1990.

168. **Jarrett, R. F., Clark, D. A., Josephs, S. F., and Onions, D. E.,** Detection of human herpesvirus-6 DNA in peripheral blood and saliva, *J. Med. Virol.,* 32, 73, 1990.

169. **Sandhoff, T., Kleim, J. P., and Schneweis, K. E.,** Latent human herpesvirus-6 DNA is sparsely distributed in peripheral blood lymphocytes of healthy adults and patients with lymphocytic disorders, *Med. Microbiol. Immunol., (Berlin),* 180, 127, 1991.

170. **Kido, S., Kondo, K., Kondo, T., Morishima, T., Takahashi, M., and Yamanishi, K.,** Detection of human herpesvirus 6 DNA in throat swabs by polymerase chain reaction, *J. Med. Virol.,* 32, 139, 1990.

171. **Qavi, H. B., Green, M. T., SeGall, G. K., and Font, R. L.,** Demonstration of HIV-1 and HHV-6 in AIDS-associated retinitis, *Curr. Eye Res.,* 8, 379, 1989.

172. **Qavi, H. B., Green, M. T., SeGall, G. K., Hollinger, F. B., and Lewis, D. E.,** The incidence of HIV-1 and HHV-6 in corneal buttons, *Curr. Eye Res.,* 10, 97, 1991.

173. **Torelli, G., Marasca, R., Luppi, M., Selleri, L., Ferrari, S., Narni, F., Mariano, M. T., Federico, M., Ceccherini-nelli, L., Bendinelli, M., Montagnani, G., Montorsi, M., and Artusi, T.,** Human herpesvirus-6 in human lymphomas: identification of specific sequences in Hodgkin's lymphomas by polymerase chain reaction, *Blood,* 77, 2251, 1991.

174. **Luka, J., Pirruccello, S. J., and Kersey, J. H.,** HHV-6 genome in T-cell acute lymphoblastic leukaemia, *Lancet,* ii, 1277, 1991.

175. **Buchwald, D., Cheney, P. R., Peterson, D. L., Henry, B., Wormsley, S. B., Geiger, A., Ablashi, D. V., Salahuddin, S. Z., Saxinger, C., Biddle, R., Kikinis, K., Jolesz, F. A., Folks, T., Balachandran, N., Peter, J. B., Gallo, R. C., and Komaroff, A. L.,** *Ann. Intern. Med.,* 116, 103, 1992.

176. **Aubin, J. T., Collandre, H., Candotti, D., Ingrand, D., Rouzioux, C., Burgard, M., Richard, S., Huraux, J. M., and Agut, H.,** Several groups among human herpesvirus 6 strains can be distinguished by Southern blotting and polymerase chain reaction, *J. Clin. Microbiol.,* 29, 367, 1991.

177. **Collandre, H., Aubin, J. T., Agut, H., Bechet, J. M., and Montagnier, L.,** Detection of HHV-6 by the polymerase chain reaction, *J. Virol. Meth.,* 31, 171, 1991.

Chapter 14

Detection of Human Polyomaviruses by PCR

Patricia E. Gibson

TABLE OF CONTENTS

I. INTRODUCTION

The *Polyomavirus* genus is one of two genera belonging to the family Papovaviridae. These are species-specific, small DNA viruses that infect a wide range of vertebrates. The polyomaviruses are host specific and produce tumors in experimental animals. This latter property has generated great interest in these viruses although no naturally occurring tumors have been attributed to them. The viruses persist in their natural hosts following acute infection, but do not appear to produce a recognizable disease. There are two human polyomaviruses, BK (BKV) and JC (JCV). Most workers are interested in JCV because it is the etiological agent responsible for the rare demyelinating disease, progressive multifocal leukoencephalopathy (PML). This chapter deals with the use of PCR to investigate acute, chronic, and latent infection with JCV and BKV.

The polyomaviruses contain a circular double-stranded DNA molecule of approximately 3.4×10^6 Da molecular weight, with a supercoiled structure in the virion. JCV (Mad-1 isolate) has been completely sequenced and consists of 5130 bp.[1] Several BKV isolates have been completely sequenced (DUN, MM, AS) and found to have a similar number of base pairs, with small variations in size between isolates.[2-4]

The molecular organization of the RNA transcripts and resulting proteins of the polyomaviruses is well understood (Figure 14–1). Transcripts are produced from both strands of the DNA: two mRNAs are transcribed off one strand of the early region in a counterclockwise direction, and three are transcribed clockwise off the opposite strand in the late region. The two early proteins produced are the small t and large T antigens, which are not structural components of the virion. They appear to be important for replication and transformation. The three late proteins, VP1, VP2, and VP3, are structural components of the virion. The major capsid protein, VP1, is responsible for serological differences between JCV and BKV and also for serological differences between strains of BKV.

Serological studies have shown that BKV and JCV infect a large proportion of the human population.[5,6] Primary infection with BKV probably occurs early in childhood, while primary infection with JCV may occur later, in adolescence and early adulthood. The nature of the primary infection is not known but may be an acute respiratory disease or cystitis. Polyomavirus probably remains latent in the body following primary infection and may reactivate later in life following changes in the immune system. BKV and JCV have been isolated from the urine of patients who have conditions associated with impaired immunity or who require treatment with cytotoxic and immunosuppressive drugs because of renal or bone marrow transplantation, malignant disease, and both acquired and congenital immune deficiency diseases. BKV and JCV have been isolated from the urine of pregnant women, but both are rarely excreted by normal individuals. JCV can be isolated from the brains of patients with PML and is responsible for this progressive neurological disease. The virus has been observed in abnormal oligodendroglial cells of the brain. Most patients with PML have an underlying disease responsible for the abnormal immune responses.

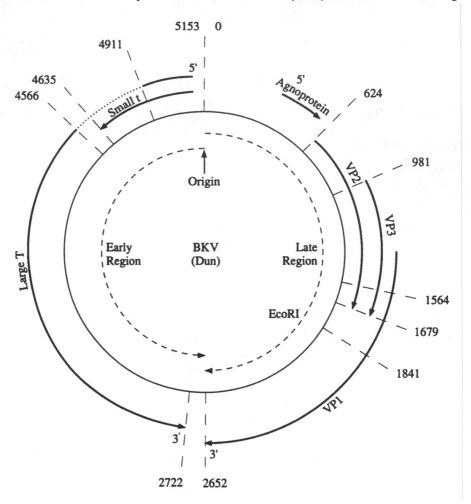

Figure 14–1 Genetic organization and location of the protein coding regions of BK virus (Genbank sequence data for PVBDUN). JC virus has a similar structure.

II. VIRUS-DETECTION METHODS

Since the first isolation of BKV and JCV in 1971 many methods of detection have been used to identify infection with human polyomaviruses. Over the last 20 years, improvements in these methods have slowly evolved, resulting in better sensitivity and specificity of the tests.[6]

The presence of intranuclear inclusions in cells of urinary sediment, most often seen by cytologists in renal transplant recipients, is an indicator of polyomavirus excretion. Such cytological changes are not a reliable indicator of polyomavirus infection, but early observations of these did lead to the discovery of BKV. Isolation of BKV and JCV in cell culture is a lengthy procedure often taking several months. The cytopathic effect is nonspecific, but JCV has a more specific host cell range than BKV, a property that aids in specific identification.

Electron microscopy (EM) is a rapid technique used for visualizing polyomavirus particles in urine and JCV in brain sections of PML patients. BKV and JCV in urine can be differentiated by immune EM using specific antisera.

Antigen-detection methods, including immunofluorescence and ELISA, using virus-specific animal antisera have been commonly used to identify BKV and JCV in exfoliated cells in urine. Monoclonal antibodies to BK and JC viruses have also been prepared for use in immunofluorescence.

Several DNA hybridization techniques have been described for the detection of BKV and JCV in urine, and JCV in brain. These tests include Southern blot, dot blot, and slot blot with radioisotope-labeled probes on urinary sediment, PML brain material, and DNA extracted from these specimens. An *in situ* hybridization test using a biotinylated DNA probe has also been used to detect JCV in paraffin-embedded brain tissues. The most recent method of DNA detection has, of course, been amplification by PCR, which has been used to detect the human polyomavirus in both urine and brain tissue.

III. PCR APPLICATION

The first report of the development of a PCR for BKV and JCV and its application to their detection in urine and brain tissue was by Arthur and colleagues.[7] These workers selected oligonucleotide primer sequences from the conserved early T antigen (Ag) region of the polyomavirus genome common to both BKV and JCV. They were able to amplify these viral sequences from urine of bone marrow and renal transplant recipients. They also found polyomavirus sequences in the urine of 3 of 30 healthy individuals. Amplified DNA from the two viruses was specifically identified either by digestion with restriction endonucleases, and recognized as an appropriate size fragment by gel electrophoresis, or by slot or Southern blot hybridization of the amplicons with specific internal oligonucleotide probes. They were also able to detect JCV in the paraffin-embedded brain tissue of a patient with PML. Resuspended urine sediment was tested, and similar amplification was observed with unextracted samples as with DNA extracted from the sedimented cells. Marshall and colleagues,[8] using the same BKV and JCV common T antigen primers described by Arthur and co-workers,[7] were able to amplify both BKV and JCV DNA from urine sediments of 44% of posttransplant patients. They used Southern blot hybridization of the amplified DNA with radiolabeled specific internal probes to confirm the presence of BKV and/or JCV. This same group also used PCR to detect JCV DNA in formalin-fixed, paraffin-embedded brain tissue sections from 24 patients with PML.[9] Again, they used the T Ag region primers as described by Arthur and colleagues,[7] plus another primer pair which amplified a 207-bp sequence from the late VP1 antigen gene region of JCV and which did not amplify BKV DNA. Confirmation in both cases was by dot blot hybridization with specific internal radiolabeled probes.

Their results suggested that prolonged tissue fixation may inhibit PCR amplification, leading to false-negative results. Another group[10] used PCR to amplify a sequence of over 700 bp within the control region of the BKV and JCV genomes from urines that had already been confirmed as BKV and JCV positive with other primers. They then directly sequenced the DNA amplified from the control region and compared the different amplicons for size and restriction enzyme sites. They used primers common to both BKV and JCV and were interested in investigating the heterogeneity that occurs in the noncoding region, near the origin of DNA replication. By sequencing PCR products directly from urine, they were able to exclude variation occurring during the long growth period in cell culture.

The devastating encephalopathy of the central nervous system caused by JCV occurs only rarely in people with impaired immunity. Rapid neurological deterioration, resulting in death, occurs within a year of onset. In recent years there has been an increase in the number of PML cases reported, directly related to the AIDS epidemic. Definitive diagnosis depends on examination of brain material by histological and virological techniques. Diagnosis of PML by examination of cerebrospinal fluid (CSF) would be a much less invasive procedure than brain biopsy, and since PCR is applicable to very small amounts of clinical sample it is an attractive alternative to examination of biopsies. Such a rapid, sensitive, and noninvasive procedure would greatly assist in the assessment and management of this disease. A few reports of PCR for the detection of JCV DNA in CSF from PML patients have already been made.[11-13] Primers were chosen from the T antigen region for these investigations.

IV. STUDIES AT THE VRL CPHL

In the Virus Reference Laboratory at Colindale, London, I tested several sets of primers for amplification of regions coding for the T antigen and the VP1 structural protein. Some of these primer pairs were common to both BKV and JCV, allowing identification of infections with both viruses, while other primer pairs were more specific and amplified only one viral DNA. Those that worked most satisfactorily under our conditions of amplification are listed in Table 14–1. This consisted of an initial denaturation step of 94°C for 2 min and then 35 c of 55°C for 1 min, 72°C for 1 min, and 91°C for 1 min. The amplification was completed with a step of 55°C for 1 min and finally 72°C for 4 min.

The T Ag primers amplify a target sequence of 386 nucleotides from the T antigen coding region of the genome of both BK and JC polyomavirus. The nested T primers amplify a sequence of 214 nucleotides within the 386-nucleotide amplicon, and are used as a secondary confirmatory test on the primary reaction product. The two primer pairs from the VP1 region are specific for either virus DNA,

Table 14–1   Primer pairs and specific probes used to amplify and identify polyomaviral DNA sequences

Primers	Position/JCV[b]	Position/BKV[a]	Sequence (5′–3′)	Specific Probes[c] (5′–3′)
T Ag-1[d]	3889–4274	4026–4411	CTGGGTTAAAGTCATGCTCC	AAACAAAATATTATGACCCCCAAAACCATG — JCV
T Ag-2			GGTAGAAGACCCTAAAGAC	GAATATAATATTATGCCCAGCACACATGTG — BKV
JC VP1–1[e]	1662–1842	—	CAGATACATTTGAAAGTGAC	ACCCAATCTAAATGAGGATCTAACCTGTGG
JC VP1–2			CCATTAGAGTGCACATTCATC	
BK VP1–1[e]		1777–1960	GCTGAAAATGACTTTAGCAG	CCCTCCCCAATTTAAATGAGGACCTAACCT
BK VP1–2			ACCCTGCATGAAGGTTAAGC	
Nested T-1[c]	3975–4188	4112–4325	AAGTATTCCTTATT(C,A)ACACC	
Nested T-2			CTGTGTATAC(C,T)ACTAAAGAA	

[a] BKV PVB DUN GenBank sequence data.

[b] JCV PLYCG GenBank sequence data.

[c] Primers and probes developed at the VRL by J. P. Clewley and P. E. Gibson.

[d] Sequence data provided by J. Snook, Oxford.

[e] Sequence data provided by P. Watkins and M. Jones, London.

one amplifying a 181-nucleotide target sequence of the JCV genome, the other a 184-nucleotide sequence of the BKV genome.

Using these primers, human polyomavirus DNA was detected in urine, in CSF, and in DNA extracted from brain tissue. The urine specimens had to be pretreated to remove inhibitors of *Taq* polymerase: 1 ml of urine was spun in a microfuge for 1 min; the supernatant was discarded, and the pellet was resuspended in 100 µl water and heated at 95°C for 10 min; this suspension was spun in a microfuge for 20 s, the supernatant was removed and stored at 4 or –30°C prior to testing by PCR; 10 µl were tested in each 50-µl amplification reaction.

DNA was extracted from the brain biopsy material by proteinase K digestion, followed by phenol chloroform extraction and ethanol precipitation. The precipitated DNA was resuspended in water, and 10 µl were used in each 50-µl PCR reaction. DNA was similarly extracted from CSF and tested both prior to and following extraction to determine if inhibition of amplification occurred without extraction. There appeared to be no inhibition of amplification by untreated CSF, and in fact, there was less amplification when the DNA was extracted. This was observed both when JC virus was seeded into normal CSF and when DNA was amplified from CSF taken from patients with confirmed PML.

In all these experiments one water control was used for each four specimens and was treated in the same manner during sample preparations. Great care was taken to avoid cross contamination in handling the specimens, setting up the reactions, and testing the reaction products. Separate rooms were also used for reagent preparation, DNA extraction, amplification, and analysis. Using control DNA and virus dilutions, the PCR was able to amplify as little as 10 ag of starting target, so that a visible band was detectable on gels following electrophoresis of 10 µl of the reaction product.

Thirty-eight urine samples from pregnant women, all of whom had cytological evidence of polyomavirus infection, were tested using the T Ag primers. The reaction products were then hybridized with specific probes for either BKV or JCV and were also tested in secondary PCRs using nested primers. JCV DNA only was detected in 31 urines; JCV and BKV DNA in 5; BKV DNA only in 1; in one urine no polyomavirus DNA detected. These results confirmed the previous studies on polyomavirus excretion in this group of pregnant women.[14]

Urine samples from 100 children admitted to a general pediatric ward were also tested using T Ag primers in the primary reaction and nested T primers in the secondary reaction. Polyomavirus DNA was detected in the urine from four children. This was shown by hybridization to be BKV in all cases. These four children were all 5 years old or under. No JCV DNA was detected.

We investigated the presence of JCV in a large panel of CSF specimens, many of which were from patients with a confirmed diagnosis of PML.[15] Of the 60 specimens tested, 18 were from 13 patients in whom PML had been confirmed by histological or virological investigations following brain biopsy or at autopsy; 42 specimens were from patients with other neurological illnesses. No JCV DNA was detected in CSF from this latter group. However, JCV DNA was detected in the CSF of 10 of the 13 patients with PML (Table 14–2). In six of these cases JCV DNA was detected with both T Ag and JC VP1 primers, and in three it was detected with T Ag primers only. In these nine cases, the nested T primers gave amplification in the secondary reaction, and in another specimen DNA amplified only in the secondary reaction with the nested T primers. No JCV DNA was detected in the CSF of the remaining three patients. JCV DNA was also detected in the extracted brain tissue of two patients with PML. This is not unexpected since virus is readily detected in brain tissue by other less sensitive techniques. In all cases the BK VP1 primers gave a negative reaction. This investigation indicates that detection of JCV DNA in CSF establishes a diagnosis of PML, but that a negative result does not necessarily exclude it.

In our laboratory, L. Jin and colleagues[16,17] have also investigated DNA VP1 gene variation of BKV by directly sequencing the PCR amplicons. Using several primer pairs, she amplified regions of the genome between 1564 and 2652 nucleotides (DUN VP1). She was able to establish that the sequence differences between BK isolates, in this very small area of the VP1 gene, correlated with the antigenic variability previously observed between strains.

V. CONCLUSIONS

BKV and JCV infections can be identified by laboratory techniques that are reliable and less prone to laboratory error than PCR. However, the very slow growth and narrow host cell range makes isolation in tissue culture an impractical procedure. Immune EM is also not practical for large numbers of specimens. In the diagnosis of PML, the rapid identification of JC virus in brain biopsy material, or in

Table 14–2 **Presence of JCV DNA detected by PCR in 18 cerebrospinal fluid specimens from 13 patients with a confirmed diagnosis of progressive multifocal leukoencephalopathy**

Patients (No. of Specimens)	Primers (No. of Specimens Positive)		
	T Ag	Nested T	JC VP1
1 (1)	+	+	+
2 (1)	+	+	+
3 (1)	+	+	+
4 (1)	+	+	+
5 (1)	+	+	+
6 (2)	+ (1)	+ (1)	+ (2)
7 (1)	+	+	–
8 (3)	+ (1)	+ (1)	–
9 (2)	+ (1)	+ (1)	–
10 (1)	–	+	–
11 (1)	–	–	–
12 (2)	–	–	–
13 (1)	–	–	–

CSF, would greatly assist in the management of this rapidly progressing fatal disease. Therefore, the use of PCR in identification of these viruses is worthwhile not only because the test can be rapidly performed, but also because small clinical samples can be accommodated. The use of PCR for sequence investigation also increases the ease and speed at which strains or genetic variants can be identified. Future work may help to identify the primary infection of these viruses; assist in epidemiological studies; and investigate as-yet unidentified roles both human polyomaviruses may play in human disease.

REFERENCES

1. **Frisque, R. J., Bream, G. L., and Cannella, M. T.,** Human polyomavirus JC virus genome, *J. Virol.,* 51, 458, 1984.
2. **Seif, I., Khoury, G., and Dhar, R.,** The genome of human papovavirus BKV, *Cell,* 18, 963, 1979.
3. **Yang, R. C. A. and Wu, P. R.,** BK Virus DNA: complete nucleotide sequence of a human tumour virus, *Science,* 206, 456, 1979.
4. **Travis, J. E., Walker, D. L., Gardner, S. D., and Frisque, R. J.,** Nucleotide sequence of the human polyomavirus AS virus, an antigenic variant of BK virus, *J. Virol.,* 63, 901, 1989.
5. **Gardner, S. D.,** Implications of papovaviruses in human diseases, in *Comparative Diagnosis of Viral Diseases,* Vol. 1. *Human and Related Viruses, Part A,* Kurstak, E., Ed., Academic Press, New York, 1977, 41–84.
6. **McCance, D. J. and Gardner, S. D.,** Papovaviruses: papillomaviruses and polyomaviruses, in *Principles and Practice of Clinical Virology,* Zuckerman, A. J., Banatvala, J. E., and Pattison, J. R., Eds., John Wiley & Sons, New York, 1987, 479–506.
7. **Arthur, R. R., Dagostin, S., and Shah, K. V.,** Detection of BK virus and JC virus in urine and brain tissue by the polymerase chain reaction, *J. Clin. Microbiol.,* 27, 1174, 1989.
8. **Marshall, W. F., Telenti, A., Proper, J., Aksamit, A. J., and Smith, T. F.,** Survey of urine from transplant recipients for polyomaviruses JC and BK using the polymerase chain reaction, *Mol. Cell. Probes,* 5, 125, 1991.
9. **Telenti, A., Aksamit, A. J., Proper, J., and Smith, T. F.,** Detection of JC virus DNA by polymerase chain reaction in patients with progressive multifocal leukoencephalopathy, *J. Infect. Dis.,* 162, 858, 1990.

10. **Flaegstad, T., Sundsfjord, A., Arthur, R. A., Pedersen, M., Traavik, T., and Subramani, S.,** Amplification and sequencing of the control regions of BK and JC virus from human urine by polymerase chain reaction, *Virology,* 180, 553, 1991.

11. **Henson, J., Rosenblum, M., Armstrong, D., and Furneaux, H.,** Amplification of JC virus DNA from brain and cerebrospinal fluid of patients with progressive multifocal leukoencephalopathy, *Neurology,* 41, 1967, 1991.

12. **Brouqui, P., Bollet, C., Delmont, J., and Bourgeade, A.,** Diagnosis of progressive multifocal leucoencephalopathy by PCR detection of JC virus from CSF, *Lancet,* 339, 1182, 1992.

13. **Telenti, A., Marshall, W. F., Aksamit, A. J., Smilack, J. D., and Smith, T. F.,** Detection of JC virus by polymerase chain reaction in cerebrospinal fluid from two patients with progressive multifocal leukoencephalopathy, *Eur. J. Clin. Microbiol. Infect. Dis.,* 11, 253, 1992.

14. **Coleman, D. V., Wolfendale, M. R., Daniel, R. A., Dhanjal, N. K., Gardner, S. D., Gibson, P. E., and Field, A. M.,** A prospective study of human polyomavirus infection in pregnancy, *J. Infect. Dis.,* 142, 1, 1980.

15. **Gibson, P. E., Knowles, W. A., Hand, J. F., and Brown, D. W. G.,** Detection of JC virus in the cerebrospinal fluid of patients with progressive multifocal leukoencephalopathy, *J. Med. Virol.,* 39, 278, 1993.

16. **Jin, L., Gibson, P. E., Knowles, W. A., and Clewley, J. P.,** BK virus antigenic variants: sequence analysis with the capsid VP1 epitope, *J. Med. Virol.,* 39, 50, 1993.

17. **Jin, L., Gibson, P. E., Booth, J. C., and Clewley, J. P.,** Genomic typing of BK virus in clinical specimens by direct sequencing of polymerase chain reaction products, *J. Med. Virol.,* 41, 11, 1993.

Chapter 15

Investigation of Human Parvovirus B19 Infection Using PCR

Jonathan P. Clewley and Bernard J. Cohen

TABLE OF CONTENTS

I. INTRODUCTION

Human parvovirus B19 is a member of the genus *Parvovirus* (family Parvoviridae) and is a small (23 nm), round, nonenveloped virus.[1] It is very stable and may resist heating to 80°C for 72 h.[2] Its genome is a single-stranded DNA molecule of 5.6 kb. A virion may contain one DNA molecule of either polarity, such that on chemical extraction of the DNA in the laboratory under annealing conditions, a population of predominantly double-stranded molecules is formed. The terminal 383 nucleotides at each end of the genome are identical inverted repeats, the distal 365 nucleotides of which are imperfect palindromes that fold to form hairpin structures.[3,4] The right-hand half genome codes for a minor 85-kDa VP1 capsid protein and a major 58-kDa VP2 protein, which is encoded within the VP1 gene. The left-hand half of the genome codes for a 77-kDa nonstructural (NS) protein, which is required for DNA replication. There are other open reading frames (ORFs) that may be transcribed and translated into NS proteins in the infected cell, although their function is not yet clear. As far as is understood, there is only one antigenic type of B19, but this is something that requires further investigation. PCR sequencing of the major antigenic domains identified by Sato and Brown and colleagues[5,6] may help clarify this question. It is a ubiquitous virus which, in the U.K. at least, appears to cause peaks of infection in late winter, spring, and early summer. There may be an epidemic cycle of 4 to 6 years. Transmission occurs through respiratory droplets and across the placenta from mother to fetus, and can also be spread by the parenteral route via contaminated factor VIII. Although B19 DNA in blood donations can be detected by dot blot hybridization, the presence of DNA in blood products needs to be investigated by a more sensitive assay such as PCR. The transmission rate is high (20 to 50%) in school and household outbreaks of erythema infectiosum (EI; see below). Indeed, the association of B19 with EI was first observed in a school. During the incubation period there may be a "flu-like" illness associated with viremia 7 to 10 days after infection. Symptoms of rash and arthropathy occur 14 to 20 days after infection and are associated with an antibody response (see Figure 15–1).

In common with other parvoviruses, B19 requires actively dividing cells for its productive replication. It replicates in erythroid progenitors of the bone marrow, and of the liver in the developing fetus. It infects only humans, and is the only known human pathogenic parvovirus (the adeno-associated viruses are thought to be benign). Infection is often asymptomatic (20 to 50%) and, if not, usually manifests as EI (fifth disease) in children. In adults, however, symptoms can be more severe, and there may be an acute

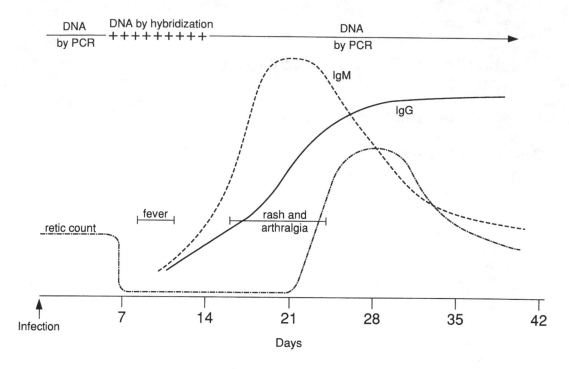

Figure 15–1 The course of infection with parvovirus B19. +++, viremia; retic., reticulocyte. (From Clewley, J. P., in *Frontiers in Virology I: Diagnosis of Human Viruses by Polymerase Chain Reaction Technology,* Becker, Y. and Darai, G., Eds., Springer-Verlag, Berlin, 1992, 285. With permission.)

arthritis, which occurs more often in women than in men. During infection, erythropoiesis is temporarily arrested and reticulocytes disappear from the peripheral blood, a condition that is life threatening in those with an underlying hemolytic disorder (e.g., sickle cell anemia). If a nonimmune pregnant woman is infected, the virus may cross the placenta and cause fetal anemia, hydrops fetalis, and intrauterine death. PCR provides a tool for investigating fetal infections. However, while it may be relatively straightforward to apply PCR to a postmortem specimen from a fetal death, PCR is a more difficult and controversial technique when applied to an ongoing pregnancy. B19 can cause persistent anemia in immunocompromised patients, and this has increasingly been seen in individuals who are immunodeficient because of HIV infection. Therapy with immunoglobulin containing anti-B19 offers a treatment for those with persistent infection, and PCR is very useful in their diagnosis and clinical management.

II. LABORATORY DIAGNOSIS OF B19 INFECTIONS

The most useful test for acute B19 infection is a B19-specific IgM antibody capture RIA[7] or ELISA.[8] During the viremic stage, the virus itself can be detected by direct DNA hybridization, by antigen capture RIA or ELISA, by immunoblotting, or by electron microscopy. Tissue samples can be examined by *in situ* hybridization. There is no routine cell culture system.

III. PCR DETECTION OF B19

The theoretical and practical aspects of PCR for B19 diagnosis have been reviewed.[9,10] The use of PCR for B19 DNA detection was first described in 1989 by Salimans and colleagues[11] and by Clewley.[12] These workers used both ethidium bromide staining and subsequent radioactive detection to identify the PCR amplicons. Using a single-round PCR, femtogram quantities (about 100 copies) of B19 DNA were detectable with the most sensitive ³²P-labeled probes. B19 PCR has subsequently been refined in several laboratories.[13–21] Durigon and co-workers[14] analyzed 16 different primer pairs for B19 PCR and found that the most sensitive detected 350 to 3500 DNA copies in a single round of PCR and that nested amplification was required for additional sensitivity. They were unable to explain the differing sensitivities between these primer pairs. Neither sequence variation between B19 genomes nor amplicon size was apparently responsible.

A. SERUM SPECIMENS

From these studies, and from observations in this laboratory, it seems that it may not be possible to approach single-copy detection of B19 DNA extracted from clinical specimens using one round of PCR. There are several possible explanations for this. One is that inhibitors of *Taq* polymerase are co-purifying with the extracted B19 DNA. Proteases present in serum that degrade *Taq* polymerase may be responsible for some inhibitory effects, as suggested by Frickhofen and Young,[22] who found that controlled heating of serum for 45 s at 75°C allowed it to be added directly to the PCR reagent mixture. Another method for overcoming the effects of inhibitors is by extraction with the ion exchange resin Chelex®.*[22] This type of method is of particular use for the extraction of DNA for PCR from oropharyngeal secretions that contain inhibitors of PCR.[22,24] Cassinotti and colleagues[15] found that a single round of PCR could detect only about 300 copies of a recombinant B19 plasmid template. Therefore, for their analysis of clinical samples they used nested PCR, which they estimated to have a sensitivity of 10 to 100 copies. The extraction method they used was to heat serum for 10 min at 95°C, centrifuge it, and use an aliquot of the supernatant in the reaction. It may be that shorter heating times would achieve greater sensitivity.[22] Koch and Adler[16] were able to add 1 to 3 µl of serum directly to their PCR, and estimated that fewer than ten B19 genomes were detectable by radioactive hybridization following a single-round PCR. Patou and colleagues[20] also tried adding serum directly to the PCR, but found this to be insensitive compared to a detergent extraction method. Sevall[17] added boiled, clarified clinical serum samples to his PCR and estimated that only 1 to 10 pg could be detected by ethidium bromide staining, and 0.1 pg by autoradiography. It is probable that the quality of serum specimens will have differing effects on a PCR, depending on when they were collected, how often they have been frozen and thawed, how much fat they contain, etc., making meaningful comparison between published accounts difficult. Heparin is known to inhibit *Taq,* and blood collected in heparinized tubes should not be used for PCR unless unavoidable. Treatment with heparinase may overcome this inhibitory effect. Schwarz and colleagues[21] extracted DNA by proteinase K digestion and phenol-chloroform extraction and found that one fetal serum sample, which was positive for B19 DNA by dot blot hybridization, was negative when tested by PCR. They suggest that heparin caused this false-negative result. Excessive EDTA can also inhibit PCR, presumably by chelating magnesium ions. We have not been able to obtain satisfactory, reproducible results by adding serum to PCR without first extracting DNA chemically. This is done by absorption of the DNA to silica in the presence of the chaotropic agent guanidinium isothiocyanate.[25,26]

A second reason why single-copy detection of B19 DNA from serum with one round of PCR may be difficult is that the viral genome may anneal back together so efficiently that very little of the intended amplicon is synthesized during the first few cycles of PCR, so that subsequent amplification is very inefficient. Experiments using higher melting temperatures and a more thermostable enzyme than *Taq* may help increase the sensitivity of single-round B19 PCR assays. A high melting temperature (95 to 98°C) for the first few (5 to 10) cycles could be followed by a lower (90°C) temperature that will melt only the amplicon.

The natural history of infection shown in Figure 15–1 indicates that B19 DNA is detectable by PCR from shortly after infection until more than 1 month later. (B19 DNA may be detectable in an immunocompromised individual for the entire course of the infection, which may be several years.) Patou and colleagues have shown that there is a statistically significant association between the amount of B19 IgM in serum and the PCR result, and an inverse relationship with the amount of B19 IgG.[20] This was significant only on a population basis, as testing of individual sera failed to predict the timing of infection. Some serum samples collected within 11 days of the onset of symptoms were found not to contain B19 DNA in the presence of specific IgM.[20] It has also been shown that in normal individuals the DNA may be detectable by PCR for up to 3 months after the initial infection, in the presence of anti-B19 specific IgM and IgG.[12] However, this is not the case for every specimen that has been examined, as a survey of the literature illustrates[12,15–17] (Table 15–1).

---

* Bio-Rad, CA.

Briefly, it appears possible to find every combination of the presence of IgM, IgG, and DNA. Does the apparent absence of B19 DNA in some anti-B19 specific IgM individuals represent a real phenomenon, or is it a laboratory artifact due to unidentified inhibitors of PCR and to reaction failures? It has previously been argued[12] that the detection of B19 DNA months after the original infection is due to a stochastic decline from very high levels of viremia (10^{10} to 10^{12} particles per milliliter of serum). It may be that low levels of virus replication are occurring in some erythroid precursor cells, or that virus is being sequestered and degraded in macrophages, and that DNA from these sources gives a positive PCR signal. As viral DNA is usually extracted from serum or plasma containing variable and unknown numbers of intact and degraded white cells, statistical sampling errors would explain observed discrepancies between serological and PCR results for those cases after the peak of viremia (more than 2 weeks; see Table 15–1 and Figure 15–1). However it is likely that some false-negative amplifications contribute to the reported results. A useful control would be to add any suspected "false-negative" serum to a known positive to attempt to show that it was not itself inhibitory.

Table 15–1 shows that the incidence of B19 DNA-positive samples in the absence of specific B19 antibodies ranges from 1 to 18.5% in four independent studies. This is a misleading estimate of the general prevalence of B19 since the numbers of specimens in some of those studies were low and because, in all cases, specimens were selected for the likelihood of their containing B19 virus. Of the five patients found by Koch and Adler[16] to have B19 DNA by PCR in the absence of specific IgM, one was immunocompromised and the other four had sickle cell disease and aplastic crisis. These four patients may have been tested before a specific anti-B19 IgM response developed. The incidence of B19 DNA in the presence of either IgM of IgG ranged from 31.6 to 62.5%.

The experience of this laboratory using IgG and IgM antigen-capture RIAs is that almost all IgM-positive samples are also IgG positive. It is possible that the estimate of the incidence of DNA-positive samples in the presence of either specific IgM or IgG alone is biased by insufficiently sensitive serological assays. The incidence of IgM (indicative of a recent infection) in sera containing no DNA ranged from 18.5 to 60.4%. Erdman and colleagues[18] examined specimens from 128 cases of EI and found that 120 had B19 DNA and specific IgM and IgG. Of the remainder, five were positive for specific IgM. The reason for the PCR-negative results in these five cases is unclear but may be caused by inhibitors or amplification failures. It has been suggested that genetic heterogeneity explains the failure of some probes to react with B19 DNA.[16] However, compared to many viruses, B19 displays little genetic heterogeneity[4] and this explanation is unconvincing to us and is not supported by the data of Frickhofen and Young[13] and Durigon and colleagues.[14]

Besides serum, B19 DNA has been detected by PCR in peripheral blood cells, including lymphocyte lysates from a patient with aplastic crisis; in throat swabs; in various fetal tissues and in fetal serum, ascitic, pericardial, and amniotic fluids; placentas; chorionic villi; synovial tissue and fluid; cerebrospinal fluid (CSF); and blood products.[2,12,15,16,19,21,27–30] The detection of B19 DNA by PCR in any tissue that can become contaminated by blood during collection must be interpreted with caution, as viremia can be very intense. This applies to the detection of B19 DNA in fetal tissues following spontaneous abortion and to the detection of B19 DNA in placentas.[12] Only *in situ* hybridization[31] or *in situ* PCR (see Chapter 3 be Staskus et al.) can provide information on the cellular location of B19 virus.

The intensity of B19 viremia (e.g., 10^{10} to 10^{12} genomes per milliliter) means that a nonradioactive dot hybridization test very often has sufficient sensitivity as a routine laboratory test for the detection of B19 DNA.[32] This may be a colorimetric detection test or, preferably for specimens that may themselves by

Table 15–1 **B19 DNA by PCR in populations with evidence of recent parvovirus infection**[a]

DNA	IgM	IgG	Cassinotti et al.[15]	Clewley[12]	Sevall[17]	Koch and Adler[16]
+	–	–	5/136 (11%)	1/96 (1%)	2/53 (3.8%)	5/27 (18.5%)
+	+	–	7/136 (5.1%)	0/96 (0%)[b]	3/53 (5.6%)	13/27 (48.15%)
+	+	+	20/136 (14.7%)	60/96 (62.5%)	11/53 (20/7%)	4/27 (14.8%)
+	–	+	16/136 (11.8%)	N/A[c]	5/53 (9/4%)	N/A[c]
–	+	–	27/136 (19.8%)	0/96 (0%)[b]	2/53 (3.8%)	5/27 (18.5%)
–	+	+	51/139 (37.5%)	35/96 (36.5%)%	30/53 (56.6%)	N/A[c]

[a] Excluding all samples DNA –ve, IgM –ve, and IgG –ve; and those only IgG +ve; [b] assuming all IgM +ves are IgG +ve; [c] N/A, data not available.

colored, a chemiluminescent one. A nonradioactive test is able to detect about 10^7 genomes per milliliter, whereas the sensitivity of a test using a ^{32}P-labeled probe can be greater than this.[33] However, even a radioactive detection system may be insufficiently sensitive to detect B19 DNA in persistent infections or in blood products such as factor VIII. Therefore, a PCR test is desirable. At present PCR is more labor intensive than dot hybridization, but if it can be made contamination free and as user friendly as an ELISA, and if it can also give a quantitative result, it will probably be readily adopted by most B19 routine diagnostic laboratories. One issue that such laboratories will have to resolve is the diagnostic significance of low-level viremia. As mentioned above, several studies have shown that B19 DNA can be detected in the serum of healthy individuals up to 1 to 3 months after the resolution of acute infection.[12,15,16,20]

B. BLOOD PRODUCTS

B19 virus has been recovered from blood donations given by healthy individuals who had yet to develop symptoms. Indeed such donations have been the major source of supply of native virus for diagnostic tests and molecular cloning; only recently have recombinant proteins become available for the development of serological assays. Some laboratories undertake screening of blood donations in order to find a plentiful supply of virus. During one such exercise using dot blot hybridization it was found that 1 in 24,000 donations contained B19 virus.[34] Despite the increasing availability of recombinant antigens, reference B19, laboratories are still likely to require native virus, and blood transfusion centers may want to screen donations for the presence of B19 in order to exclude the virus from blood products. McOmish and colleagues[19] have developed a system for screening donations for B19 DNA by PCR that fulfills both requirements. They achieved this by pooling samples from donations prior to DNA extraction and PCR. Positive pools were then subdivided for further screening to identify the viremic donors. B19 DNA was detected in 6 out of 20,000 donations in concentrations ranging from 2.4×10^4 to 5×10^{10} genomes per milliliter. The lower DNA concentrations would not have been detected by dot hybridization. They also detected B19 DNA in 18 out of 27 separate batches of non-heat-treated factor VIII and IX concentrates. This confirmed previous observations that B19 can be transmitted by clotting factor preparations.[35,36] They concluded that they would be able to develop a screening protocol for B19 DNA by PCR in blood donations that would greatly reduce the incidence of virus infection hemophiliacs and other susceptible individuals with clotting disorders. Zakrzewska and colleagues[30] examined 25 clotting factor concentrates by both hybridization and PCR for the presence of B19 DNA; 9 out of 25 of the products were positive by PCR, but B19 DNA was detectable in only 2 of these by hybridization. These included untreated, chloroform-treated, and dry-heated products, but the numbers studied were too small for the authors to be able to draw any conclusions about the effect of different treatments on the survival of B19. Nevertheless it is clear from retrospective studies that B19 infection can be transmitted by treated as well as untreated concentrates.[2,37,38] Although screening of blood products by PCR for B19 DNA would allow contaminated batches to be identified, a means of assaying the infectivity of the virus in the products is also necessary. The absence of a routine culture system for B19 makes this difficult. In laboratories that have access to bone marrow or fetal livers, sufficient erythroid progenitors may be obtained to assess B19 infectivity.[39,40]

C. INTRAUTERINE INFECTIONS

Perhaps the most contentious and controversial use of B19 PCR is for the investigation of intrauterine infections. Products of conception or specific fetal organs (fresh or formalin fixed) from an abortion or fetal death can be tested by hybridization or by PCR for the presence of B19 DNA. Hybridization is usually sufficiently sensitive for this purpose, as viremia in a hydropic fetus is often intense. However if the tissues are necrotic DNA will probably be degraded and be more difficult to detect by direct hybridization. In such situations PCR is appropriate. Also, we have recently shown that a discrete-sized amplicon of B19 DNA can be obtained by PCR whereas only degraded DNA is demonstrable by Southern blotting.[41] We have found the level of B19 DNA is usually highest in fetal liver (where the concentration of erythroid progenitor target cells is greatest), but other fetal organs and the placenta may also be positive.

Prenatal diagnosis of intrauterine B19 infection may be attempted by examining fetal blood or amniotic fluid. PCR investigation of fetal infection where pregnancy is continuing is more problematic, however, as illustrated by the study of Török and colleagues,[28] who investigated 56 pregnancies at risk for B19 infection. Fifteen women were positive for both B19 IgM and IgG, and 15 of their fetuses were positive for B19 DNA by PCR. Of these 15 fetuses, 10 became healthy full-term infants. Twenty four women were B19 IgG-positive and IgM negative, indicating prior B19 infection. B19 DNA was found by PCR in 4 maternal serum specimens in this group and in 3 of the corresponding fetal specimens. Seventeen women were specific IgM and IgG negative, indicating that they had not been previously exposed to B19. However, B19 DNA was found by PCR in serum in three of these women, and in two of the corresponding fetuses. A fourth woman from this group had a positive PCR result only for her fetus. As there was no serological evidence of B19 infection in these women it is difficult to be certain of the significance of the PCR findings in them.

Cassinotti and colleagues[15] also reported unusual findings in a case of B19 infection in pregnancy. B19 DNA was detected by PCR in amniotic fluid, chorionic villi, and maternal blood from a fetal loss in the 27th week of pregnancy. It was not reported whether B19 DNA detection by hybridization was attempted. The maternal blood contained anti-B19 IgG both at the time of the spontaneous abortion and also 2 years earlier, indicating a past infection. Consistent with this was the absence of detectable anti-B19 IgM at the time of abortion. There was clinical evidence that a 5-year-old daughter had been infected with B19 at this time, and the authors suggest that a secondary infection (reinfection or viral reactivation) had occurred in the mother. However, secondary B19 infection has been observed only following experimental infection in human volunteers.[42] It has not been documented in immunocompetent individuals exposed to natural infection, but as reinfection with other viruses (e.g., rubella) has been associated with embryopathy, the possibility of reinfection with B19 merits further study.

Of most benefit for studies of the outcome of congenital B19 and its clinical management will be the wider availability of sensitive and specific IgM and IgG tests. At present these are often done as "in-house" tests, and thus confined to a few reference centers.

D. MONITORING PERSISTENT B19

Chronic B19 infection of immunocompromised patients, resulting in anemia, has been reported.[43,44] This condition is difficult to diagnose serologically as B19-specific IgM and IgG antibodies may be lacking or present at low levels. As these patients respond to treatment with immunoglobulin, which clears virus, it is important that an early diagnosis be attempted. This can be achieved by detection of viral DNA either by hybridization or by PCR. Following immunoglobulin therapy, the amount of any virus remaining is low, and PCR is the preferred technique to demonstrate viral clearance and the efficacy of treatment. Also, patients should be monitored by PCR (perhaps at monthly intervals) to provide an early warning of B19 recurrence, which can be successfully retreated.[45–47] A recent case report[48] described B19 DNA persistence for 4 years in an immunologically healthy woman in the presence of an apparently normal B19 IgM and IgG response. During the time B19 DNA was detectable, the woman experienced recurrent episodes of paresthesia. PCR was necessary to detect the level of DNA present as it would not have been found by hybridization. Cases such as these, where B19 is diagnosed by PCR alone, illustrate the need for a PCR confirmatory test. Perhaps alternative amplification systems such as the ligase chain reaction, Qβ amplification, or transcription-based assays will be able to fill this role.

E. PATHOGENESIS OF PERSISTENT ARTHROPATHIES

Joint symptoms (arthropathy) are a common complication of B19 infection in adults and may persist for months or even years.[49–51] Initial studies using DNA hybridization failed to show the presence of any B19 DNA persisting in these cases.[50] Patients with these symptoms may be diagnosed as having rheumatoid arthritis (RA), although rheumatoid factor is typically absent.[52] However, it was recently reported that B19 DNA was found in synovial biopsies of 15 of 20 patients with RA,[29] and B19 DNA was detected in the peripheral mononuclear cells (PBL) of three of these patients. It was also detected in the synovia (4 of 24) and PBL (4 of 17) of patients with other arthropathies, and in the PBL of 5 of 15 healthy controls. These five healthy controls were all anti-B19 IgG seropositive, whereas only five of the 15 synovial-biopsy DNA-positive patients were anti-B19 seropositive. It is difficult to explain the absence of anti-B19 from such a large proportion of RA patients with B19 PCR-positive synovial biopsies. Moreover, in view of the high (15%) prevalence of B19 DNA in the PBL of anti-B19 IgG-positive, healthy controls, the results of this study must be treated with caution and need to be confirmed.

F. INVESTIGATION OF DOUBTFUL IgM SEROLOGY

The detection of anti-B19 IgM is the most useful test for the diagnosis of a recent B19 infection since it remains positive for only 2 to 3 months after infection.[7] However, a small proportion of samples obtained within a few days of onset of symptoms may be IgM negative because they are collected too early. In these circumstances, tests for B19 virus by methods including PCR may be positive. For example, Schwarz and colleagues[21] examined anti-B19 IgM-negative sera from six patients collected during the incubation period of virus infection and showed that they could detect B19 DNA by PCR. However, these sera had all been selected on the basis of a positive dot blot hybridization result. In other cases the results of IgM serology may be doubtful, and PCR can be used to evaluate their significance. This is particularly likely to occur in patients with diseases like RA and systemic lupus erythematosus (SLE), where the presence of autoantibodies (rheumatoid factor; anti-DNA) can cause false-positive reactions. Even with IgM assays based on antibody capture, which are less prone to interference by these factors than assays using other test formats, there is a risk that low-level IgM reactions could be misinterpreted.[52] The date of onset of symptoms needs to be taken into account, and elevated B19 IgM reactions, sometimes over a period of months or years, may be misconstrued as evidence of persistent infection. In this situation PCR is helpful because a negative result would exclude B19 persistence. It has been pointed out by Frickhofen and Young[13] that PCR also enables samples that cannot be tested for IgM, such as paraffin-embedded tissues, to be examined for evidence of B19 infection.

G. PRIMERS

Many different primers have been used for the detection of B19 DNA in the studies described in the literature.[10–22,27–30,43,44,52] The sequences of two almost-full-length B19 genomes are very similar (98.9%), and restriction mapping studies show relatively few differences between isolates.[4,54,55] Thus there is only limited genomic genetic heterogeneity, and most well-designed primers would be expected to work equally well. For reasons that are not understood (see above), this does not seem to be the case. For instance, a 100-fold difference in sensitivity of various primers has been observed.[13,14] Examples of sets of diagnostic nested and single-round amplification primers are described by Clewley,[10] Frickhofen and Young,[13] and Durigon et al.[14] Some suitable diagnostic primers are shown in Table 15–2.

Primers in the conserved region of the NS gene[54] are a good choice for attempting to develop a generic parvovirus PCR that might be capable of detecting as-yet-uncharacterized parvoviruses.[12,18] One candidate uncharacterized parvovirus is the small, round virus(es) associated with gastroenteritis.[56–58] Primers located near the termini of the genome can be used to amplify almost-full-length B19 DNA for cloning and other recombinant DNA manipulative purposes, using appropriate conditions.[10,59,60]

H. OTHER APPLICATIONS

The study of genome variation of viruses has been achieved by PCR amplification of antigenically important regions and by subsequent sequencing of the PCR amplicons.[61] Until recently, it was unclear which region of the B19 genome might be the most suitable for this. The mapping of the binding sites of neutralizing and monoclonal antibodies on B19 capsids has defined an immunodominant region,[5,6] which is a good target for similar B19 PCR amplicon-sequencing studies. Finally, a PCR-generated digoxigenin or other labeled probe can be used as an alternative to random primed or nick-translated cloned DNA probes in hybridization tests.[62,63]

Table 15–2 Primers used for amplification of B19 DNA[a]

Primer		Sequence 5' → 3'	5' Nucleotide Position	Gene
PV1[b]	Sense	GGT AAG AAA AAT ACA CTG T	1390	NS
PV2	Antisense	TTG CCC GCC TAA AAT GGC TTT	1608	NS
PV3	Sense	ATG GGC CGC CAA GTA CAG GAA A	1415	NS
PV4	Antisense	TCA TTA AAT GGA AAG TTT TCA TT	1520	NS
H[c]	Sense	GGG CCG CCA AGT ACA GGA	1417	NS
C	Antisense	AGG TGT GTA GAA GGC TTC TT	2160	NS
F	Sense	AAT GAA AAC TTT CCA TTT AAT GA	1498	NS
I	Antisense	TCC TGA ACT GGT CCC GGG GAT GGG	2088	NS
Z[d]	Sense	GGA ACA GAC TTA GAG CTT ATT C	2537	VP
Y	Antisense	GCT TGT GTA AGT CTT CAC TAG	2774	VP
V	Probe	ACC CAT CCT CTC TGT TTG ACT AGT TGT CTC G	2607	VP

[a] Nucleotide numbering is taken from Shade et al.[54]; GenBank accession no. M13178; [b] References 7, 9, and 15; [c] References 7 and 36; [d] Reference 7.

IV. CONCLUSIONS

The most important test that a B19 diagnostic laboratory needs is one for the detection of anti-B19 IgM. Other useful tests include the detection of anti-B19 IgG and B19 DNA by dot blot hybridization. PCR is also able to contribute to the diagnosis of B19 infections, particularly those in immunocompromised patients, where it can be used to monitor the progress of therapy. It will be interesting to sequence PCR products from sequential specimens taken from patients who have persistent B19 infection, to study the evolution of the virus *in vivo*. B19 PCR has also been applied to the diagnosis of intrauterine infections, although some of the results obtained are as yet difficult to interpret. PCR may help elucidate the etiological role of B19 virus in arthropathies, but claims that it is associated with rheumatoid arthritis need to be substantiated. Adeno-associated virus DNA integrates into cellular chromosomal DNA,[64] and although there is as yet no evidence that B19 also does this, the possibility can be investigated by PCR.

Many animal parvoviruses cause gastroenteritis by replicating in rapidly dividing gut epithelial cells. Parvovirus-like particles have been found in cases of human gastroenteritis, although evidence that they are causally involved is not strong. PCR could be used to see if B19 is involved with any gastroenteritis outbreaks, for example, those associated with the so-called small, round, unstructured viruses.[57] Moreover, a "generic" PCR based on conserved NS gene sequences can be used to screen for other, as-yet-uncharacterized human parvoviruses.

Serum is the clinical sample most commonly investigated for evidence of B19 infection, though other specimens may be expected to contain the virus. These include saliva, throat swabs, and urine, all of which can be collected noninvasively. It is expected that PCR will be adapted to examine these specimens, to determine whether they are useful for diagnostic purposes. Of particular promise is saliva, which can serve as a substitute for serum in serological tests.[65]

REFERENCES

1. **Pattison, J. R.,** Parvoviruses, medical and biological aspects, in *Virology,* 2nd ed., Fields, B. N. and Knipe, D. M., Eds., Raven Press, New York, 1990, chap. 63.
2. **Lyon, D. J., Chapman, C. S., Martin, C., Brown, K. E., Clewley, J. P., Flower, A. J. E., and Mitchell, V. E.,** Symptomatic parvovirus B19 infection and heat treated factor IX concentrate, *Lancet,* i, 1085, 1089.
3. **Deiss, V., Tratschin, J.-D., Weitz, M., and Siegl, G.,** Cloning of the human parvovirus B19 genome and structural analysis of its palindromic termini, *Virology,* 175, 247, 1990.
4. **Mori, J., Beattie, P., Melton, D. W., Cohen, B. J., and Clewley, J. P.,** Structure and mapping of the DNA of human parvovirus B19, *J. Gen. Virol.,* 68, 2797, 1987.
5. **Sato, H., Hirata, J., Kuroda, N., Shiraki, H., Maeda, Y., and Okochi, K.,** Identification and mapping of neutralizing epitopes of human parvovirus B19 by using human antibodies, *J. Virol.,* 65, 5485, 1991.
6. **Brown, C. S., Jensen, T., Melonen, R. H., Puijk, W., Sugamura, K., Sato, H., and Spaan, W. J. M.,** Localization of an immunodominant domain on baculovirus produced parvovirus B19 capsids: correlation to a major surface region on the native virus particle, *J. Virol.,* 66, 6989, 1992.
7. **Cohen, B. J.,** Laboratory tests for the diagnosis of infection with B19 virus, in *Parvoviruses and Human Disease,* Pattison, J. R., Ed., CRC Press, Boca Raton, FL, 1986, 69.
8. **Anderson, L. J., Tsou, C., Parker, R. A., Chorba, T. L., Wulff, H., Tattersall, P., and Mortimer, P. P.,** Detection of antibodies and antigens of human parvovirus B19 by enzyme-linked immunosorbent assay, *J. Clin. Microbiol.,* 24, 522, 1986.
9. **Clewley, J. P.,** PCR diagnosis of parvovirus B19, in *Frontiers of Virology 1: Diagnosis of Human Viruses by Polymerase Chain Reaction Technology,* Becker, Y. and Darai, G., Eds., Springer-Verlag, Berlin, 1992, 285.
10. **Clewley, J. P.,** PCR detection of parvovirus B19, in *Diagnostic Molecular Microbiology,* Persing, D. H., Ed., ASM Press, Washington, D.C., 1993, 367.
11. **Salimans, M. M. M., Holsappel, S., Van de Rijke, F. M., Jiwa, N. M., Raap, A. K., and Weiland, H. T.,** Rapid detection of human parvovirus B19 DNA by dot-hybridization and the polymerase chain reaction, *J. Virol. Meth.,* 23, 19, 1989.
12. **Clewley, J. P.,** Polymerase chain reaction assay of parvovirus B19 DNA in clinical specimens, *J. Clin. Microbiol.,* 27, 2647, 1989.

13. **Frickhofen, N. and Young, N. S.,** Polymerase chain reaction for detection of parvovirus B19 in immunodeficient patients with anemia, *Behring Institute Mitteilungen,* 85, 46, 1990.

14. **Durigon, E. L., Erdman, D. D., Gary, G. W., Pallansch, M. P., Török, T. J., and Anderson, L. J.,** Multiple primer pairs for polymerase chain reaction (PCR) amplification of human parvovirus B19 DNA, *J. Virol. Meth.,* 44, 155, 1993.

15. **Cassinotti, P., Weitz, M., and Siegl, G.,** Human parvovirus B19 infections: routine diagnosis by a new nested polymerase chain reaction assay, *J. of Med. Virol.,* 40, 228, 1993.

16. **Koch, W. C. and Adler, S. P.,** Detection of human parvovirus B19 by using the polymerase chain reaction, *J. Clin. Microbiol.,* 28, 65, 1990.

17. **Sevall, J. S.,** Detection of parvovirus B19 by dot-blot and polymerase chain reaction, *Mol. Cell. Probes,* 4, 237, 1990.

18. **Erdman, D. D., Usher, J. M., Tsou, C., Caul, E. O., Gary, G. W., Kajigaya, S., Young, N. S., and Anderson, L. J.,** Human parvovirus B19 specific IgG, IgA, and IgM antibodies and DNA in serum specimens from persons with erythema infectiosum, *J. Med. Virol.,* 35, 110, 1991.

19. **McOmish, F., Yap, P. L., Jordan, A., Hart, H., Cohen, B. J., and Simmonds, P.,** Detection of parvovirus B19 in blood donations: a model system for screening by polymerases chain reaction, *J. Clin. Microbiol.,* 31, 323, 1993.

20. **Patou, G., Pillay, D., Myint, S., and Pattison, J.,** Characterization of a nested polymerase chain reaction assay for the detection of parvovirus B19, *J. Clin. Microbiol.,* 31, 540, 1993.

21. **Schwarz, T. F., Jäger, G., Holzgreve, W., and Roggendorf, M.,** Diagnosis of human parvovirus B19 infections by polymerase chain reaction, *Scan. J. Infect. Dis.,* 24, 691, 1992.

22. **Frickhofen, N. and Young, N. S.,** A rapid method of sample preparation for detection of DNA viruses in human serum by polymerase chain reaction, *J. Virol. Meth.,* 35, 65, 1991.

23. **Welsh, P. S., Metzger, D. A., and Higuchi, R.,** Chelex 100 as a medium for simple extraction of DNA for PCR-based typing from forensic material, *BioTechniques,* 10, 506, 1991.

24. **Ochert, A.,** personal communication, 1993.

25. **Boom, R., Sol, C. J. A., Salimans, M. M. M., Larsen, C. L., Wertheim-van Dillen, P. M. E., and Van der Noorda, J.,** Rapid and simple method for purification of nucleic acids, *J. Clin. Microbiol.,* 28, 495, 1990.

26. **Boom, R., Sol, C. J. A., Salimans, M. M. M., Heijtink, R., Wertheim-van Dillen, P. M. E., and Van der Noorda, J.,** Rapid purification of hepatitis B virus DNA from serum, *J. Clin. Microbiol.,* 29, 1804, 1991.

27. **Salimans, M. M. M., Van der Rijke, F. M., Raap, A. K., and Van Elsacker-Niele, A. M. W.,** Detection of parvovirus B19 DNA in fetal tissues by *in situ* hybridization and polymerase chain reaction, *J. Clin. Pathol.,* 42, 525, 1989.

28. **Török, T. J., Wang, Q.-Y., Gary, G. W., Yang, C.-F., Finch, T. M., and Anderson, L. J.,** Prenatal diagnosis of intrauterine parvovirus B19 infection by the polymerase chain reaction technique, *Clin. Infect. Dis.,* 14, 149, 1992.

29. **Saal, J. G., Steidle, M., Einsele, H., Müller, C. A., Fritz, P., and Zacher, J.,** Persistence of B19 parvovirus in synovial membranes of patients with rheumatoid arthritis, *Rheumatol. Int.,* 12, 147, 1992.

30. **Zakrzewska, K., Azzi, A., Patou, G., Morfini, M., Rafanelli, D., and Pattison, J. R.,** Human parvovirus B19 in clotting factor concentrates: B19 DNA detection by the nested polymerase chain reaction, *Br. J. Haematol.,* 81, 407, 1992.

31. **Porter, H. J., Quantrill, A. M., and Fleming, K. A.,** B19 parvovirus infection of myocardial cells, *Lancet,* i, 535, 1988.

32. **Mori, J., Field, A. M., Clewley, J. P., and Cohen, B. J.,** Dot blot hybridization assay of B19 virus DNA in clinical specimens, *J. Clin. Microbiol.,* 27, 459, 1989.

33. **Clewley, J. P.,** The application of DNA hybridization to the understanding of human parvovirus disease, in *DNA Probes for Infectious Diseases,* Tenover, F. C., Ed., CRC Press, Boca Raton, FL, 1989, 211.

34. **Cohen, B. J., Field, A. M., Gudnadottir, S., Beard, S., and Barbara, J. A. J.,** Blood donor screening for parvovirus B19, *J. Virol. Meth.,* 30, 233, 1990.

35. **Mortimer, P. P.,** Transmission of serum parvovirus-like virus by clotting-factor concentrates, *Lancet,* ii, 482, 1983.

36. **Morifini, M., Longo, G., Rossi Ferrini, P., Azzi, A., Zakrewska, C., Ciappi, S., and Kolumban, P.,** Hypoplastic anemia in a hemeophiliac first infused with a solvent/detergent treated factor VIII concentrate: the role of human B19 parvovirus, *Am. J. Hematol.,* 39, 149, 1992.

37. **Corsi, O. B., Azzi, A., Morfini, M., Fanci, R., and Ferrani, P. R.,** Human parvovirus infection in haemophiliacs first infused with treated clotting factor concentrates, *J. Med. Virol.,* 25, 165, 1988.

38. **Azzi, A., Zakrewska, K., Gentilomi, A., Musiani, M., and Zerbini, M.,** Detection of B19 parvovirus infections by a dot-blot hybridization assay using a digoxigenin labelled probe, *J. Virol. Meth.,* 27, 125, 1990.

39. **Shimomura, S., Komatsu, N., Frickhofen, N., Anderson, S., Kajigaya, S., and Young, N. S.,** First continuous propagation of B19 parvovirus in a cell line, *Blood,* 79, 18, 1992.

40. **Morey, A. L., Patou, G., Myint, S., and Fleming, K. A.,** *In vitro* culture for the detection of infectious parvovirus B19 and B19-specific antibodies using foetal haematopoietic precursor cells, *J. Gen. Virol.,* 73, 3313, 1992.

41. **Leruez, M., Cohen, B. J., and Clewley, J. P.,** unpublished observations, 1992.

42. **Anderson, M. J., Higgins, P. G., Davis, L. R., Willman, J. S., Jones, S. E., Kidd, I. M., Pattison, J. R., and Tyrell, D. A. J.,** Experimental parvoviral infection in humans, *J. Infect. Dis.,* 152, 257, 1985.

43. **Kurtzman, G. J., Ozawa, K., Cohen, B., Hanson, G., Oseas, L., and Young, N. S.,** Chronic bone marrow failure due to persistent B19 parvovirus infection, *N. Engl. J. Med.,* 317, 287, 1987.

44. **Kurtzman, G. J., Cohen, B. J., Myers, P., Amunullah, A., and Young, N. S.,** Persistent B19 infection as a cause of severe chronic anaemia in children with acute lymphocytic leukemia, *Lancet,* ii, 1159, 1988.

45. **Koch, W. C., Massey, G., Russell, C. E., and Adler, S. P.,** Manifestations and treatment of human parvovirus B19 infection in immunocompromised patients, *J. Pediatr.,* 116, 355, 1990.

46. **Mitchell, S. A., Welch, J. M., Weston-Smith, S., Nicholson, F., and Bradbeer, C. S.,** Parvovirus infection and anaemia in a patient with AIDS: case report, *Genitourinary Med.,* 66, 95, 1990.

47. **Kurtzman, G., Frickhofen, N., Kimball, J., Jenkins, D. W., Niehuis, A. W., and Young, N. S.,** Pure red-cell aplasia of 10 years' duration due to persistent parvovirus B19 infection and its cure with immunoglobulin therapy, *N. Engl. J. Med.,* 321, 519, 1989.

48. **Faden, H., Gary, W. G., and Anderson, L. J.,** Chronic parvovirus infection in a presumably immunologically healthy woman, *Clin. Infect. Dis.,* 15, 595, 1992.

49. **White, D. G., Woolf, A. D., Mortimer, P. P., Cohen, B. J., Blake, D. R., and Bacon, P. A.,** Human parvovirus arthropathy, *Lancet,* i, 419, 1985.

50. **Cohen, B. J., Buckley, M. M., Clewley, J. P., Jones, V. E., Puttick, A. H., and Jacoby, R. K.,** Human parvovirus infection in early rheumatoid and inflammatory arthritis, *Ann. Rheumatic Dis.,* 45, 832, 1986.

51. **Frickhofen, N. and Young, N. S.,** Persistent parvovirus B19 infections in humans, *Microbial Pathogenesis,* 7, 319, 1989.

52. **Naides, S. J., Scharosch, L. L., Foto, F., and Howard, E. J.,** Rheumatological manifestations of human parvovirus B19 infection in adults, *Arthritis and Rheumatism,* 33, 1297, 1990.

53. **Brown, D. W. G.,** Viral diagnosis by antibody capture assay, *Public Health Virology 12 Reports,* Mortimer, P. P., Ed., PHLS Press, London, 1986, chap. 7.

54. **Shade, R. O., Blundell, M. C., Cotmore, S. F., Tattersall, P., and Astell, C. R.,** Nucleotide sequence and genome organization of human parvovirus B19 isolated from the serum of a child during aplastic crisis, *J. Virol.,* 58, 921, 1986.

55. **Blundell, M. C. and Astell, C. R.,** A GC-box motif upstream of the B19 parvovirus unique promoter is important for *in vitro* transcription, *J. Virol.,* 63, 4814, 1989.

56. **Paver, W. K. and Clarke, S. K. R.,** Comparison of human fetal and serum parvolike viruses, *J. Clin. Microbiol.,* 4, 76, 1976.

57. **Appleton, H.,** Small round viruses: classification and role in food-borne infections, in *Novel Diarrhoea Viruses, Ciba Foundation Symposium 128,* Bock, G. and Whelan, J., Eds., 1987, 108.

58. **Turton, J., Appleton, H., and Clewley, J. P.,** Similarities in nucleotide sequence between serum and faecal human parvovirus DNA, *Epidemiol. Infect.,* 105, 197, 1990.

59. **Ohler, L. D. and Rose, E. A.,** Optimization of long-distance PCR using a transposon-based model system, *PCR Meth. Applic.,* 2, 51, 1992.

60. **Mori, J., Leruez, M., and Clewley, J. P.,** unpublished observations, 1992.

61. **Ou, C.-Y., Ciesielski, C. A., Myers, G., Bandea, C. I., Luo, C.-C., Korber, B. T. M., Mullins, J. I., Schochetman, G., Berkelman, R. L., Economou, A. N., Witte, J. J., Furman, L. J., Satten, G. A., MacInnes, K. A., Curran, J. W., and Jaffe, H. W.,** Molecular epidemiology of HIV transmission in a dental practice, *Science,* 256, 1165, 1992.

62. **Yun, Z. B. and Hornsleth, A.,** Production of digoxigenin-labelled parvovirus DNA probe by PCR, *Res. Virol.,* 113, 926, 1991.

63. **Hilfenhaus, S., Cohen, B. J., and Clewley, J. P.,** unpublished observations, 1992.

64. **Kotin, R. M., Siniscalco, M., Samulski, R. J., Zhu, X., Hunter, L., Laughlin, C. A., McLaughlin, S., Muzyczka, N., Rocchi, M., and Berns, K. I.,** Site-specific integration by adeno-associated virus, *Proc. Natl. Acad. Sci. U.S.A.,* 87, 2211, 1990.

65. **Mortimer, P. P. and Parry, J. V.,** Non-invasive virological diagnosis: are saliva and urine specimens adequate substitutes for blood?, *Rev. Med. Virol.,* 1, 73, 1991.

INDEX